Bocconi & Springer Series

Mathematics, Statistics, Finance and Economics

Volume 12

The **Bocconi & Springer Series** aims to publish research monographs and advanced textbooks covering a wide variety of topics in the fields of mathematics, statistics, finance, economics and financial economics. Concerning textbooks, the focus is to provide an educational core at a typical Master's degree level, publishing books and also offering extra material that can be used by teachers, students and researchers.

The series is born in cooperation with Bocconi University Press, the publishing house of the famous academy, the first Italian university to grant a degree in economics, and which today enjoys international recognition in business, economics, and law.

The series is managed by an international scientific Editorial Board. Each member of the Board is a top level researcher in his field, well-known at a local and global scale. Some of the Board Editors are also Springer authors and/or Bocconi high level representatives. They all have in common a unique passion for higher, specific education, and for books.

Volumes of the series are indexed in Web of Science - Thomson Reuters.

Manuscripts should be submitted electronically to Springer's mathematics editorial department: francesca.bonadei@springer.com

THE SERIES IS INDEXED IN SCOPUS

Donatien Hainaut

Continuous Time Processes for Finance

Switching, Self-exciting, Fractional and other Recent Dynamics

BOCCONI
UNIVERSITY
PRESS

Donatien Hainaut
ISBA
Université Catholique de Louvain
Louvain-La_Neuve, Belgium

ISSN 2039-1471   ISSN 2039-148X   (electronic)
Bocconi & Springer Series
ISBN 978-3-031-06360-2   ISBN 978-3-031-06361-9   (eBook)
https://doi.org/10.1007/978-3-031-06361-9

Mathematics Subject Classification: 60-xx, 62-xx

This Springer imprint is published by the registered company Springer Nature Switzerland AG
The registered company address is: Gewerbestrasse 11, 6330 Cham, Switzerland

Preface

Quantitative finance no longer needs an introduction. Mathematical modeling has opened new perspectives both for trading and for risk management. Sometimes seen as a thread contributing to financial crises such as the credit crunch of 2008, quantitative finance offers foremost opportunities to improve our understanding of the dynamics of financial markets. Forty years after the publication of the celebrated Black and Scholes formula, mathematical modeling has opened astonishing new perspectives. This book aims to present some of them, with a strong emphasis on computational and econometric aspects.

Structure of the Book

The first chapter is an introduction to switching models, which replicate economic cycles observed in financial markets. In this framework, parameters of a basic model are modulated by a hidden Markov chain representative of the economic state. We focus on the modulated geometric diffusion with jumps at transitions of economic regimes. As risk management applications require the disposal of econometric methods for estimating parameters from time series, we review the Hamilton filter. This is used both for statistical inference and for detecting the most likely state of the hidden Markov chain. We next price options by Fast Fourier Transform (FFT) and discuss the conditions under which an equivalent measure is eligible as a risk neutral measure. The next section reviews the switching multifractal model, in which the volatility of asset log-returns is stochastic and ruled by multiple Markov chains. We show that the multifractal dynamic may be reformulated as a switching regime process with a large number of regimes and that it outperforms econometric models such as GARCH, in term of goodness of fit. The chapter concludes with an introduction to the fluid embedding technique. We use this to infer the Laplace transform of hitting times of a switching diffusion with transition jumps. A multivariate extension of switching diffusions is proposed in Chap. 2.

Log-likelihood maximization is the standard approach for estimating a model. Nevertheless, this task is difficult to carry out when the calculation of the likelihood is computationally intensive. As this is the case for most of processes presented in this book, the second chapter presents an alternative method based on a Bayesian learning paradigm. This estimation procedure, called Markov Chain Monte Carlo (MCMC), is a powerful framework providing information about the density of parameter estimates. The chapter starts by reviewing the main features of Markov chains and presents the Metropolis–Hastings algorithm adapted to perform Bayesian inference. This approach is the cornerstone of the third chapter and is repeatedly used throughout the book. As an illustration, the MCMC algorithm is adapted for estimating a multivariate switching regime diffusion that is a natural extension of the model studied in the first chapter.

The estimation by MCMC methods requires the existence of a closed form expression for the likelihood. When unavailable, the likelihood can nevertheless be approximated by simulations. Chapter 3 develops such approaches based on what is called "particle filtering." Particle filtering is a technique to compute the most likely sample path of a hidden process in nested models. A nested model is a visible price process ruled by one or several hidden dynamic factors. As an illustration, we consider the Heston model, in which the stock price is driven by a geometric diffusion with a mean reverting stochastic variance. After recalling its main features, we develop an algorithm to guess the sample path of variance from observed log-returns. Next, we combine this filter with an MCMC method for estimating Heston parameters from time series. Particle filtering is used throughout the book. In Chap. 4, this serves to filter the intensity of a self-exciting jump process. In Chap. 9, particle filtering is used to estimate a rough Heston model in which the variance depends on a rough fractional Brownian motion. Finally, in Chap. 10, this technique serves to filter the stochastic clock triggering illiquidity periods.

Looking to time series of stock log-returns reveals that large shocks often occur during short periods of time and that shocks in, for instance, foreign markets trigger sudden price adjustments in the domestic market. The first phenomenon is called "clustering of jumps" and is mainly caused by an onslaught of bad news. The second one is known as mutual contagion. The switching processes of Chap. 1 and the models with stochastic volatility seen in Chap. 3 fail to explain clustering and contagion between shocks. Chapter 4 explores an alternative approach based on self-exciting jump processes that circumvents this drawback. In this framework, the probability of observing a new shock increases as soon as a jump in price occurs. We start with a presentation of the mathematical features of the self-exciting jump-diffusion for stock prices. The following sections focus on the econometric estimation of this process. We detect jumps with a "peak-over-threshold" method and infer jump parameters by log-likelihood. Next, we learn how to sample jumps with Ogata's algorithm in order to adapt the particle filter of Chap. 2 to estimate the sample path of jump intensity. The remainder of the chapter is devoted to option pricing and changes of measure.

In the self-exciting processes of Chap. 4, the occurrence of a shock depends on previous shocks. In the SEJD specification, the intensity of jumps, which is akin to the instantaneous probability of a shock, increases as soon as a shock occurs. The influence of this jump on the intensity then decays with time according to an exponential memory function, also called a kernel. If the memory kernel is not exponential decaying, we lose most of the analytical tractability offered by the SEJD, mainly because the process is no longer a semi-martingale. In Chap. 5, we show that the moment-generating function of such a process nevertheless admits a closed-form expression if the memory kernel is the Fourier transform of a probability measure. We focus on cumulated jump processes and exploit the results to analyze the contagion between large positive and negative shocks in S&P 500 daily returns.

In Chaps. 4 and 5, the memory of past events plays a crucial role in determining the behavior of self-exciting processes. In a broader context, we could legitimately ask whether financial time series display long or short memory and how to model this in continuous time. We try to answer these questions in Chaps. 6–9. Chapter 6 starts by reviewing the mathematical properties of fractional Brownian motion. Unlike regular Brownian motion, fBm has dependent increments as for self-exciting processes. After having defined the stochastic integral with the Wick product, we develop two applications. The first concerns the pricing of European options on an asset ruled by a geometric fractional Brownian motion. The second proposes a fractional interest rate model. Concepts seen in this chapter will help us to understand the rough Heston model in Chap. 9, which is a fractional extension of the stochastic volatility model of Chap. 3.

Chapter 7 introduces the main properties of Gaussian fields, which naturally extend the notion of fractional Brownian motion. A Gaussian field is a family of normal random variables indexed by time, with a covariance structure. We focus on a market model based on Gaussian fields defined through the spectral measure of their autocovariance function, in a similar way to the memory kernels of Chap. 5. After a concise description of the model's properties, we emphasize the importance of the covariance in the valuation of exotic derivatives. For this purpose, we derive closed-form expressions for calendar exchange spread and Asian calendar exchange spread options. The numerical examples reveal that a misspecification of the autocovariance function generates significant deviations of option prices.

In most popular approaches to pricing interest derivatives, the short-term rate is a mean reverting process. As underlined in Chap. 7, this is equivalent to assuming that interest rates are a random field with an exponential decaying autocovariance function. In Chap. 8, we replace this memory kernel by a Mittag-Leffler decaying function. This function is sub-exponential and is a generalization of the exponential. This is also closely related to differential fractional calculus and will appear again in the results of Chap. 10 about illiquidity modeling. The main consequence of replacing the exponential memory by a Mittag-Leffler function is that the interest rate remembers its sample path over a longer period. The main drawback is that the process is no longer a semi-martingale. As in Chap. 5, we nevertheless take advantage of an integral representation of the memory kernel to convert the model into an infinite dimensional Markov process. Approximating this process allows us to infer the dynamics of bond prices and many of their features.

In Chaps. 6 or 8, we study processes depending on the convolution of a memory kernel and their past sample path. Chapter 9 extends this approach and focuses on Affine Volterra processes that are solutions of stochastic convolution equations with affine coefficients. This family includes a broad range of dynamics such as Ornstein–Uhlenbeck, Heston, and the Rough Heston processes. In this last model, the volatility behaves like the mean reverting fractional motion studied in Chap. 6. After a presentation of Volterra processes, we adapt the particle filter of Chap. 3 to fit the Rough Heston model to the S&P 500 with the MCMC algorithm of Chap. 2.

The previous chapters have initiated our journey into the world of processes that are not martingales. We introduce in Chap. 10 a new category of non-martingale processes perfectly adapted for modeling illiquidity. At high frequency or in small cap markets, transactions are sparse. Time-series of stock prices in such conditions display characteristic periods in which they stay motionless. We replicate this phenomenon by a sub-diffusion, which is a standard Brownian motion observed on a different scale of time. We next show that the density of a sub-diffusion is described in terms of a Fractional Fokker–Planck (FFP) equation. We evaluate European options in this setting and discuss the estimation of model parameters with a particle MCMC algorithm such as detailed in Chap. 3.

Chapter 11 extends the results of Chap. 10 in several directions. Firstly, we consider a fractional jump-diffusion with stochastic volatility for the asset return. Secondly, we consider a wider family of fractional dynamics. In the previous chapter, fractional processes are built by replacing the time scale by a random clock that is the inverse of an α-stable Lévy process. Here, we enlarge the family of clocks to all invertible Lévy subordinators. In this framework, we show that European option prices are solutions of a fractional forward differential equation. The fractional derivative in this Dupire equation is a convolution-type derivative. We focus on fractional dynamics based on inverted Poisson and α-stable subordinators and propose a numerical method to evaluate options in these two cases.

Reading Guide

This book targets the following audience: master's students in quantitative finance, or in the actuarial, statistical, or mathematical sciences, PhD students, scholars, and professional quantitative analysts. Its reading requires a basic to a good knowledge of stochastic calculus and statistical inference. The chapters are all interconnected, but some of them may be read independently if a few sections are skipped.

- Chapter 1 is an introduction to switching regime processes. The fast Fourier transform (FFT) algorithms of Sect. 1.6 are used in Chaps. 2, 4, 5, 8, and 9.
- Chapter 2 presents the MCMC algorithm and the illustration is linked to Chap. 1. This is a prerequisite for the reading of Chap. 3.
- Chapter 3 is about Bayesian estimation by particle filtering. This technique is again used in Chaps. 4, 9, and 10.

- Chapter 4 studies a diffusion model with self-exciting jumps. This is a prerequisite of Chap. 5. The FFT and MCMC algorithms of Chaps. 1 and 3 are used for estimating the model.
- Chapter 5 analyzes the features of non-Markov self-exciting processes. The Bochner spectral theorem is used in Chap. 7, but the reading of Chap. 5 is not an absolute prerequisite. The FFT algorithm of Chap. 1 is used in the illustration.
- Chapter 6 introduces fractional Brownian motion. Its reading is recommended but not compulsory before Chap. 9.
- Chapter 7 explores models based on Gaussian fields. It is indirectly related to Chap. 5 but may be read independently.
- Chapter 8 introduces a Lévy interest rate model with a long memory. It is connected to Chap. 5 but may be read independently. The FFT algorithm of Chap. 1 is used in the illustration.
- Chapter 9 is about Volterra's equations. The reading of Chap. 6 is recommended. The FFT and MCMC algorithms of Chaps. 1 and 3 are used to estimate the rough Heston model.
- Chapter 10 explores illiquidity modeling with stochastic clocks. This is an absolute prerequisite of Chap. 11. The MCMC algorithm of Chap. 3 is used for estimating the model.
- Chapter 11 extends results of Chap. 10, the reading of which is a prerequisite.

Acknowledgments

This book grew out of courses and research articles written during the last 10 years. I would like to thank all my students and colleagues who provided useful feedback through their comments and suggestions. I also thank my wife Vinita and my daughter Constance who sustained me in ways that I never knew that I needed. Finally, yet importantly, I want to thank Catriona Byrne, who accepted this book for publication by Springer.

Louvain-La_Neuve, Belgium   Donatien Hainaut

Contents

About the Author

Donatien Hainaut is Professor of Quantitative Finance and Actuarial Sciences at UCLouvain, where he manages of the master's program in data science and statistical orientation. Prior to this, he held several positions as associate professor at Rennes School of Business and the ENSAE in Paris. He also has several field experiences having worked as risk officer, quantitative analyst, and ALM officer. He is a qualified actuary and holds a PhD in the area of assets and liability management. His current research focuses on contagion mechanism in stochastic processes, fractional processes, and their application in insurance and finance.

Notation

$\mathrm{A}f$	Infinitesimal generator of f
$B^H(t)$	Fractional Brownian motion with a Hurst index H
$\left.\begin{array}{l} C(t,k) \\ P(t,k) \end{array}\right\}$	Call price, maturity t, and strike price $K = S_0 e^k$
$(\delta_t)_{t \geq 0}$	Continuous-time Markov chain
$\left.\begin{array}{l} \{\mathrm{F}_t\}_{t \geq 0} \\ \{\mathrm{G}_t\}_{t \geq 0} \\ \{\mathrm{H}_t\}_{t \geq 0} \end{array}\right\}$	Filtrations
e_j	Null vector, except the jth element, which is equal to 1
$\mathbb{E}(\cdot)$	Expectation
$\mathbb{V}(\cdot)$	Variance
$\mathbb{E}^{\mathbb{Q}}(\cdot)$	Expectation under the risk neutral measure
$E_\alpha(x)$	Mittag-Leffler function
FFT	Fast Fourier transform
DFT	Discrete Fourier transform
$f(\boldsymbol{x}\lvert\Theta)$	Likelihood of a vector of observations with a distribution f and parameters in Θ
$\mathrm{FC}(\omega)$	Fourier transform of $C(t,k)$
i.i.d.	Identically independent distributed
J_{ij} or J	Random variable for jumps
$\nu(z)$	Jump or Lévy density
mgf	Moment-generating function
$N(\mu,\sigma)$	Normal random distribution with mean μ and standard deviation σ
$N_{i,j}(t)$	Transition jump process
Ω	Probability space
$\mathbb{P}$	Real probability measure
$\mathbb{Q}$	Risk neutral probability measure
Q_0	Matrix of instantaneous transition probabilities of a Markov chain
$\mathbb{S}^d$	Space of symmetric positive definite matrices of dimension d
L^2	Space of square integrable functions

$[X_t, Y_t]$	Quadratic covariation of processes X_t and Y_t
$\langle M \rangle_t$	The angular bracket of M_t
$\widehat{\theta}$	Estimator of θ
$\frac{\partial^\alpha f}{\partial t^\alpha}$	Fractional derivative of order $\alpha \in (0, 1]$
$(X_t)_{t \geq 0}$	Stochastic process
W_t	Brownian motion
pdf	Probability density function
cdf	Cumulative distribution function

Chapter 1
Switching Models: Properties and Estimation

There is considerable empirical evidence suggesting that the random walk model for changes in stock prices is not appropriate. One of the reasons is that this model fails to account for economic cycles because increments are independent and identically distributed. A reliable solution for modeling economic cycles consists in modulating the parameters of a basis process, e.g., a Brownian motion by a hidden Markov chain. This approach has received a lot of attention in the recent econometric literature. In this chapter, we aim to introduce its main features in continuous time. The first sections focus on regime switching diffusions with transition jumps. We present their properties and an estimation procedure to fit this process to stock log-returns. Pricing of options is also discussed. We next introduce the switching multifractal model of Calvet and Fisher [2]. This is a switching diffusion with a large number of regimes that are structured in order to limit the number of parameters. This chapter partly serves as introduction to Chap. 2 in which a multivariate extension is estimated by a Monte Carlo Markov Chain method.

1.1 A Hidden Markov Chain

The economic conjuncture varies considerably over time, and certain events such as economic slowdown trigger sudden changes in the dynamics of financial markets. In order to model this behavior, we may assume that the economy is categorized into a finite number of states or regimes. The process representative of the economic conjuncture may therefore be modeled by a continuous Markov chain.

We consider a complete probability space Ω equipped with a probability measure $\mathbb{P}$. Here $\mathbb{P}$ denotes the real-world probability measure. Throughout this chapter, the economy is categorized into n states or regimes, indexed by a set of integers $\mathcal{N} := \{1, 2, \cdots, n\}$. All states are assumed attainable.

© The Author(s), under exclusive license to Springer Nature Switzerland AG 2022
D. Hainaut, *Continuous Time Processes for Finance*,
Bocconi & Springer Series 12, https://doi.org/10.1007/978-3-031-06361-9_1

The information about the economic regime over time is carried by a hidden Markov chain that we denote by $(\delta_t)_{t\geq 0}$. It takes values from the canonical state space $E = \{e_1, \ldots, e_n\}$, where $e_j = (0, \ldots, 1, \ldots, 0)^\top$ (the null vector of dimension n, except the jth element, which is equal to 1).

The natural filtration generated by $(\delta_t)_{t\geq 0}$ is denoted by $\{\mathcal{G}_t\}_{t\geq 0}$. The generator of δ_t is an $n \times n$ matrix $Q_0 := [q_{i,j}]_{i,j\in\mathcal{N}}$, with

$$q_{i,j} \geq 0 \quad \text{for } i \neq j \text{ and } \sum_{j=1}^{n} q_{i,j} = 0 \,.$$

The $q_{i,j}$ for $j \neq i$ may be interpreted as the instantaneous probability of transition from states i to j:

$$\mathbb{P}\left(\delta_{t+\Delta} = e_j \mid \delta_t = e_i\right) \approx q_{i,j}\Delta \text{ when } \Delta \to 0 \,.$$

The diagonal elements of Q_0 are all negative: $q_{i,i} = -\sum_{j\neq i} q_{i,j} < 0$. The probability of staying in regime i satisfies the following relation:

$$\mathbb{P}\left(\delta_{t+\Delta} = e_i \mid \delta_t = e_i\right) \approx 1 + q_{i,i}\Delta \text{ when } \Delta \to 0 \,.$$

The matrix of transition probabilities over the time interval $[t, s]$ is the following matrix exponential:

$$P(t, s) = \exp\left(Q_0(s - t)\right) \,,$$

where the exponential corresponds to the following expansion:

$$\exp(Q_0(s - t)) = \sum_{i=0}^{\infty} \frac{(Q_0(s - t))^i}{i!} \,. \tag{1.1}$$

It can be proved that this series converges to a finite matrix. The (i, j)th element of the matrix $P(t, s)$, denoted $p_{i,j}(t, s)$, is the probability of switching from state i at time t to state j at time s. Note that one can prove that for every matrix of transition probabilities, there exists a matrix of instantaneous probabilities Q_0.

We consider an irreducible Markov chain. A Markov chain is said to be *irreducible* if for all $i \neq j$, there is a finite time h such that $p_{i,j}(t, t + h) > 0$. In other words, every economic regime is reachable from another one. For such a chain, there is no state in which the economy can stay trapped. For an irreducible Markov chain, the following limit always exists:

$$\lim_{t\to\infty} \mathbb{P}(\delta_t = e_i) = \pi_i > 0 \,,$$

for all $i \in \mathcal{N}$. The vector $\pi = (\pi_i)_{i \in \mathcal{N}}$ forms what is called the steady or stationary distribution. If the chain starts with its stationary distribution, the marginal distribution of all states at any time will always be the stationary distribution.

To each pair of distinct states (i, j) in the state space of the Markov chain δ_t, we define a point process $N_{i,j}(t)$ as follows:

$$N_{i,j}(t) := \sum_{0 < s \leq t} \mathbf{1}_{\{\delta_{s-}=e_i\}} \mathbf{1}_{\{\delta_s=e_j\}}, \tag{1.2}$$

where $\mathbf{1}$ is the indicator function. $N_{i,j}(t)$ counts the number of transitions from state i to state j up to time t. We also define the following intensity process:

$$\lambda_{i,j}(t) := q_{i,j} \mathbf{1}_{\{\delta_{t-}=e_i\}}. \tag{1.3}$$

If we compensate the counting process $N_{i,j}(t)$ by the integral of $\lambda_{i,j}(\cdot)$, it can be proved that the resulting process

$$M_{i,j}(t) := N_{i,j}(t) - \int_0^t \lambda_{i,j}(s)\mathrm{d}s \tag{1.4}$$

is a martingale, i.e., $\mathbb{E}\left(M_{i,j}(t) \mid \mathcal{G}_s\right) = M_{i,j}(s)$ for $s \leq t$. These martingales are mutually orthogonal and purely discontinuous. We will use them to construct a risk neutral measure.

1.2   A Modulated Asset Model

We propose a price process S_t for a financial asset, with jumps induced by change of economic regimes. This process is defined on a filtration $\{\mathcal{H}_t\}_{t \geq 0}$ of Ω that carries the information about the asset price. Remember that the information about the Markov Chain $(\delta_t)_{t \geq 0}$ is contained in another filtration $\{\mathcal{G}_t\}_{t \geq 0}$. The augmented filtration that gathers information about all processes is denoted by $\{\mathcal{F}_t\}_{t \geq 0}$ and is such that $\mathcal{G}_t \cup \mathcal{H}_t \subset \mathcal{F}_t$. The instantaneous return of this asset is the sum of a drift, a Brownian motion W_t, and a compensated jump process:

$$\frac{\mathrm{d}S_t}{S_{t-}} = \mu_t \mathrm{d}t + \sigma_t \mathrm{d}W_t \tag{1.5}$$

$$+ \sum_{i=1}^{n} \sum_{i \neq j}^{n} \left(\left(e^{J_{i,j}} - 1 \right) \mathrm{d}N_{i,j}(t) - \lambda_{i,j}(t) \mathbb{E}\left(e^{J_{i,j}} - 1 \right) \mathrm{d}t \right).$$

The drift rate μ_t and the Brownian volatility σ_t are modulated by the Markov chain δ_t:

$$\mu_t = \delta_t^\top \boldsymbol{\mu}, \qquad \sigma_t = \delta_t^\top \boldsymbol{\sigma},$$

where $\boldsymbol{\mu} = (\mu_1, \ldots, \mu_n)^\top \in \mathbb{R}^n$ and $\boldsymbol{\sigma} = (\sigma_1, \ldots, \sigma_n)^\top \in \mathbb{R}_+^n$. In this approach, the economic regime drives the instantaneous average growth rate and volatility of the stock price. The jump part allows a jump to be introduced in the stock return when the economy changes regime. If δ_t switches from state i to j ($\mathrm{d}N_{i,j}(t) = 1$), we observe a random jump of size: $e^{J_{i,j}} - 1$. For a pair (i, j), the $J_{i,j}$ are independent identically distributed (i.i.d.) random variables on $\mathbb{R}$. The probability density function (pdf) of $J_{i,j}$ is denoted by $v_{i,j}(\cdot)$. The particular form of the jump size is chosen in order to guarantee that negative shocks are not bigger than the current stock price. Otherwise, the probability of having a negative stock price would not be null. We call this *switching jump-diffusion (SJD) model*. Note that the independence between $N_{i,j}(t)$ and $J_{i,j}$ ensures that

$$\mathbb{E}\left(\left(e^{J_{i,j}} - 1\right) \mathrm{d}N_{i,j}(t)\right) = \mathbb{E}\left(e^{J_{i,j}} - 1\right) \lambda_{i,j}(t)\mathrm{d}t.$$

Therefore, the jump term is canceled on average (we say "compensated") by the last term in Eq. (1.5). The expected instantaneous return is therefore equal to $\mathbb{E}\left(\frac{\mathrm{d}S_t}{S_t} \mid \mathcal{F}_t\right) = \mu_t \, \mathrm{d}t$.

According to Itô's lemma for jump processes, any function $f(t, S_t, \delta_t)$ that is respectively C^1 and C^2 with respect to time and stock price admits the following differential for $\delta_t = e_i$:

$$\mathrm{d}f = \left(\frac{\partial f}{\partial t} + \mu_t S_t \frac{\partial f}{\partial S} + \frac{1}{2}\sigma_t^2 S_t^2 \frac{\partial^2 f}{\partial S^2}\right) \mathrm{d}t \tag{1.6}$$

$$+ \sigma_t S_t \frac{\partial f}{\partial S} \mathrm{d}W_t - S_t \frac{\partial f}{\partial S} \sum_{i=1}^{n} \sum_{i \neq j}^{n} \lambda_{i,j}(t) \mathbb{E}\left(e^{J_{i,j}} - 1\right) \mathrm{d}t$$

$$+ \sum_{i=1}^{n} \sum_{i \neq j}^{n} \left(f\left(t, S_t e^{J_{i,j}}, e_j\right) - f(t, S_t, \delta_t)\right) \mathrm{d}N_{i,j}(t).$$

We refer the reader to Jeanblanc et al. [17] for an introduction to stochastic calculus. The infinitesimal generator of f is the operator $\mathscr{A}f = \mathbb{E}\left(\frac{\mathrm{d}f}{\mathrm{d}t} \mid \mathcal{F}_t\right)$. When $\delta_t = e_i$, this is given by

$$\mathscr{A}f(t, S, e_i) = \left(\frac{\partial f}{\partial t} + \mu_i S \frac{\partial f}{\partial S} + \frac{1}{2}\sigma_i^2 S^2 \frac{\partial^2 f}{\partial S^2}\right)$$

$$+ \sum_{j \neq i}^{n} q_{i,j} \int_{\mathbb{R}} \left(f(t, Se^z, e_j) - f(t, S, e_i) - (e^z - 1) S \frac{\partial f}{\partial S}\right) v_{i,j}(z)\mathrm{d}z.$$

Applying Itô's lemma for switching process to $\ln S_t$ leads to the following representation of stock prices

$$d \ln S_t = \left(\mu_t - \frac{1}{2}\sigma_t^2 \right) dt + \sigma_t dW_t \tag{1.7}$$

$$+ \sum_{i=1}^{n} \sum_{i \neq j}^{n} \left(J_{i,j} dN_{i,j}(t) - \lambda_{i,j}(t) \mathbb{E}\left(e^{J_{i,j}} - 1 \right) dt \right).$$

By direct integration, we infer that S_t is an exponential process:

$$S_t = S_0 \exp \left(\int_0^t \left(\mu_s - \frac{1}{2}\sigma_s^2 \right) ds + \int_0^t \sigma_s dW_s \right) \tag{1.8}$$

$$\times \exp \left(\sum_{i=1}^{n} \sum_{i \neq j}^{n} \left(\int_0^t J_{i,j}(s) \, dN_{i,j}(s) - \int_0^t \lambda_{i,j}(s) \mathbb{E}\left(e^{J_{i,j}(s)} - 1 \right) ds \right) \right).$$

Notice that in this last equation we use an integral representation of the jump process, that is

$$\int_0^t J_{i,j}(s) \, dN_{i,j}(s) = \sum_{k=1}^{N_{i,j}(t)} J_{i,j}(t_k),$$

where $J_{i,j}(t_k)$ is the kth jump caused by a transition of δ_t from i to j, at time t_k.

At this stage, we have not discussed the choice of the statistical distribution yet. If we do not consider any jumps during regime transitions, $J_{i,j} = 0$, the stock price process is a switching geometric Brownian motion (SGBM). An alternative consists in considering constant jumps, but this assumption is rather unrealistic. In theory, any statistical distribution defined on $\mathbb{R}$ is admissible. However, very few present sufficient analytical tractability for both options pricing and econometric estimation. A common practice consists in assuming that jumps are i.i.d. copies of a double exponential distribution, as in Kou [18] and Kou and Wang [19, 20]. A double exponential distributed random variable J may take positive or negative values. Its probability density function (defined on $\mathbb{R}$) is given by

$$\nu(z) = p\rho^+ e^{-\rho^+ z} 1_{\{z \geq 0\}} - (1-p)\rho^- e^{-\rho^- z} 1_{\{z < 0\}}, \tag{1.9}$$

while the associated cumulative distribution function is

$$\mathbb{P}[J \leq z] = (1-p)e^{-\rho^- z} 1_{\{z \leq 0\}} + \left[(1-p) + p\left(1 - e^{-\rho^+ z} \right) \right] 1_{\{z > 0\}}.$$

This distribution depends on three parameters: $\rho^+ \in \mathbb{R}^+$, $\rho^- \in \mathbb{R}^-$, and $p \in (0, 1)$, where p (resp., $(1 - p)$) denotes the probability of observing an upward (resp., downward) exponential jump, and $\frac{1}{\rho^+}$ (*resp.*, $\frac{1}{\rho^-}$) gives the size of an average positive (*resp.*, negative) jump. When only unidirectional jumps are considered, everything remains valid with $p = 1$ or $p = 0$, for positive and negative exponential jumps. The expected value of the size of jump (J) is the weighted sum of these average sizes: $\mathbb{E}(J) = p\frac{1}{\rho^+} + (1-p)\frac{1}{\rho^-}$. The moment generating function of J is given by

$$\psi(\omega) = \mathbb{E}\left(e^{\omega J}\right) = p\frac{\rho^+}{\rho^+ - \omega} + (1-p)\,\frac{\rho^-}{\rho^- - \omega}\,,$$

and therefore,

$$\mathbb{E}\left(e^J - 1\right) = \psi(\omega) - 1 = p\frac{1}{\rho^+ - 1} + (1-p)\left(\frac{1}{\rho^- - 1}\right).$$

In the following sections, we assume that random variables $J_{i,j}$ for $i \neq j$ have a double exponential distribution defined by a triplet of parameters $\left(\rho_{i,j}^+, \rho_{i,j}^-, p_{i,j}\right)$.

We conclude this section by studying the moment generating function of the log-return of S_t. We introduce some new notation. First, the drift of $d \ln S_t$ is a process denoted by $\tilde{\mu}_t$:

$$\tilde{\mu}_t = \delta_t^\top \tilde{\boldsymbol{\mu}}\,, \tag{1.10}$$

where $\tilde{\boldsymbol{\mu}}$ is a vector $(\tilde{\mu}_1, \ldots, \tilde{\mu}_n)^\top$ of $\mathbb{R}^n$ with

$$\tilde{\mu}_i = \mu_i - \frac{1}{2}\sigma_i^2 - \sum_{i \neq j}^n q_{i,j}\mathbb{E}\left(e^{J_{i,j}} - 1\right)$$

$$= \mu_i - \frac{1}{2}\sigma_i^2 - \sum_{i \neq j}^n q_{i,j}\left(\psi_{i,j}(1) - 1\right),$$

for $i \in \mathcal{N}$. The log-return $X_t := \ln\frac{S_t}{S_0}$ is given by

$$X_t = \int_0^t \tilde{\mu}_s ds + \int_0^t \sigma_s dW_s + \sum_{i=1}^n \sum_{i \neq j}^n \int_0^t J_{i,j}(s)\,dN_{i,j}(s)\,. \tag{1.11}$$

For a function $f(t, X_t, \delta_t)$ satisfying the smoothness conditions of Itô's lemma, its infinitesimal generator is

$$\mathscr{A}f(t, x, e_i) = \left(\frac{\partial f}{\partial t} + \frac{\partial f}{\partial X}\tilde{\mu}_i + \frac{1}{2}\frac{\partial^2 f}{\partial X^2}\sigma_i^2\right) \tag{1.12}$$

$$+ \sum_{j \neq i}^n \lambda_{i,j}(t)\int \left(f(t, x + z, e_j) - f(t, x, e_i)\right)\nu_{i,j}(z)dz\,.$$

We use these results to infer the moment generating function of X_s, which is used later for European option pricing:

Proposition 1.1 *The mgf of X_s for $s \geq t$ with $\omega \in \mathbb{C}_-$ is given by the following expression:*

$$\mathbb{E}\left(e^{\omega X_s} \mid \mathcal{F}_t\right) = \left(\frac{S_t}{S_0}\right)^{\omega} \exp\left(A(\omega, t, s, \delta_t)\right), \tag{1.13}$$

where $A(\omega, t, s, \cdot)$ is such that the vector of functions

$$\tilde{A}(\omega, t, s) = \left(\tilde{A}(\omega, t, s, e_i)\right)_{i \in \mathcal{N}}^{\top} := \left(e^{A(\omega, t, s, e_1)}, \ldots, e^{A(\omega, t, s, e_n)}\right)^{\top}$$

is a solution of the ODE system:

$$0 = \frac{\partial}{\partial t} \tilde{A}(\omega, t, s, e_i) + \left(\omega \tilde{\mu}_i + \omega^2 \frac{\sigma_i^2}{2}\right) \tilde{A}(\omega, t, s, e_i) \tag{1.14}$$

$$+ \sum_{\substack{j \neq i}}^{n} q_{i,j} \left(\tilde{A}(\omega, t, s, e_j) \psi_{i,j}(\omega) - \tilde{A}(\omega, t, s, e_i)\right)$$

with the terminal boundary condition:

$$\tilde{A}(\omega, s, s, e_i) = 1 \quad i \in \mathcal{N}.$$

Proof Let us define $f(t, X_t, \delta_t) = \mathbb{E}\left(e^{\omega X_s} \mid \mathcal{F}_t\right)$. The infinitesimal generator of this function is the operator defined by

$$\mathcal{A}f = \lim_{h \to 0} \frac{\mathbb{E}\left(f\left(t+h, X_{t+h}, \delta_{t+h}\right) - f(t, X_t, \delta_t) \mid \mathcal{F}_t\right)}{h}$$

and is equal to Eq. (1.12). Nevertheless, this limit is null by definition of f. Therefore, we infer that if $\delta_t = e_i$, the function f is the solution of the equation:

$$0 = \frac{\partial f}{\partial t} + \frac{\partial f}{\partial X} \tilde{\mu}_i + \frac{\partial^2 f}{\partial X^2} \frac{\sigma_i^2}{2} \tag{1.15}$$

$$+ \sum_{\substack{j \neq i}}^{n} q_{i,j}(t) \int \left(f(t, x + z, e_j) - f(t, x, e_i)\right) v_{i,j}(z) \mathrm{d}z.$$

Let us further assume that f is an exponential affine function of X_t:

$$f(t, X_t, e_i) = \exp\left(A(\omega, t, s, e_i) + B(\omega, t, s) X_t\right),$$

where $A(\omega, t, s, e_i)$ (for $i = 1, \ldots, n$) and $B(\omega, t, s)$ are time-dependent functions with terminal conditions $A(\omega, s, s, e_i) = 0$ and $B(\omega, s, s) = \omega$. The partial derivatives of f with respect to the state variables are given by

$$\frac{\partial f}{\partial t} = \left(\frac{\partial}{\partial t} A(\omega, t, s, e_i) + \frac{\partial}{\partial t} B(\omega, t, s) X_t \right) f,$$

$$\frac{\partial f}{\partial X} = B(\omega, t, s) f, \qquad \frac{\partial^2 f}{\partial X^2} = B(\omega, t, s)^2 f.$$

The last term in Eq. (1.15) can be written as follows:

$$\sum_{j \neq i}^{n} q_{i,j} \int \left(f(t, x + z, e_j) - f(t, x, e_i) \right) v_{i,j}(z) dz$$

$$= e^{B(\omega, t, s) X_t} \sum_{j \neq i}^{n} q_{i,j} \int \left(e^{A(\omega, t, s, e_j) + B(\omega, t, s) z} - e^{A(\omega, t, s, e_i)} \right) v_{i,j}(z) dz$$

$$= e^{B(\omega, t, s) X_t} \sum_{j \neq i}^{n} q_{i,j} \left(e^{A(\omega, t, s, e_j)} \psi_{i,j}(B(\omega, t, s)) - e^{A(\omega, t, s, e_i)} \right).$$

Injecting these expressions into Eq. (1.15) leads to the following relation:

$$0 = \left(\frac{\partial A}{\partial t} + \frac{\partial B}{\partial t} X_t \right) e^{A(\omega, t, s, e_i)} + B \tilde{\mu}_i e^{A(\omega, t, s, e_i)}$$

$$+ B^2 \frac{\sigma_i^2}{2} e^{A(\omega, t, s, e_i)} + \sum_{j \neq i}^{n} q_{i,j} \left(e^{A(\omega, t, s, e_j)} \psi_{i,j}(B(t, s)) - e^{A(\omega, t, s, e_i)} \right),$$

from which we infer that $B(\omega, t, s) = \omega$. Regrouping terms allows us to conclude that $A(\omega, t, s, e_i)$ for $i = 1, \ldots, n$ are solutions of a system of ODEs:

$$0 = \frac{\partial A}{\partial t} e^{A(\omega, t, s, e_i)} + \omega \tilde{\mu}_i e^{A(\omega, t, s, e_i)}$$

$$+ \omega^2 \frac{\sigma_i^2}{2} e^{A(\omega, t, s, e_i)} + \sum_{j \neq k}^{n} q_{i,j} \left(e^{A(\omega, t, s, e_j)} \psi_{i,j}(\omega) - e^{A(\omega, t, s, e_i)} \right).$$

If we define $\tilde{A}(t, s) = (e^{A(\omega, t, s, e_i)})_{i \in \mathcal{N}}$, this last equation is rewritten as follows:

$$0 = \frac{\partial}{\partial t} \tilde{A}(\omega, t, s, e_i) + \left(\omega \tilde{\mu}_i + \omega^2 \frac{\sigma_i^2}{2} \right) \tilde{A}(\omega, t, s, e_i)$$

$$+ \sum_{j \neq k}^{n} q_{i,j} \left(\tilde{A}(\omega, t, s, e_j) \psi_{i,j}, j(\omega) - \tilde{A}(\omega, t, s, e_i) \right).$$

$\square$

The moment generating function of X_t may be inverted numerically by a discrete Fourier transform (DFT) to estimate parameters from a time series of stock returns. However, an alternative, more efficient approach to econometric estimation of the model will be presented in the next section. First, we show that this moment generating function admits a closed form expression if there is no transition jump.

Proposition 1.2 *Let us define an $n \times n$ matrix:*

$$D(\omega) = Q_0^\top + \text{diag} \begin{pmatrix} \omega \left(\mu_1 - \frac{\sigma_1^2}{2} \right) + \omega^2 \frac{\sigma_1^2}{2} \\ \vdots \\ \omega \left(\mu_n - \frac{\sigma_n^2}{2} \right) + \omega^2 \frac{\sigma_n^2}{2} \end{pmatrix}. \tag{1.16}$$

If $J_{i,j} = 0$, $(S_t)_{t \geq 0}$ is a switching regime diffusion, and the mgf of X_s for $s \geq t$ with $\omega \in \mathbb{C}_-$ is given by the following expression:

$$\mathbb{E} \left(e^{\omega X_s} \mid \mathcal{F}_t \right) = \mathbb{E} \left(\left(\frac{S_s}{S_0} \right)^\omega \mid \mathcal{F}_t \right) \tag{1.17}$$

$$= \left(\frac{S_t}{S_0} \right)^\omega \mathbf{1}_n^\top \exp \left(D(\omega)(s - t) \right) \delta_t ,$$

where $\mathbf{1}_n = (1, \ldots, 1)^\top$ is an n-vector of ones.

Proof This result relies on properties of the matrix exponential. If D is an $n \times n$ matrix, the matrix exponential of D is the infinite sum $\sum_{k=0}^{\infty} \frac{(D)^k}{k!}$. From this definition, we immediately infer that for any $t \in \mathbb{R}^+$

$$\frac{\mathrm{d}}{\mathrm{d}t} \exp(t D) = D \exp(t D) = \exp(t D) D . \tag{1.18}$$

On the other hand, the matrix exponential of a diagonal matrix is the matrix of exponentials. If $u = (u_1, \ldots, u_n)^\top$ is a vector of dimension n, then

$$\exp \left(\text{diag}(u) \right) = \text{diag} \left(\left(e^{u_1}, \ldots, e^{u_n} \right) \right) .$$

In the notation of Proposition 1.1, we have

$$\tilde{A}(\omega, t, s, \delta_t) = \mathbf{1}_n^\top \exp\left(D(\omega)(s - t)\right) \delta_t \,.$$

Differentiating with respect to time, from the property (1.18), we infer that

$$\frac{\partial \tilde{A}(\omega, t, s, e_i)}{\partial t} = -\mathbf{1}_n^\top \left(D(\omega) \exp\left(D(\omega)(t - s)\right) e_i\right)$$

$$= -\sum_j^n q_{i,j} \tilde{A}(\omega, t, s, e_j) - \left(\omega\left(\mu_i - \frac{\sigma_i^2}{2}\right) + \omega^2 \frac{\sigma_i^2}{2}\right) \tilde{A}(\omega, t, s, e_i) \,.$$

On the other hand, we have that $q_{i,i} = -\sum_{j \neq i}^n q_{i,j}$. We can then rewrite this last expression as follows:

$$\frac{\partial \tilde{A}(\omega, t, s, e_i)}{\partial t} = -\left(\omega\left(\mu_i - \frac{\sigma_i^2}{2}\right) + \omega^2 \frac{\sigma_i^2}{2}\right) \tilde{A}(\omega, t, s, e_i)$$

$$-\sum_{j \neq i}^n q_{i,j} \left(\tilde{A}(\omega, t, s, e_j) - \tilde{A}(\omega, t, s, e_i)\right) \,,$$

and we see that Eq. (1.14) of Proposition 1.1 is satisfied since $\tilde{\mu}_i = \mu_i - \frac{\sigma_i^2}{2}$ in this case.   □

1.3 A Modified Hamilton Filter for Estimation

When a model is used for option pricing, the best parameter estimates are those that replicate derivative prices. The set of parameters obtained in this manner are said to be "risk neutral" and in theory exclude arbitrages. In risk management, models are needed for simulating future asset prices. In this context, the parameter estimates are those explaining at best the evolution of historical prices. In this case, the parameters are appraised under the real probability measure $\mathbb{P}$. We will come back to options valuation later. In this section, we instead propose an enhanced version of the Hamilton filter (Hamilton [14]) to calibrate under $\mathbb{P}$ the SJD model with double exponential transition jumps. This procedure requires the following result:

Proposition 1.3 *Let us consider a double exponential random variable J, with parameters $\left(\rho^+, \rho^-, p\right)$ and pdf as defined in Eq. (1.9). The probability density function of the sum $J + \sigma_t W_\Delta$, where Δ is a time interval and when $\delta_t = e_i$, is* *equal to*

$$g(z \mid \delta_t = e_i) = p\rho^+ \exp\left(\frac{1}{2}\left(\rho^+\right)^2 \sigma_i^2 \Delta - \rho^+ z\right) \Phi\left(\frac{z - \rho^+ \sigma_i^2 \Delta}{\sqrt{\Delta}\sigma_i}\right) \quad (1.19)$$

$$- (1-p)\,\rho^- \exp\left(\frac{1}{2}\left(\rho^-\right)^2 \sigma_i^2 \Delta - \rho^- z\right)\left(1 - \Phi\left(\frac{z - \rho^- \sigma_i^2 \Delta}{\sqrt{\Delta}\sigma_i}\right)\right)$$

for $z \in \mathbb{R}$ and where $\Phi(\cdot)$ is the cdf of a standard normal random variable $N(0, 1)$.

Proof $g(z \mid \delta_t = e_i)$ is the convolution of densities of J and $\sigma_i W_\Delta$ (denoted by $\tilde{f}$),

$$g(z \mid \delta_t = e_i) = \int_{-\infty}^{+\infty} v(u)\tilde{f}(z - u)\mathrm{d}u$$

$$= p\rho^+ \int_0^{+\infty} e^{-\rho^+ u}\,\frac{1}{\sqrt{2\pi\,\Delta}\sigma_i}\exp\left(-\frac{1}{2}\frac{(z-u)^2}{\sigma_i^2 \Delta}\right)\mathrm{d}u$$

$$- (1-p)\,\rho^- \int_{-\infty}^0 e^{-\rho^- u}\,\frac{1}{\sqrt{2\pi\,\Delta}\sigma_i}\exp\left(-\frac{1}{2}\frac{(z-u)^2}{\sigma_i^2 \Delta}\right)\mathrm{d}u,$$

which can be rewritten as follows:

$$g(z \mid \delta_t = e_i) = \frac{p\rho^+}{\sqrt{2\pi\,\Delta}\sigma_i} \int_0^{+\infty} \exp\left(-\frac{1}{2}\frac{(z-u)^2 + 2\rho^+ \sigma_i^2 \Delta u}{\sigma_i^2 \Delta}\right)\mathrm{d}u \quad (1.20)$$

$$- \frac{(1-p)\,\rho^-}{\sqrt{2\pi\,\Delta}\sigma_i} \int_{-\infty}^0 \exp\left(-\frac{1}{2}\frac{(z-u)^2 + 2\rho^- \sigma_i^2 \Delta u}{\sigma_i^2 \Delta}\right)\mathrm{d}u.$$

Given that

$$(z-u)^2 + 2\rho^+ \sigma_i^2 \Delta u = \left((z-u) - \rho^+ \sigma_i^2 \Delta\right)^2 - \left(\rho^+ \sigma_i^2 \Delta\right)^2 + 2\rho^+ \sigma_i^2 \Delta z,$$

the first integral in Eq. (1.20) becomes

$$\frac{1}{\sqrt{2\pi\,\Delta}\sigma_i} \int_0^{+\infty} \exp\left(-\frac{1}{2}\frac{(z-u)^2 + 2\rho^+ \sigma_i^2 \Delta u}{\sigma_i^2 \Delta}\right)\mathrm{d}u \quad (1.21)$$

$$= \frac{\exp\left(\frac{1}{2}\left(\rho^+\right)^2 \sigma_i^2 \Delta - \rho^+ z\right)}{\sqrt{2\pi\,\Delta}\sigma_i} \int_0^\infty \exp\left(-\frac{1}{2}\frac{\left((z-u) - \rho^+ \sigma_i^2 \Delta\right)^2}{\sigma_i^2 \Delta}\right)\mathrm{d}u.$$

Using the substitution

$$v = \left((z - u) - \rho^+ \sigma_i^2 \Delta\right)$$

implies that $u = \left((z - v) - \rho^+ \sigma_i^2 \Delta\right)$ and $du = -dv$. Moreover, if $u = 0$, then $v = z - \rho^+ \sigma_i^2 \Delta$ and when $u = +\infty$, $v = -\infty$. As a consequence, the integral in Eq. (1.21) becomes

$$\frac{1}{\sqrt{2\pi}\Delta\sigma_i} \int_0^\infty \exp\left(-\frac{1}{2}\frac{\left((z - u) - \rho^+ \sigma_i^2 \Delta\right)^2}{\sigma_i^2 \Delta}\right) du$$

$$= \frac{1}{\sqrt{2\pi}\Delta\sigma_i} \int_{-\infty}^{z - \rho^+ \sigma_i^2 \Delta_t} \exp\left(-\frac{1}{2}\frac{v^2}{\Delta\sigma_i^2}\right) dv$$

$$= \Phi\left(\frac{z - \rho^+ \sigma_i^2 \Delta}{\sqrt{\Delta}\sigma_i}\right),$$

where $\Phi(\cdot)$ is the cdf of an $N(0, 1)$ random variable. On the other hand, the second integral in Eq. (1.20) is equal to

$$\frac{1}{\sqrt{2\pi}\Delta\sigma_i} \int_{-\infty}^0 \exp\left(-\frac{1}{2}\frac{(z - u)^2 + 2\rho^- \sigma_i^2 \Delta u}{\Delta\sigma_i^2}\right) du \qquad (1.22)$$

$$= \frac{\exp\left(\frac{1}{2}(\rho^-)^2 \sigma_i^2 \Delta - \rho^- z\right)}{\sqrt{2\pi}\Delta\sigma_i} \int_{-\infty}^0 \exp\left(-\frac{1}{2}\frac{\left((z - u) - \rho^- \sigma_i^2 \Delta\right)^2}{\Delta\sigma_i^2}\right) du .$$

Analogously, using the substitution $v = \left((z - u) - \rho^- \sigma_i^2 \Delta\right)$ implies that $u = \left((z - v) - \rho^- \sigma_i^2 \Delta\right)$ and $du = -dv$. Moreover, if $u = 0$, then $v = z - \rho^- \sigma_i^2 \Delta$, and when $u = -\infty$, $v = +\infty$. Therefore, the integral in Eq. (1.22) turns out to equal

$$\frac{1}{\sqrt{2\pi}\Delta\sigma_i} \int_{-\infty}^0 \exp\left(-\frac{1}{2}\frac{\left((z - u) - \rho^- \sigma_i^2 \Delta\right)^2}{\Delta\sigma_i^2}\right) du$$

$$= \frac{-1}{\sqrt{2\pi}\Delta\sigma_i} \int_{+\infty}^{z - \rho^- \sigma_i^2 \Delta} \exp\left(-\frac{1}{2}\frac{v^2}{\Delta\sigma_i^2}\right) dv$$

$$= \frac{1}{\sqrt{2\pi}\Delta\sigma_i} \int_{z-\rho^-\sigma_i^2\Delta}^{+\infty} \exp\left(-\frac{1}{2}\frac{v^2}{\Delta\sigma_i^2}\right) dv$$

$$= \left(1 - \Phi\left(\frac{z - \rho^-\sigma_i^2\Delta}{\sqrt{\Delta}\sigma_i}\right)\right).$$

Combining these expressions (1.20)–(1.22) leads to the result. $\qquad\square$

In the rest of this section, we denote by $x = \{x_1, \ldots, x_T\}$ the time series of log-returns of a financial asset, measured at times $t_1, \ldots, t_T$ equally spaced by Δ (which is not necessarily equal to the Δ_t involved in the definition of Q_0 for δ_t):

$$x_i = \ln\left(\frac{S_{t_{i-1}+\Delta}}{S_{t_{i-1}}}\right) \quad i = 1, \ldots, T.$$

We assume that the Markov chain δ_t only changes regime at times t_i for $i = 1, \ldots, T$. The Markov chain is also hidden in the sense that we only observe stock prices. The available information is contained in the filtration $\mathcal{H}_t$. If the economy stays in the jth state over the period of time $[t_{i-1}, t_i]$, the log-return is normally distributed $X_i \sim N(\tilde{\mu}_j\Delta, \sigma_j\sqrt{\Delta})$. If the system switches from regime i to j, the density of the log-return is equal to $g_{i,j}(z \mid \delta_t = e_i)$ given by Eq. (1.19) and parameters $\left(\rho_{i,j}^+, \rho_{i,j}^-, p_{i,j}\right)$ for the distribution of $J_{i,j}$. We denote by

$$\Theta = \left\{\bar{\mu}, Q_0, \bar{\sigma}, \left(\rho_{i,j}^+, \rho_{i,j}^-, p_{i,j}\right)_{i\neq j\in\mathcal{N}}\right\}$$

the set of parameters of the SJD model. Δ_t and the number of states, n, are not considered as parameters and are chosen a priori. We denote by $x_{1:k} = \{x_1, \ldots, x_k\}$ the time series of observations up to time t_k. Using Bayes' rule, we reformulate the log-likelihood of observed returns as follows:

$$\ln f(x \mid \Theta) = \ln f(x_1|\Theta) + \ln f(x_2|\Theta, x_1) \qquad (1.23)$$
$$+ \ln f(x_3|\Theta, x_{1:2}) + \cdots + \ln f(x_T|\Theta, x_{1:T-1}),$$

where $f(x_k|\Theta, x_{1:k-1})$ is the density function of the return on the kth period, for parameters Θ, and conditional on previous observations $x_1, \ldots, x_{k-1}$. The parameters are estimated by maximizing this log-likelihood function. Therefore, we concentrate on the terms in the right-hand side of this log-likelihood. Conditioning upon the state of δ_k allows us to infer that $f(x_k|\Theta, x_{1:k-1})$ is equal to

$$f(x_k|\Theta, x_{1:k-1}) = \sum_{i=1}^{n}\sum_{j=1}^{n} p_i(t_{k-1}|\Theta, x_{1:k-1}) \, p_{i,j}(t_{k-1}, t_k|\Theta)$$

$$\times f(x_k|\Theta, \delta_{t_k} = e_j, \delta_{t_{k-1}} = e_i),$$

where:

- $f(x_k | \Theta, \delta_{t_k} = e_j, \delta_{t_{k-1}} = e_i)$ is either:

  – The Gaussian density of the return in state i, $N(\tilde{\mu}_i \Delta, \sigma_i \sqrt{\Delta})$, or
  – $g_{i,j}(z - \tilde{\mu}_i \Delta \mid \delta_t = e_i)$ with $g_{i,j}(. \mid \delta_t = e_i)$ the probability density function of the sum $J_{i,j} + \sigma_i W_\Delta$ as given by Eq. (1.19).

- $p_{i,j}(t_{k-1}, t_k | \Theta)$ is the probability of transition from state i at time t_{k-1} to state j at time t_k for the set of parameters Θ.
- $p_i(t_{k-1} | \Theta, x_{1:k-1})$ is the probability of presence in state i at time t_{k-1}, conditional on all observations up to t_{k-1}.

Again using Bayes' rule, the probability $p_i(t_{k-1} | \Theta, x_{1:k-1})$ is recursively computed with $f(x_{k-1} | \Theta, x_{1:k-2})$ as follows:

$$p_i(t_{k-1} | \Theta, x_{1:k-1}) = \frac{\begin{array}{c} \sum_{j=1}^n p_j(t_{k-2} | \Theta, x_{1:k-2}) \, p_{j,i}(t_{k-2}, t_{k-1} | \Theta) \\ \times f(x_{k-1} | \Theta, \delta_{t_{k-1}} = e_i, \delta_{t_{k-2}} = e_j) \end{array}}{f(x_{k-1} | \Theta, x_{1:k-2})}.$$

In order to initiate the recursion, we need to determine $f(x_1 | \Theta)$. If the Markov chain has been running for a sufficiently long period of time, we assume that the probability of presence in a given state is equal to its stationary probability, denoted $\pi_i(\Theta)$ for $i = 1, \ldots, n$. Then, we infer that

$$f(x_1 | \Theta) = \sum_{i=1}^n \sum_{j=1}^n \pi_i(\Theta) \, p_{i,j}(t_0, t_1 | \Theta) \, f\left(x_1 | \Theta, \delta_{t_1} = e_j, \delta_{t_0} = e_i\right).$$

Therefore, the log-likelihood as defined by Eq. (1.23) is evaluated by recursion and maximized numerically to estimate the parameters. We use the parameter estimates, denoted $\widehat{\Theta}$, to filter the states through which the Markov chain transits by the relation:

$$\mathbb{E}\left(\delta_{t_k}^\top \begin{pmatrix} 1 \\ \vdots \\ n \end{pmatrix} \middle| \mathcal{H}_{t_k} \right) = \sum_{i=1}^n p_i(t_k | \widehat{\Theta}, x_1, \ldots x_k) \, i.$$

The estimate of the log-likelihood, $\ln f(x_1, \ldots, x_T | \widehat{\Theta})$, is used to evaluate two other important criteria to judge the overall quality of the model and the optimal number of regimes. The first one is the *Akaike information criterion (AIC)*. If m is the number of parameters in the model, then the AIC value of the model is defined by

$$\mathrm{AIC} = 2m - 2 \ln f(x | \widehat{\Theta}).$$

Given a set of candidate models, the preferred model is the one with the lowest AIC. The AIC rewards goodness of fit (assessed by the likelihood function), but it also penalizes models with a large number of parameters. The second criterion is the *Bayesian information criterion (BIC)*:

$$BIC = \ln(n)m - 2\ln f(\boldsymbol{x}|\widehat{\Theta}).$$

Like the AIC, the model with the lowest BIC is preferred, and the BIC also penalizes models with too many parameters. However, the penalty term is larger in BIC than in AIC.

Using the asymptotic property of the log-likelihood, we can perform a hypothesis test in order to compare two models with different numbers of regimes. This test is done with a log-likelihood ratio test (LR) of the two models against each other.

Let us consider two models M_r and M_s with, respectively, r and s parameters and such that $M_s \subset M_r$. Let $\widehat{\Theta^r}$ and $\widehat{\Theta^s}$ be the parameter estimates for models M_r and M_s, respectively. The LR statistic for testing M_s against M_r is equal to

$$LR\left(\widehat{\Theta^r}, \widehat{\Theta^s}\right) = 2\left(\ln f\left(\boldsymbol{x}|\widehat{\Theta^r}\right) - \ln f\left(\boldsymbol{x}|\widehat{\Theta^r}\right)\right).$$

This difference is approximately χ^2 distributed with $r - s$ degrees of freedom.

1.4 Numerical Illustration

To illustrate this chapter, we fit the SJD and SGBM models with the modified Hamilton filter to the time series of the S&P 500 stock index from 31/1/2001 to 31/1/2020 (4779 observations). Over this window of time, the S&P 500 alternates between periods of stable growth and moderate or deep crisis. In 2003, we observe a depreciation of the index due to the second Gulf war. In 2008, the S&P 500 collapses during the subprime crisis (Fig. 1.1).

We estimate four models: the SGBM and SJD with two and three regimes. Table 1.1 reports some statistics of goodness of fit. The first and second columns present the log-likelihoods (Log. Like.) and the number of parameters (# par.) of each model, whereas the last two columns contain the AIC and BIC. The highest likelihood and the lowest AIC/BIC are obtained by the switching jump- diffusion and three regimes. This model counts 30 parameters that are shown in Tables 1.2 and 1.3. The first regime corresponds to a deep financial turmoil: the volatility is above 40%, whereas the return is below -40%! The third state may be identified with economic growth: the average stock return climbs up to 24%, while the volatility falls to 8%. The second regime is an intermediate state in which the economy is in light recession. Transition jumps from state 2 or 3 to regime 1 are mainly negative.

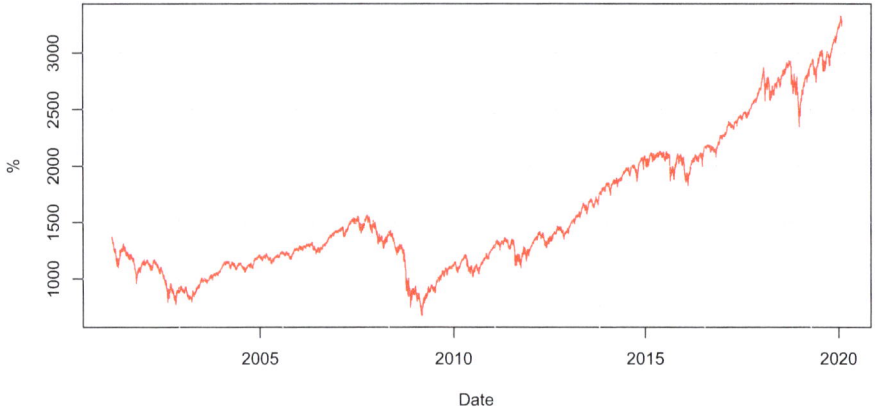

Fig. 1.1 S&P 500 daily values from 21/1/2001 to 21/1/2020

Table 1.1 Comparison of SJD and SGBM models with two and three regimes

	Log. like.	# par.	AIC	BIC
SGBM 2D	15,464	6	−30,916	−30,877
SGBM 3D	15,687	12	−31,350	−31,272
SJD 2D	15,497	12	−30,982	−30,943
SJD 3D	15,706	30	−31,387	−31,310

Table 1.2 3D SJD model: matrix of transition probabilities over 1 day

$p_{i,j}\,(t, t+1\,\text{day})$	1	2	3
1	0.97	0.03	0.00
2	0.01	0.96	0.03
3	0.00	0.02	0.98

Table 1.3 Parameter estimates of the 3D SJD model

3D SJD model					
μ_1	−42.89%	μ_2	−4.10%	μ_3	24.35%
σ_1	43.23%	σ_2	18.12%	σ_3	8.54%
$p_{1,1}$	–	$p_{1,2}$	0.82	$p_{1,3}$	0.60
$p_{2,1}$	0.09	$p_{2,2}$	–	$p_{2,3}$	1.00
$p_{3,1}$	0.32	$p_{3,2}$	0.01	$p_{3,3}$	–
$\rho_{1,1}^{+}$	–	$\rho_{1,2}^{+}$	94.39	$\rho_{1,3}^{+}$	172.00
$\rho_{2,1}^{+}$	535.71	$\rho_{2,2}^{+}$	–	$\rho_{2,3}^{+}$	15.15
$\rho_{3,1}^{+}$	301.58	$\rho_{3,2}^{+}$	4440.06	$\rho_{3,3}^{+}$	–
$\rho_{1,1}^{-}$	–	$\rho_{1,2}^{-}$	−220.40	$\rho_{1,3}^{-}$	−122.13
$\rho_{2,1}^{-}$	−66.99	$\rho_{2,2}^{-}$	–	$\rho_{2,3}^{-}$	−600.74
$\rho_{3,1}^{-}$	−83.72	$\rho_{3,2}^{-}$	−98.42	$\rho_{3,3}^{-}$	–

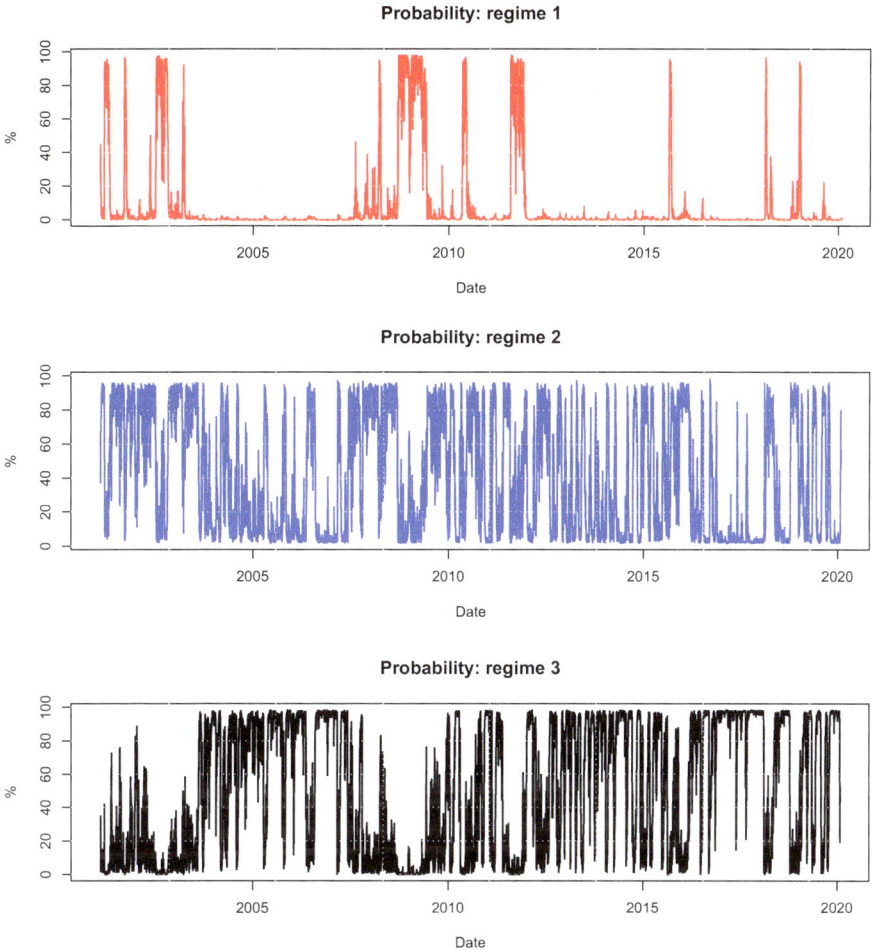

Fig. 1.2 These graphs show the probability of presence in each regime over the period 2001–2020

Figure 1.2 shows the probability of presence in each regime over the period 2001–2020. We clearly observe that the hidden Markov chain is in regime 1 during the major economic crisis of these last two decades. In 2003, the second Gulf war drives down the S&P 500 over several months. The chain is also in the first regime during the subprime crisis of 2008 and the second dip of the double dip recession in 2011. In 2015, the chain briefly enters into regime 1 during the stock market selloff caused by turbulences on the Chinese market. We also observe the first regime in 2018 when oil price falls due to overproduction.

1.5 Change of Measure

In this section, we present a family of measure changes for point processes $Z_{i,j}(t) := \sum_{k=1}^{N_{i,j}(t)} J_{i,j}(k)$. Let $v_{i,j}^b(\cdot)$ be a probability density defined on the same domain as $v_{i,j}(\cdot)$, the pdf of the jumps J under $\mathbb{P}$. We define the following log-ratio:

$$\phi_{i,j}(h,u) := \ln\left(h\,\frac{v_{i,j}^b(u)}{v_{i,j}(u)}\right), \tag{1.24}$$

where $h \in \mathbb{R}^+$ and u is in the support of $v_{i,j}(\cdot)$. For any $h_{i,j} \in \mathbb{R}^+$ such that $h_{i,j} > -1$ for $i, j \in N$, the compensated jump process

$$M_{i,j}(t) = \sum_{k=1}^{N_{i,j}(t)} \left(e^{\phi_{i,j}(h_{i,j},J_{i,j}(k))} - 1\right)$$

$$- \int_0^t \lambda_{i,j}(s)\mathbb{E}\left(e^{\phi_{i,j}(h_{i,j},J_{i,j}(k))} - 1 \mid \mathcal{F}_0\right) ds \tag{1.25}$$

is a martingale by construction. But the expectation in the integral of Eq. (1.25) is equal to

$$\mathbb{E}\left(e^{\phi_{i,j}(h_{i,j},J_{i,j})} - 1 \mid \mathcal{F}_0\right) = h_{i,j}\int_{-\infty}^{\infty}\left(\frac{v_{i,j}^b(u)}{v_{i,j}(u)}\right)v_{i,j}(u)\mathrm{d}u - 1,$$

which allows us to simplify the definition of $M_{i,j}(t)$ as follows:

$$M_{i,j}(t) = \sum_{k=1}^{N_{i,j}(t)}\left(e^{\phi_{i,j}(h_{i,j},J_{i,j}(k))} - 1\right) - \int_0^t \lambda_{i,j}(s)\left(h_{i,j} - 1\right)ds. \tag{1.26}$$

We will later see that $h_{i,j}$ is involved in the definition of a risk neutral measure. Note that one could consider the generalization in which $h_{i,j}$ is an $\mathcal{F}_t$-adapted process but, in this case, the process δ_t would no longer be a Markov chain under the equivalent measures.

In the next proposition, we will define an interesting family of equivalent measures and the law of the point process $Z_{i,j}(t)$ under the new measure. This result follows from Girsanov's theorem for semi-martingales, but for completeness we sketch the proof in this specific case.

Proposition 1.4 *Let $Z_{i,j}(t)$, $i, j \in N$, be the point processes defined by*

$$Z_{i,j}(t) := \sum_{k=1}^{N_{i,j}(t)} J_{i,j}(k).$$

If $h_{i,j} > -1$, the processes $L_{i,j}(t)$, $i, j \in \mathcal{N}$, defined as follows:

$$L_{i,j}(t) = \exp\left(\int_0^t \phi_{i,j}\left(h_{i,j}, J_{i,j}(s)\right) dN_{i,j}(s) - \int_0^t \lambda_{i,j}(s)\left(h_{i,j} - 1\right) ds \right),$$

are Radon–Nikodym derivatives $\frac{d\mathbb{P}^b}{d\mathbb{P}}$ from the real measure $\mathbb{P}$ to a new probability measure $\mathbb{P}^b$. Under $\mathbb{P}^b$, $Z_{i,j}(t)$ is still a point process, but its dynamics are modified as follows:

$$Z_{i,j}(t) = \sum_{k=1}^{N_{i,j}^h(t)} J_{i,j}^b(k), \tag{1.27}$$

where $N_{i,j}^h(t)$ is a counting process of intensity $\lambda_{i,j}(t)h_{i,j}$ and $J_{i,j}^b$ are i.i.d. jumps with pdf $v_{i,j}^b(u)$. According to the definition (1.3) of $\lambda_{i,j}(t)$, the matrix of transition probabilities of $(\delta_t)_{t\geq 0}$ under $\mathbb{Q}$ is equal to $Q_0^{\mathbb{Q}} = \left(q_{i,j}h_{i,j}\right)_{i,j\in\mathcal{N}}$.

Proof From Eq. (1.26), $M_{i,j}(t)$ is a martingale satisfying the SDE:

$$dM_{i,j}(t) = \left(e^{\phi_{i,j}(h_{i,j}, J_{i,j})} - 1 \right) dN_{i,j}(t) - \lambda_{i,j}(t)\left(h_{i,j} - 1\right) dt.$$

We can construct a martingale $L_{i,j}(t)$ with geometric dynamics given by

$$dL_{i,j}(t) = L_{i,j}(t)\, dM_{i,j}(t)$$
$$= L_{i,j}(t)\left(e^{\phi_{i,j}(h_{i,j}, J_{i,j})} - 1 \right) dN_{i,j}(t) - L_{i,j}(t)\lambda_{i,j}(t)\left(h_{i,j} - 1\right) dt.$$

If we apply Itô's lemma to the function $\ln L_{i,j}(t)$, the differential of $\ln L_{i,j}(t)$ is equal to

$$d \ln L_{i,j}(t) = \phi_{i,j}\left(h_{i,j}, J_{i,j}\right) dN_{i,j}(t) - \lambda_{i,j}(t)\left(h_{i,j} - 1\right) dt,$$

from which we infer the expression of $L_{i,j}(t)$ by direct integration. The expectation of $Z_{i,j}(t)$ under the measure $\mathbb{P}^b$, defined by the Radon–Nikodym derivative $L_{i,j}(t)$, is given by

$$\mathbb{E}^{P^b}\left(e^{uZ_{i,j}(t)} | \mathcal{F}_0 \right) \tag{1.28}$$
$$= \mathbb{E}\left(e^{\int_0^t (uJ_{i,j}(s) + \phi_{i,j}(h_{i,j}, J_{i,j}))dN_{i,j}(s) - \int_0^t \lambda_{i,j}(s)(h_{i,j} - 1)ds} | \mathcal{F}_0 \right).$$

If the filtrations of $N_{i,j}(t)$ and $\lambda_{i,j}(t)$ are momentaneously denoted by $\mathcal{G}_t^{i,j} \subset \mathcal{F}_t$ and $\mathcal{H}_t^{i,j} \subset \mathcal{F}_t$, using nested expectations allows us to rewrite the expectation (1.28) as follows:

$$\mathbb{E}\left(e^{\int_0^t (uJ_{i,j}(s)+\phi_{i,j}(h_{i,j},J_{i,j}))\mathrm{d}N_{i,j}(s)-\int_0^t \lambda_{i,j}(s)(h_{i,j}-1)\mathrm{d}s}|\mathcal{F}_0\right) \quad\quad (1.29)$$

$$= \mathbb{E}\left(e^{-\int_0^t \lambda_{i,j}(s)(h_{i,j}-1)\mathrm{d}s}\mathbb{E}\left(\prod_{k=1}^{N_{i,j}(t)} Y_{i,j}|\mathcal{H}_t^{i,j}\vee\mathcal{F}_0\right)|\mathcal{F}_0\right),$$

where $Y_{i,j} = \mathbb{E}\left(e^{uJ_{i,j}+\phi_{i,j}(h_{i,j},J_{i,j})}|\mathcal{G}_t^{i,j}\vee\mathcal{H}_t^{i,j}\vee\mathcal{F}_0\right)$. By definition of $\phi(\cdot)$, we have that

$$Y_{i,j} = \int h_{i,j}e^{uz}v_{i,j}^b(z)\mathrm{d}z = h_{i,j}\mathbb{E}\left(e^{uJ_{i,j}^b}\right).$$

Furthermore, conditionally to $\mathcal{H}_t^{i,j}\vee\mathcal{H}_0$, $N_{i,j}(t)$ is an inhomogeneous Poisson process with the following moment generating function:

$$\mathbb{E}\left(\prod_{k=1}^{N_{i,j}(t)} h_{i,j}\mathbb{E}\left(e^{uJ_{i,j}^b}\right)|\mathcal{H}_t^{i,j}\vee\mathcal{F}_0\right)$$

$$= \mathbb{E}\left(e^{N_{i,j}(t)\ln\left[h_{i,j}\mathbb{E}\left(e^{uJ_{i,j}^b}\right)\right]}|\mathcal{H}_t^{i,j}\vee\mathcal{F}_0\right)$$

$$= \exp\left(\int_0^t \lambda_{i,j}(s)\left(h_{i,j}\mathbb{E}\left(e^{uJ_{i,j}^b}\right)-1\right)\mathrm{d}s\right).$$

We infer from this last equation that the expectation (1.29) is equal to

$$\mathbb{E}\left(e^{\int_0^t (uJ_{i,j}(s)+\phi_{i,j}(h_{i,j},J_{i,j}))\mathrm{d}N_{i,j}(s)-\int_0^t \lambda_{i,j}(s)(h_{i,j}-1)\mathrm{d}s}|\mathcal{F}_0\right)$$

$$= \mathbb{E}\left(e^{-\int_0^t \lambda_{i,j}(s)(h_{i,j}-1)\mathrm{d}s+\int_0^t \lambda_{i,j}(s)\left(h_{i,j}\mathbb{E}\left(e^{uJ_{i,j}^b}\right)-1\right)\mathrm{d}s}|\mathcal{F}_0\right)$$

$$= \mathbb{E}\left(\exp\left(\int_0^t h_{i,j}\lambda_{i,j}(s)\left(\mathbb{E}\left(e^{uJ_{i,j}^b}\right)-1\right)\mathrm{d}s\right)|\mathcal{F}_0\right),$$

which turns out to be the moment generating function of $Z_{i,j}(t)$ under the equivalent measure $\mathbb{P}^b$.   □

To avoid arbitrage opportunities, financial derivatives are priced under an equivalent risk neutral measure under which discounted (non-dividend paying) asset prices are martingales. In the remainder of this section, we consider a financial market composed of two assets: a risk-free cash account and a stock. The interest rate depends on δ_t and is defined as $r_t = \delta_t^\top \, r$, where $r = (r_1, \ldots, r_n)^\top \in \mathbb{R}^n$. The stock price S_t is defined by Eq. (1.8). By construction, the risk neutral measure is not unique. We consider Radon–Nikodym derivatives of the following form:

$$L_t = \prod_{i,j=1}^{n} \exp\left(\int_0^t \phi_{i,j}\left(h_{i,j}, J_{i,j}(s)\right) \mathrm{d}N_{i,j}(s) - \int_0^t \lambda_{i,j}(s)\left(h_{i,j} - 1\right) \mathrm{d}s \right)$$

$$\times \exp\left(-\frac{1}{2}\int_0^t \beta_s^2 \mathrm{d}s - \int_0^t \beta_s \mathrm{d}W_s \right), \tag{1.30}$$

where $(\beta_t)_{t\geq 0}$ is an $\mathcal{F}_t$-measurable process. The first factor in the definition of this Radon–Nikodym derivative implies that

$$\mathrm{d}W_t^\beta = \mathrm{d}W_t + \beta_t \mathrm{d}t$$

is a Brownian motion under the new equivalent measure. The next proposition establishes the dynamics of S_t under such a new martingale measure, which will be denoted by $\mathbb{Q}$.

Proposition 1.5 *The dynamics of the asset price under the equivalent measure $\mathbb{Q}$ defined by the Radon–Nikodym derivative (1.30) equals*

$$\frac{\mathrm{d}S_t}{S_t} = (\mu_t - \sigma_t \beta_t)\,\mathrm{d}t + \sigma_t \mathrm{d}W_t^\beta \tag{1.31}$$

$$+ \sum_{i=1}^{n}\sum_{j\neq i}^{n} \lambda_{i,j}(t) \left(h_{i,j}\mathbb{E}\left(\mathrm{e}^{J_{i,j}^b} - 1\right)\mathrm{d}t - \mathbb{E}\left(\mathrm{e}^{J_{i,j}} - 1\right)\mathrm{d}t \right)$$

$$+ \sum_{i=1}^{n}\sum_{j\neq i}^{n} \left(\left(\mathrm{e}^{J_{i,j}^b} - 1\right)\mathrm{d}N_{i,j}^h(t) - h_{i,j}\lambda_{i,j}(t)\mathbb{E}\left(\mathrm{e}^{J_{i,j}^b} - 1\right)\mathrm{d}t \right).$$

Proof We temporarily denote $\ln S_t/S_0$ by Y_t. From Eq. (1.7), we infer that

$$\mathrm{d}Y_t = \left(\mu_t - \frac{1}{2}\sigma_t^2 - \sigma_t \beta_t \right)\mathrm{d}t + \sigma_t(\mathrm{d}W_t + \beta_t \mathrm{d}t)$$

$$+ \sum_{i=1}^{n}\sum_{j\neq i}^{n} \lambda_{i,j}(t)\left(h_{i,j}\mathbb{E}\left(\mathrm{e}^{J_{i,j}^b} - 1\right)\mathrm{d}t - \mathbb{E}\left(\mathrm{e}^{J_{i,j}} - 1\right)\mathrm{d}t \right)$$

$$+ \sum_{i=1}^{n}\sum_{j\neq i}^{n} \left(\mathrm{d}Z_{i,j}(t) - h_{i,j}\lambda_{i,j}(t)\mathbb{E}\left(\mathrm{e}^{J_{i,j}^b} - 1\right)\mathrm{d}t \right).$$

Applying Itô's lemma to the function $f(Y_t) = \mathrm{e}^{Y_t}$ leads to the dynamics (1.31) under $\mathbb{Q}$.   □

Given that under the risk neutral measure all assets earn on average the risk-free rate, one easily obtains the condition that ensures that L_t defines a pricing measure:

Corollary 1.6 *An equivalent measure defined by the Radon–Nikodym derivative (1.30) is a risk neutral measure if and only if $(\beta_t)_{t \geq 0}$, $h_{i,j} > -1$ for $i, j \in N$ and $v_{i,j}^b(\cdot)$ satisfy the following constraint:*

$$\delta_t^\top r = (\mu_t - \sigma_t \beta_t) + \sum_{i=1}^{n} \sum_{j \neq i}^{n} \left(h_{i,j} \mathbb{E} \left(e^{J_{i,j}^b} - 1 \right) - \mathbb{E} \left(e^{J_{i,j}} - 1 \right) \right) . \quad (1.32)$$

When there is no transition jump, the asset price is ruled by a switching geometric Brownian motion (SGBM). In this particular case, the non-arbitrage condition (1.32) implies that $\beta_t = \delta_t^\top \boldsymbol{\beta}$, where $\boldsymbol{\beta} = (\beta_1, \dots \beta_n)^\top$ is a vector with $\beta_i = \frac{\mu_i - r_i}{\sigma_i}$.

1.6 European Options Pricing

Let us consider European call and put options of maturity T, written upon an underlying price process $(S_t)_{t \geq 0}$ as given by (1.8). In the following, we express their payoff and strike as functions of the log-return $X_T = \ln(\frac{S_T}{S_0})$ and of the log-strike $k = \ln\left(\frac{K}{S_0}\right)$. We assume that the Markov chain process $(\delta_t)_{t \geq 0}$ is observable. For the sake of simplicity, the risk-free rate is assumed constant. The available information is carried by the filtration $(\mathcal{F}_t)_{t \geq 0}$ (we assume that the Markov chain is visible). The prices at time t of call and put options, denoted by $C(t, k, \delta_t)$ and $P(t, k, \delta_t)$, are functions of the log-strike k. It is well known that prices are equal to their expected discounted payoffs under the risk neutral measure (denoted by $\mathbb{E}^{\mathbb{Q}}$), and therefore if the risk neutral density at time $t \leq T$ of the log-return $\ln \frac{S_T}{S_0} | \mathcal{F}_t$ is denoted by $f_{t,T}(x, \delta_t)$:

$$C(t, k, \delta_t) = \mathbb{E}^{\mathbb{Q}} \left(e^{-r(T-t)} \left(S_0 e^{X_T} - K \right)_+ | \mathcal{F}_t \right) \quad (1.33)$$

$$= S_0 \int_k^{+\infty} e^{-r(T-t)} \left(e^x - e^k \right) f_{t,T}(x, \delta_t) \, dx ,$$

$$P(t, k, \delta_t) = \mathbb{E}^{\mathbb{Q}} \left(e^{-r(T-t)} \left(K - S_0 e^{X_T} \right)_+ | \mathcal{F}_t \right)$$

$$= S_0 \int_{-\infty}^{k} e^{-r(T-t)} \left(e^k - e^x \right) f_{t,T}(x, \delta_t) \, dx ,$$

where $r \in \mathbb{R}^+$ is assumed to be the constant risk-free rate. Option prices are computable by two methods, both using the Fast Fourier algorithm. The first one, proposed by Carr and Madan [4], directly estimates option prices. The second one approaches the density of the log-return. Option values are next calculated by computing the integrals in Eq. (1.33). We review both methods in the next paragraphs.

As $C(\cdot)$ (resp., $P(\cdot)$) tends to S_t (resp., $-S_t$) when $k \to -\infty$ (resp., $k \to +\infty$), $C(\cdot)$ and $P(\cdot)$ are not square integrable with respect to k and their Fourier transforms are not defined. For this reason, we consider the modified call and put prices, denoted by $c(k) = e^{\epsilon k} C(t, k, \delta_t)$, $p(k) = e^{\epsilon k} P(t, k, \delta_t)$, for which the Fourier transform exists for some damping factor ϵ ($\epsilon > 1$ for the call and $\epsilon < -1$ for the put). The Fourier transforms of $c(k)$ and $p(k)$ are defined as follows:

$$\mathcal{F}C(\omega) = \int_{-\infty}^{\infty} e^{i\omega k}\, c(k)\, \mathrm{d}k\,,$$

$$\mathcal{F}P(\omega) = \int_{-\infty}^{\infty} e^{i\omega k}\, p(k)\, \mathrm{d}k\,.$$

We denote by $\Upsilon_{t,s}(\omega) = \mathbb{E}^{\mathbb{Q}}\left(e^{\omega X_s} \mid \mathcal{F}_t\right)$ the moment generating function of Proposition 1.1 with parameters under $\mathbb{Q}$. The Fourier transform of $c(k)$ is equal to

$$\mathcal{F}C(\omega) = S_t e^{-r(T-t)} \int_{-\infty}^{\infty} \int_{k}^{+\infty} e^{(i\omega+\epsilon)k}\left(e^x - e^k\right) f_{t,T}(x, \delta_t)\, \mathrm{d}x\, \mathrm{d}k\,. \tag{1.34}$$

Inverting the order of integration, we infer that:

$$\int_{-\infty}^{\infty} \int_{k}^{+\infty} e^{(i\omega+\epsilon)k}\, e^x\, f_{t,T}(x, \delta_t)\, \mathrm{d}x\, \mathrm{d}k \tag{1.35}$$

$$= \int_{-\infty}^{\infty} e^x\, f_{t,T}(x, \delta_t) \int_{-\infty}^{x} e^{(i\omega+\epsilon)k}\, \mathrm{d}k\, \mathrm{d}x$$

$$= \frac{\Upsilon_{t,T}(i\omega + \epsilon + 1)}{(i\omega + \epsilon)}\,.$$

Similarly, we have that

$$\int_{-\infty}^{\infty} \int_{k}^{+\infty} e^{(i\omega+\epsilon+1)k}\, f_{t,T}(x, \delta_t)\, \mathrm{d}x\, \mathrm{d}k \tag{1.36}$$

$$= \int_{-\infty}^{\infty} f_{t,T}(x, \delta_t) \int_{-\infty}^{x} e^{(i\omega+\epsilon+1)k}\, \mathrm{d}k\, \mathrm{d}x$$

$$= \frac{\Upsilon_{t,T}(i\omega + \epsilon + 1)}{(i\omega + \epsilon + 1)}\,.$$

The Fourier transform (1.34) of $c(k)$ admits therefore a simple representation, as a function of $\Upsilon_{t,s}$:

$$\mathcal{F}C(\omega) = \frac{S_t e^{-r(T-t)}}{(i\omega + \epsilon)^2 + (i\omega + \epsilon)} \Upsilon_{t,T}(i\omega + \epsilon + 1),$$

with ϵ positive. We can check that $\mathcal{F}P(\omega) = \mathcal{F}C(\omega)$, but ϵ must be negative. The values of call options are then obtained by inverting the Fourier transform:

$$C(t, k, \delta_t) = \frac{S_t e^{-\epsilon k - r(T-t)}}{\pi} \int_0^\infty e^{-i\omega k} \frac{\Upsilon_{t,T}(i\omega + \epsilon + 1)}{(i\omega + \epsilon)^2 + (i\omega + \epsilon)} d\omega. \tag{1.37}$$

As the same expressions hold for puts, except that $\epsilon < 0$, we exclusively focus on call options in the remainder of this section. The naive approach consists in numerically calculating the integral in Eq. (1.37). Setting $\omega_m = \Delta_\omega(m-1)$ and letting M be the number of steps used in the Discrete Fourier Transform (DFT) as in Carr and Madan [4], an approximation of the call price is then given by

$$C(t, k, e_j) \approx \frac{S_t e^{-\epsilon k - r(T-t)}}{\pi} \tag{1.38}$$

$$\times \sum_{m=1}^{M} e^{-i\omega_m k} \varrho_m \left[\frac{\Upsilon_{t,T}(i\omega_m + \epsilon + 1)}{(i\omega_m + \epsilon)^2 + (i\omega_m + \epsilon)} \right] \Delta_\omega,$$

where $\varrho_m = \frac{1}{2} 1_{\{m=1\}} + 1_{\{m \neq 1\}}$. A judicious choice of the discretization steps in Eq. (1.38) allows us to use a Fast Fourier Transform algorithm to speed up the calculations. This point is detailed in the following proposition.

Proposition 1.7 *Let M be the number of steps used in the Discrete Fourier Transform (DFT) and $\Delta_k = \frac{2k_{\max}}{M-1}$ be the discretization step. Let us define $\varrho_m = \frac{1}{2} 1_{\{m=1\}} + 1_{\{m \neq 1\}}$, $\Delta_\omega = \frac{2\pi}{M \Delta_k}$, and $\omega_m = (m-1) \Delta_\omega$. The values of $C(t, k, \delta_t)$ at points $k_j = -\frac{M}{2} \Delta_k + (j-1) \Delta_k$ are approximated by*

$$C(k_j) \approx \frac{2 S_0 e^{-\epsilon k_j - r(T-t)}}{M \Delta_k} \tag{1.39}$$

$$\times \operatorname{Re} \left(\sum_{m=1}^{M} \varrho_m \left(\frac{\Upsilon_{t,T}(i\omega_m + \epsilon + 1)}{(i\omega_m + \epsilon)^2 + (i\omega_m + \epsilon)} (-1)^{m-1} \right) e^{-i \frac{2\pi}{M}(m-1)(j-1)} \right).$$

This last relation can be computed with a Fast Fourier Transform algorithm.

The main difficulty is the selection of parameters M and $k_{\max}$ that keep Δ_ω and Δ_k relatively small. Indeed, a high granularity of strike prices is obtained with a small

Δ_k. However, decreasing Δ_k increases Δ_ω and then the inaccuracy in the frequency domain.

An alternative is to numerically calculate the probability density function of the log-return. The characteristic function of a random variable, here $\Upsilon_{t,T}(i\omega) = \mathbb{E}^{\mathbb{Q}}\left(e^{i\,\omega X_T} \mid \mathcal{F}_t\right)$, is also the inverse Fourier transform of its probability density function (pdf):

$$f_{t,T}(x, \delta_t) = \frac{1}{2\pi} \int_{-\infty}^{+\infty} \Upsilon_{t,T}(i\omega)\, e^{-i\,x\,\omega} d\omega$$

$$= \frac{1}{\pi} \mathrm{Re} \left(\int_0^{+\infty} \Upsilon_{t,T}(i\omega) e^{-i\,x\,\omega} d\omega \right).$$

Therefore, we can retrieve the pdf by numerically computing its Fourier transform as stated in the next proposition.

Proposition 1.8 *Let M be the number of steps used in the Discrete Fourier Transform (DFT) and $\Delta_x = \frac{2x_{\max}}{M-1}$ be this step of discretization. Let us define $\Delta_\omega = \frac{2\pi}{M\,\Delta_x}$ and*

$$\omega_m = (m-1)\Delta_\omega,$$

for $m = 1, \ldots, M$. Let $\Upsilon_{t,T}(\omega) = \mathbb{E}^{\mathbb{Q}}\left(e^{\omega X_T} \mid \mathcal{F}_t\right)$ be the mgf of X_T. The values of $f_{t,T}(\cdot, \cdot)$, the pdf of $X_T\mid\mathcal{F}_t$ at points $x_k = -\frac{M}{2}\Delta_x + (k-1)\Delta_x$, are approached by the sum:

$$f(x_k, \delta_t) \approx \frac{2}{M\,\Delta_x}\mathrm{Re}\left(\sum_{m=1}^{M} \varrho_m \Upsilon_{t,T}(i\,\omega_m)(-1)^{m-1} e^{-i\frac{2\pi}{M}(m-1)(k-1)} \right), \quad (1.40)$$

where $\varrho_m = \frac{1}{2}1_{\{m=1\}} + 1_{\{m\neq1\}}$.

Proposition 1.8 can also be used to compute the pdf of log-returns under the real measure $\mathbb{P}$. To illustrate this point, we fit a switching diffusion with three regimes (SGBM) to the time series of S&P 500 daily values from 21/1/2001 to 21/1/2020. The estimation procedure is the Hamilton filter of Sect. 1.3. We do not consider any transition jumps. Parameter estimates are provided in Tables 1.4 and 1.5. The first regime corresponds to a deep recession with large negative average returns and a volatility above 40%. In the third state, stock prices grow on average by 25% a year and the volatility is low. In the second regime, the stock market is depressed, and the volatility is around 18%. Figure 1.3 shows the pdf of the one year return with $M = 64$ steps and for different initial regimes. The switching pdf still looks like a bell curve with slightly fatter tails. The peak observed in the distribution of the log-return when the process is initially in regime 1 is due to the limited number of discretization steps.

Table 1.4 3D SGBM model: matrix of transition probabilities over 1 day for the S&P 500

$p_{i,j}\,(t, t+1\,\mathrm{day})$	1	2	3
1	0.97	0.03	0.00
2	0.01	0.96	0.03
3	0.00	0.02	0.98

Table 1.5 Parameters of the 3D SGBM model fitted to the S&P 500

3D SGBM model					
μ_1	−33.21%	μ_2	−6.65%	μ_3	25.20%
σ_1	42.68%	σ_2	18.24%	σ_3	8.58%

Pdf under P

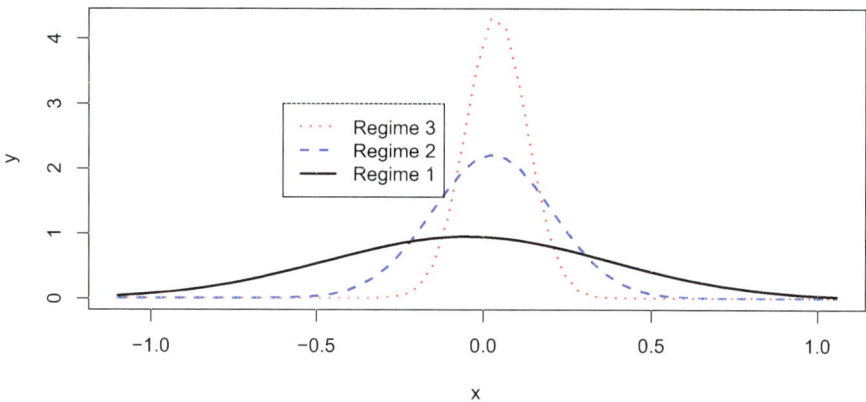

Fig. 1.3 Densities of the 1 year log-return under $\mathbb{P}$ per initial state, for the S&P 500. The model is an SGBM. Volatilities and transition probabilities are those from Tables 1.4 and 1.5

1.7 The Markov Switching Multifractal (MSM) Model

In this section, we present a continuous version of the multifractal process developed by Calvet and Fisher [2, 3]. This type of process presents many interesting features. The volatility specification is highly parsimonious and requires only a few parameters. The Markov Switching Multifractal (MSM) model is also consistent with the slowly declining autocovariograms and fat tails of financial series. In this setting, the dynamics of stock prices can be reformulated as an SGBM. The likelihood admits a closed form expression, and the model can be estimated with the procedure of Sect. 1.3.

The stock price is a process $(S_t)_{t\geq 0}$ defined on $\left(\Omega, \{\mathcal{H}_t\}_{t\geq 0}, \mathbb{P}\right)$ as the solution of the following SDE:

$$\frac{\mathrm{d}S_t}{S_t} = \mu\mathrm{d}t + \sigma_t\mathrm{d}W_t, \tag{1.41}$$

where μ, σ_t, and W_t are, respectively, a positive constant, the volatility process, and a Brownian motion. The volatility process is defined on a filtration $\mathcal{G}_t$ different from the filtration $\mathcal{H}_t$. σ_t is the product of a constant σ_0 and of the d elements of a multivariate process, $V_t = (V_1(t), V_2(t) \ldots V_d(t))^\top \in \mathbb{R}_+^d$:

$$\sigma_t = \sigma_0 \left(\prod_{i=1}^{d} V_i(t) \right)^{1/2}. \tag{1.42}$$

The components of V_t, called *multipliers*, are mutually independent jump processes. Their definition involves d Poisson processes, denoted $N_i(t)$ with an intensity γ_i. The dynamics of $V_i(t)$ depends upon V, a random variable. Calvet and Fisher [2] recommend the following Bernoulli distribution:

$$V = \begin{cases} v_0 & p_0 = \frac{1}{2} \\ 2 - v_0 & 1 - p_0 = \frac{1}{2} \end{cases}, \tag{1.43}$$

where $v_0 \in (0, 1)$. This is a symmetric distribution centered around 1. The components of the vector V_t are ruled by the dynamics:

$$dV_i(t) = (V - V_i(t-)) \, dN_i(t) \quad i = 1, \ldots, d. \tag{1.44}$$

The multiplier $V_i(t)$ is by construction a stepwise process oscillating between values $v_0 < 1$ and $2 - v_0 > 1$. The frequency of oscillations depends upon the intensity of $N_i(t)$. We assume that $\gamma_{i \in \{1,\ldots,d\}}$ depends on two parameters, denoted $\gamma_1 \in \mathbb{R}^+$ and $c \in (1, \infty)$, as follows:

$$\gamma_i \equiv \gamma_1^{c^{i-1}} \quad i \in \{1, \ldots, d\}. \tag{1.45}$$

This rule of construction guarantees that $\gamma_d \leq \cdots \leq \gamma_1 < 1$ if $\gamma_1 \in (0, 1)$ or $\gamma_d \geq \cdots \geq \gamma_1 > 1$ if $\gamma_1 \in [1, \infty)$. This means that the multiplier $V_1(t)$ (resp., $V_d(t)$) oscillates at a higher frequency than $V_d(t)$ (resp., $V_1(t)$) if $\gamma_1 \in (0, 1)$ (resp., $\gamma_1 \in [1, \infty)$). By construction,

$$\mathbb{P}(V_i(t) = v_0 \mid N_i(t) = m) = \frac{1}{2} \quad m \geq 1.$$

The main advantage of this model is its ability to capture low-frequency regime shifts in the volatility process. This is a parsimonious model: with only four parameters, $(\sigma_0, v_0, \gamma_1, c)$, σ_t takes its values in the interval

$$\sigma_t \in \left[\sigma_0 (v_0)^{\frac{d}{2}}, \; \sigma_0 (2 - v_0)^{\frac{d}{2}} \right].$$

We reformulate the MSM model as a switching geometric Brownian motion (SGBM) with $n = 2^d$ regimes. As in the first sections of this chapter, we consider a continuous Markov chain $(\delta_t)_{t \geq 0}$ on a filtration $\mathcal{G}_t$. This chain takes its values from the canonical state space $E = \{e_1, \ldots, e_n\}$, where $e_j = (0, \ldots, 1, \ldots, 0)^\top$. We also denote by $\boldsymbol{\sigma} = (\sigma_1, \ldots \sigma_n)^\top$ the vector of possible outcomes for σ_t. The volatility process is therefore equal to the scalar product

$$\sigma_t = \delta_t^\top \boldsymbol{\sigma} . \tag{1.46}$$

Each element of the state space E corresponds to an occurrence of the vector V_t, denoted $\boldsymbol{v}_1, \ldots \boldsymbol{v}_n \in \mathbb{R}_+^d$. By construction, $\boldsymbol{v}_j = (v_{j,1}, \ldots, v_{j,d})^\top$ is a vector of dimension d with $v_{j,k} = v_0$ or $v_{j,k} = 2 - v_0$ for $k = 1, \ldots, d$. The volatility of stock prices in the jth regime is given by

$$\sigma_j = \sigma_0 \sqrt{\prod_{k=1}^{d} v_{j,k}} .$$

The probability of switching from state i at time t to state j at time s, denoted $p_{i,j}(t, s)$, admits a closed form expression. Given that the multipliers are independent, $p_{i,j}(t, s)$ is the following product:

$$p_{i,j}(t, s) = \mathbb{P}\left(V_s = \boldsymbol{v}_j \mid V_t = \boldsymbol{v}_i\right) \tag{1.47}$$

$$= \prod_{k=1}^{d} \mathbb{P}\left(V_k(s) = v_{j,k} \mid V_k(t) = v_{i,k}\right) .$$

Given that for $m \geq 1$,

$$\mathbb{P}\left(V_k(s) = v_0 \mid V_k(t) = v_{i,k} \; N_k(t) = m\right) = \mathbb{P}\left(V_k(s) = v_0 \mid N_k(s - t) = m\right) ,$$

we rewrite the probabilities in Eq. (1.47) as an infinite sum

$$\mathbb{P}\left(V_k(s) = v_{j,k} \mid V_k(t) = v_{i,k}\right) = \mathbb{I}_{\{v_{j,k} = v_{i,k}\}} \mathbb{P}\left(N_k(s - t) = 0\right)$$

$$+ \sum_{m=1}^{\infty} \mathbb{P}\left(V_k(s) = v_{j,k} \mid N_k(s - t) = m\right) \mathbb{P}\left(N_k(s - t) = m\right) .$$

As $\mathbb{P}\left(V_k(s) = v_0 \mid N_k(s - t) = m\right) = \frac{1}{2}$, for $m \geq 1$, we infer for $v_{j,k} \neq v_{i,k}$ that

$$\mathbb{P}\left(V_k(s) = v_{j,k} \mid V_k(t) = v_{i,k}\right) = \sum_{m=1}^{\infty} \frac{1}{2} \frac{(\gamma_k (s - t))^m}{m!} e^{-\gamma_k(s-t)} , \tag{1.48}$$

whereas for $v_{j,k} = v_{i,k}$, we have that

$$\mathbb{P}\left(V_k(s) = v_{i,k} \mid V_k(t) = v_{i,k}\right) \tag{1.49}$$

$$= e^{-\gamma_k(s-t)} + \sum_{m=1}^{\infty} \frac{1}{2} \frac{(\gamma_k(s-t))^m}{m!} e^{-\gamma_k(s-t)}.$$

The combination of Eqs. (1.47), (1.48), and (1.49) allows us to compute the transition matrix. For a sufficiently small time Δ, the transition probability is

$$p_{i,j}(t, t+\Delta) = \prod_{k=1}^{d} \left(\frac{1}{2}\gamma_k\Delta + (1 - \gamma_k\Delta)\,\mathbb{I}_{\{v_{j,k}=v_{i,k}\}}\right) + O(\Delta^2)$$

for $i, j \in N := \{1, 2, \ldots, n\}$. The matrix Q_0 of instantaneous probabilities is in this case approached by

$$q_{i,j} \approx \frac{p_{i,j}(t, t+\Delta)}{\Delta}, \quad i \neq j,$$

$$q_{i,i} \approx \frac{p_{i,i}(t, t+\Delta) - 1}{\Delta}.$$

1.8 Numerical Illustrations

To illustrate this section, we fit the multifractal model to daily S&P 500 returns from 31/1/2001 to 31/1/2020 (4779 observations). The parameters are estimated with the Hamilton filter of Sect. 1.3. Table 1.6 reports statistics of goodness of fit. The highest likelihood and the lowest AIC/BIC are obtained with six multipliers. Notice that the number of parameters (# par.) in the second column counts all parameters of the switching diffusion. These parameters are all built with the five coefficients $(\mu, \sigma_0, v_0, \gamma_1, c)$ presented in Table 1.7. The average log-return is around 19%, while the baseline volatility, σ_0, is above 20%. v_0 increases with d, whereas γ_1 is stable whatever the number of multipliers. The parameter c is very close to one.

Figure 1.4 displays the most likely volatility $(\sigma_t)_{t \geq 0}$ over the last two decades. The peaks of volatility correspond to major crises. σ_t reaches its highest values

Table 1.6 Log-likelihood, AIC, and BIC of fractal models as functions of the number of multipliers, d

d	Log. Like.	# par.	AIC	BIC
3	15,712.96	8	−31,415	−31,383
4	15,731.41	16	−31,452	−31,420
5	15,739.80	32	−31,469	−31,437
6	15,753.19	64	−31,496	−31,464
7	15,740.33	128	−31,470	−31,438

Table 1.7 Parameter estimates as functions of the number of multipliers, d

d	μ	σ_0	v_0	γ_1	c
3	0.1924	0.2626	0.4315	1.1102	5.5259
4	0.1866	0.2181	0.4578	1.0231	5.2314
5	0.1859	0.2659	0.5139	1.0156	3.8111
6	0.1898	0.2559	0.5433	1.1483	2.1225
7	0.1868	0.2211	0.6076	1.1898	1.6418

Filtered Volatility

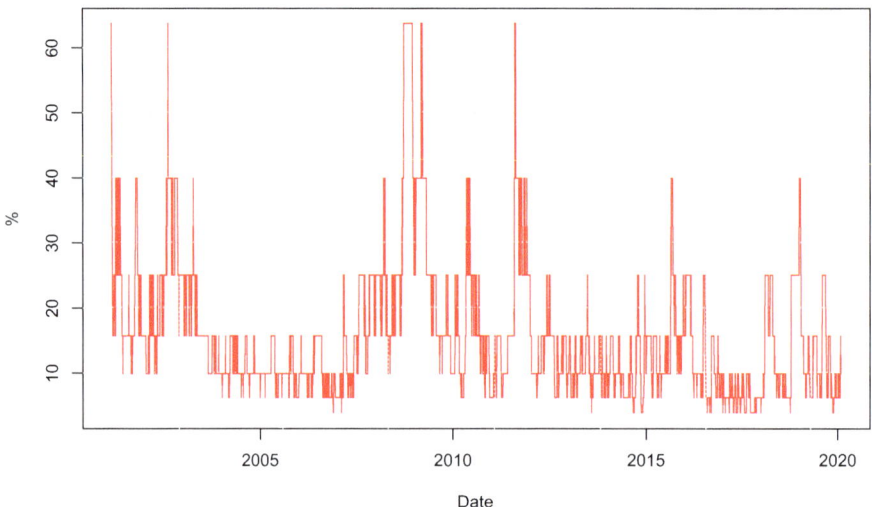

Fig. 1.4 S&P 500 filtered daily volatilities from 31/1/01 to 31/1/20. Model with $d = 6$ multipliers

in 2003, during the second Gulf war, during the subprime crisis of 2008, and in the double dip recession of 2011. Peaks are also observed in 2015–2016 due to turbulences on the Chinese market and in 2018 when oil price fell due to overproduction. To conclude this section, we compare the performance of fractal and GARCH models. The *GARCH model*, developed by Engle [6], is a discrete time approach and is a reference in the industry. In GARCH(p, q), the dynamics of daily log-returns is

$$\Delta S_t = \ln \frac{S_t}{S_{t-1}} = \mu + \sigma_t \epsilon_t \,,$$

where $\epsilon_t \sim N(0, 1)$ and μ is the daily expected return. The variance σ_t^2 depends on past realization of ϵ_t and σ_t^2 as follows:

$$\sigma_t^2 = \omega + \sum_{i=1}^{q} \alpha_i \epsilon_{t-i}^2 + \sum_{i=1}^{p} \beta_i \sigma_{t-i}^2 \,.$$

Table 1.8 Goodness of fit: GARCH models fitted to S&P 500, 31/1/01–31/1/20

	Log-likelihood	Number of parameters
Garch(1,1)	15,638	4
Garch(2,2)	15,645	6
Garch(3,3)	15,646	8

The results of the fit to the time series of S&P 500 are presented in Table 1.8. A comparison with log-likelihoods of Table 1.6 emphasizes that the switching multifractal model outperforms GARCH models.

1.9 Hitting Time of a Regime Switching Model

In this section, we introduce the fluid embedding technique of Rogers [22] that allows us to analyze the time when a switching regime model hits a floor or a ceiling. We adopt a similar approach to Jiang and Pistorius [16], and we focus on the pricing of perpetual binary options. Let us recall that X_t is the log-return of S_t as defined in Eq. (1.31),

$$\mathrm{d}X_t = \left(\delta_t^\top r - \frac{1}{2}\sigma_t^2\right)\mathrm{d}t + \sigma_t \mathrm{d}W_t^{\mathbb{Q}}$$

$$+ \sum_{i=1}^n \sum_{j\neq i} \left(\mathrm{d}Z_{i,j}(t) - q_{i,j}^{\mathbb{Q}} 1_{\{\delta_{t-}=e_i\}}\mathbb{E}\left(e^{J_{i,j}^b} - 1\right)\mathrm{d}t\right),$$

where $r = (r_1, \ldots, r_n) \in \mathbb{R}^n$ is the risk-free rate in each phase. In order to lighten our load, we assume that the transition jumps, $\left(J_{i,j}^b\right)_{i\neq j}$, have a double exponential distribution that only depends on $j \in \mathcal{N}$, with parameters ρ_j^+, ρ_j^- and p_j. The Markov chain $(\delta_t)_{t\geq 0}$ is assumed to be observable, and we recall that its matrix of instantaneous transition probabilities is $Q_0^{\mathbb{Q}}$. The elements of this matrix are $q_{i,j}^{\mathbb{Q}} = h_{i,j}q_{i,j}$ for $i, j \in \mathcal{N}$. The state space of $(\delta_t)_{t\geq 0}$ is denoted by $E_0 = \{e_1, \ldots, e_n\}$ (instead of E) in this section.

Perpetual high or low binary options have an infinite time horizon and respectively deliver a payoff equal to $\delta_\tau^\top \bar{p}$, where $\bar{p} = (\bar{p}_1, \ldots, \bar{p}_n) \in \mathbb{R}^n$, where the stopping time τ for a low and high binary option is respectively defined by $\tau = \inf\{t : X_t \leq k\}$ or $\tau = \inf\{t : X_t \geq k\}$ for a certain level $k \in \mathbb{R}^+$. The value of these binary options is equal to the expected discounted cash-flow under the risk neutral measure:

$$B^{\mathrm{high}}(X_t, \delta_t) = \mathbb{E}^{\mathbb{Q}}\left(e^{-\int_t^\tau \delta_s^\top r \mathrm{d}s}(\delta_\tau^\top \bar{p})|\mathcal{F}_t\right), \quad \tau = \inf\{t : X_t \geq k\},$$

$$B^{\mathrm{low}}(X_t, \delta_t) = \mathbb{E}^{\mathbb{Q}}\left(e^{-\int_t^\tau \delta_s^\top r \mathrm{d}s}(\delta_\tau^\top \bar{p})|\mathcal{F}_t\right), \quad \tau = \inf\{t : X_t \leq k\}.$$

Proposition 1.9 *The modulated discount factor from T to t $\leq$ T has a value equal to*

$$\mathbb{E}^{\mathbb{Q}}\left(e^{-\int_t^T \delta_s^\top r ds}\delta_T|\mathcal{F}_t\right) = \delta_t^\top e^{\left(Q_0^{\mathbb{Q}}-\text{diag}(r)\right)(T-t)}. \tag{1.50}$$

Proof Let us denote by $g(t, \delta_t)$ the left-hand term of Eq. (1.50). Using nested expectations and a first-order Taylor expansion of the exponential, we have that

$$
\begin{aligned}
0 &= \lim_{\Delta \to 0} \frac{\mathbb{E}^{\mathbb{Q}}\left(e^{-\int_t^{t+\Delta} \delta_s^\top r ds}g(t+\Delta, \delta_{t+\Delta})|\mathcal{F}_t\right) - g(t, \delta_t)}{\Delta} \\
&= \lim_{\Delta \to 0} \frac{\mathbb{E}^{\mathbb{Q}}\left((1-\delta_s^\top r\Delta)e^{-\int_t^{t+\Delta} \delta_s^\top r ds}g(t+\Delta, \delta_{t+\Delta})|\mathcal{F}_t\right) - g(t, \delta_t) + O(\Delta^2)}{\Delta}.
\end{aligned}
$$

We can rewrite this limit as

$$\mathbb{E}^{\mathbb{Q}}\left(dg(t, \delta_t)|\mathcal{F}_t\right) = \delta_t^\top r\, g(t, \delta_t)\, dt\,,$$

where the left-hand term is calculable with Itô's lemma. After expanding this last equation, we infer that $g(\cdot, \cdot)$ is a solution of the system

$$\frac{\partial g(t, e_i)}{\partial t} + \sum_{j \neq i}^n q_{i,j}^{\mathbb{Q}}\left(g(t, e_j) - g(t, e_i)\right) = r_i\, g(t, e_i) \quad i = 1, \ldots, n.$$

As $q_{i,i}^{\mathbb{Q}} = -\sum_{j \neq i}^n q_{i,j}^{\mathbb{Q}}$, we check that the function (1.50) is also a solution of this system.   □

Next, let us define the stopping time ζ, which is the first jumping time of a process N_t^r with intensity $\lambda_t^r = \int_0^t \delta_s^\top r ds$. The above expectation may then be rewritten as

$$
\begin{aligned}
\mathbb{E}^{\mathbb{Q}}\left(e^{-\int_t^T \delta_s^\top r ds}\delta_T|\mathcal{F}_t\right) &= \mathbb{E}^{\mathbb{Q}}\left(1_{\{T \leq \zeta\}}\delta_T|\mathcal{F}_t\right) \\
&= \mathbb{E}^{\mathbb{Q}}\left(\delta_T^r|\mathcal{F}_t\right),
\end{aligned}
$$

where $(\delta_t^r)_{t \geq 0}$ is a "killed" Markov chain with instantaneous probabilities, $Q^r = Q_0^{\mathbb{Q}} - \text{diag}(r)$. The sum of elements in the rows of Q^r is no longer equal to zero. The Markov chain is called *transient* for this reason and is defined on $E \cup \partial$, where ∂ is a cemetery state in which δ_t^r is a null vector. δ_t^r enters this absorbing state at time ζ when N_t^r jumps for the first time. The option prices are reformulated as $\mathbb{E}^{\mathbb{Q}}\left(\delta_\tau^r{}^\top \bar{p}|\mathcal{F}_t\right)$.

We denote by $(\gamma_t)_{t>0}$ a transient continuous time Markov chain, defined on a finite state space $E \cup \partial$. γ_t is a vector that takes values in the set of unit vectors of dimension $3n$. In the cemetery state, γ_t is the null vector. We define a new stochastic process A_t, which is the fluid embedding process of X_t, as follows:

$$A_t = A_0 + \int_0^t \gamma_s^\top \bar{m} \, ds + \int_0^t \gamma_s^\top \bar{s} \, dW_s \, ,$$

where $\bar{m}$ and $\bar{s}$ are vectors of dimension $3n$ that will be defined later. The generator of γ_t restricted to E is the following $3n \times 3n$ matrix:

$$Q_\gamma = \begin{pmatrix} -D^- & D^- & O_n \\ B^- & C & B^+ \\ O_n & D^+ & -D^+ \end{pmatrix} \, ,$$

where D^-, D^+, B^-, B^+ are $n \times n$ matrices defined by

$$D^- = \mathrm{diag}\left(-\left(\rho_j^-\right)_{j \in N} \right) \, , \quad D^+ = \mathrm{diag}\left(\left(\rho_j^+\right)_{j \in N} \right)$$

$$B^- = \left((1 - p_j) q_{i,j}^{\mathbb{Q}} 1_{\{j \neq i\}} \right)_{i,j=1,\ldots,n} \, , \quad B^+ = \left(p_j q_{i,j}^{\mathbb{Q}} 1_{\{j \neq i\}} \right)_{i,j=1,\ldots,n}$$

$$C = \mathrm{diag}\left(\left(q_{i,i}^{\mathbb{Q}} - r_i \right)_{i=1,\ldots,n} \right) \, ,$$

with O_n being an $n \times n$ null matrix. The state space E can be partitioned as $E = E^- \cup E_0 \cup E^+$. States in E^+ and E^-, respectively, correspond to positive and negative jumps, whereas states of E_0 are inherited from δ_t. The vectors $\bar{m}$ and $\bar{s}$ are defined by

$$\bar{m} = (m_i)_{i=1:3n+1} = \begin{cases} -1 & i \in \{1, \ldots, n\} \\ \tilde{\mu}_{i-n}^{\mathbb{Q}} & i \in \{n+1, \ldots, 2n\} \\ +1 & i \in \{2n+1, \ldots, 3n\} \, , \end{cases}$$

$$\bar{s} = (s_i)_{i=1:3n+1} = \begin{cases} 0 & i \in \{1, \ldots, n\} \\ \sigma_{i-n} & i \in \{n+1, \ldots, 2n\} \\ 0 & i \in \{2n+1, \ldots, 3n\} \, , \end{cases}$$

where

$$\tilde{\mu}_i^{\mathbb{Q}} = r_i - \frac{1}{2}\sigma_i^2 - \sum_{j \neq i}^n \left(q_{i,j}^{\mathbb{Q}} \mathbb{E}\left(e^{J_{i,j}^b} - 1 \right) \right) \, ,$$

for $i = 1, \ldots, n$, is the drift under $\mathbb{Q}$ of X_t, when $\delta_t = e_{i-n}$.

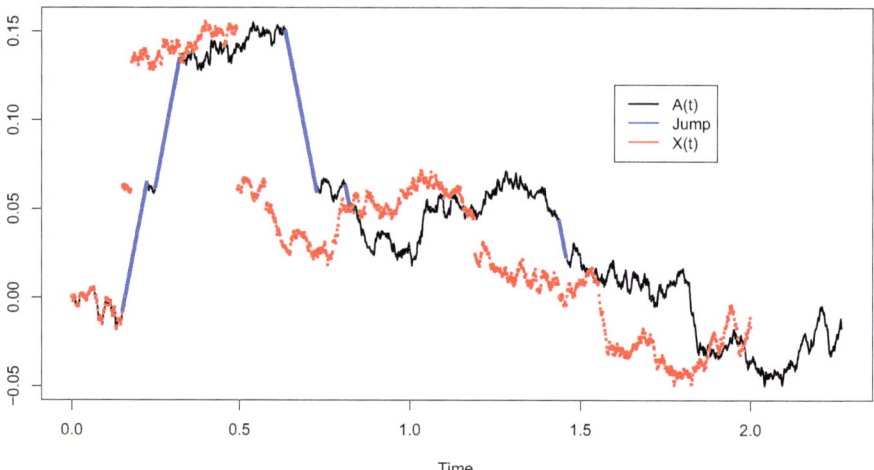

Fig. 1.5 Example of sample paths of X_t and of its fluid embedding process A_t

Figure 1.5 illustrates the fluid embedding technique by comparing the sample paths of X_t and A_t. The drift of X_t is set to $\tilde{\mu}^Q = (5\%, 0\%, 5\%)$, while the volatilities are not modulated and equal to 5%. The transition jumps share the same distribution with $p = 0.6$, $\rho_+ = 20$, and $\rho_- = -20$. The matrix Q_0^Q is such that the 1 year probability of transiting to another state is 0.30. We observe that jumps of X_t are converted into linear segments in the sample path of A_t.

Example A Markov chain with two states:

$$
Q_\gamma = \begin{pmatrix}
\rho_1^- & 0 & -\rho_1^- & 0 & 0 & 0 \\
0 & \rho_2^- & 0 & -\rho_2^- & 0 & 0 \\
0 & q_{1,2}^Q(1-p_2) & q_{1,1}^Q - r_1 & 0 & 0 & q_{1,2}^Q p_2 \\
q_{2,1}^Q(1-p_2) & 0 & 0 & q_{2,2}^Q - r_2 & q_{2,1}^Q p_2 & 0 \\
0 & 0 & \rho_1^+ & 0 & -\rho_1^+ & 0 \\
0 & 0 & 0 & \rho_2^+ & 0 & -\rho_2^+
\end{pmatrix}.
$$

We have that

$$
E^- = \{1, 2\}, \; E^0 = \{3, 4\}, \; E^+ = \{5, 6\}.
$$

Within this approach, any sample path of X_t that represents discontinuities at jump times may be converted into a continuous path of A_t. Let us define

$$
T_0(t) := \int_0^t 1_{\{\gamma_t \in E_0\}} \mathrm{d}s \,,
$$

$$
T_0^{-1}(u) := \inf\{t \geq 0 \, : \, T_0(t) > u\} \,,
$$

where $T_0(t)$ is the total time spent in E_0 by the Markov chain γ_t up to time t. By construction, $A \circ T_0^{-1}$ and $\gamma \circ T_0^{-1}$ restricted to E_0 have the same distribution as $X_{t \wedge \zeta}$ and δ_t^r. Let us define $\tilde{\tau}$ for binary low and high options by

$$\tilde{\tau} = \inf\{t \geq 0\,;\, \gamma_t \in E_0 \text{ and } A_t \leq k\}\,,$$
$$\tilde{\tau} = \inf\{t \geq 0\,;\, \gamma_t \in E_0 \text{ and } A_t \geq k\}\,.$$

Then

$$\mathbb{E}^{\mathbb{Q}}\left(e^{-\int_t^\tau \delta_s^\top r ds}(\delta_\tau^\top \bar{\boldsymbol{p}})|\mathcal{F}_t\right) = \mathbb{E}^{\mathbb{Q}}\left(1_{\tau \leq \zeta}(\delta_\tau^\top \bar{\boldsymbol{p}})|\mathcal{F}_t\right)$$
$$= \mathbb{E}^{\mathbb{Q}}\left(1_{\tilde{\tau} \leq \zeta}(\gamma_{\tilde{\tau}}^\top \bar{\boldsymbol{p}})|\mathcal{F}_t\right)$$

and $\tilde{\boldsymbol{p}}$ is an extended vector of payoffs:

$$\tilde{\boldsymbol{p}} = (\tilde{p}_i)_{i=1:3n} = \begin{cases} \bar{p}_{i-n} & i \in \{n+1, \ldots, 2n\} \\ 0 & \text{otherwise.} \end{cases}$$

We now define up-crossing and down-crossing ladders γ_t^+ and γ_t^- as follows:

$$\gamma_x^+ = \gamma_{\tau_x^+} \qquad \gamma_x^- = \gamma_{\tau_x^-}\,,$$

where

$$\tau_x^+ = \inf\{s \geq 0 : A_s > x\}\,,$$
$$\tau_x^- = \inf\{s \geq 0 : A_s < x\}\,.$$

By convention, we set τ_x^+ (resp., τ_x^-) to $+\infty$ if A_t never reaches x. Figure 1.6 shows the up-crossing ladder τ_x^+ for a sample of A_t, simulated with parameters used for Fig. 1.5. The idea behind this approach is to substitute the initial time scale by a clock function of x, the value of $(A_t)_{t \geq 0}$. While the hitting time is random when observed on the time scale, it becomes deterministic on the up-crossing ladder scale.

By construction, γ_x^+ and γ_x^- are Markov chains[1] with respective state spaces $E^0 \cup E^+$ and $E^- \cup E^0$. Their generators limited to these subspaces are $2n \times 2n$ matrices denoted by Q^+ and Q^-. The initial distributions are $n \times 2n$ matrices η^+ and η^- such that

$$\eta^+(i, j) = P(\gamma_0^+ = e_j\,,\, \tau_0^+ \leq \zeta \mid \gamma_0 = e_i) \quad e_i \in E^-\,,\, e_j \in E^0 \cup E^+\,,$$
$$\eta^-(i, j) = P(\gamma_0^- = e_j\,,\, \tau_0^- \leq \zeta \mid \gamma_0 = e_i) \quad e_i \in E^+\,,\, e_j \in E^- \cup E^0\,.$$

---

[1] As τ_x^+ and τ_x^- are stopping times, this is a consequence of the strong Markov property of γ_t.

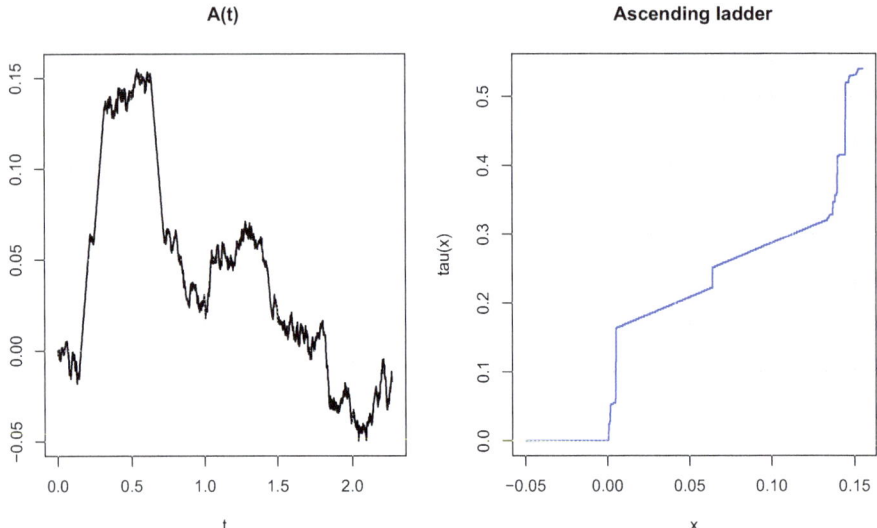

Fig. 1.6 Left plot: sample path of A_t. Right plot: τ_x^+, up-crossing ladder

Let $\Sigma = \text{diag}\left((\bar{s}_i)_{i=1:3n}\right)$ and $M = \text{diag}\left((\bar{m}_i)_{i=1:3n}\right)$ be $3n \times 3n$ matrices. The identity and null matrix of dimension n are denoted by I_n and O_n.

Proposition 1.10 $\left(\eta^+, Q^+, \eta^-, Q^-\right)$ *are solutions of the following matrix equations:*

$$\frac{1}{2}\Sigma^2 W^+ \left(Q^+\right)^2 - M W^+ Q^+ + Q_\gamma W^+ = O \,, \qquad (1.51)$$

$$\frac{1}{2}\Sigma^2 W^- \left(Q^-\right)^2 + M W^- Q^- + Q_\gamma W^- = O \,, \qquad (1.52)$$

where O is a $3n \times 2n$ null matrix and

$$W^+ = \begin{pmatrix} \eta^+ \\ I_n \ O_n \\ O_n \ I_n \end{pmatrix} \qquad W^- = \begin{pmatrix} I_n \ O_n \\ O_n \ I_n \\ \eta^- \end{pmatrix} .$$

For $X_t = x$ and $k > x$, we have that

$$B^{\text{high}}(X_t, \delta_t) = \mathbb{E}^{\mathbb{Q}} \left(1_{\{\tau_k^+ \leq \zeta\}} \left(\gamma_k^+ \bar{p}^+\right) | \mathcal{F}_t\right) \qquad (1.53)$$

$$= \gamma_t^\top W^+ \exp\left(Q^+(k - x)\right) \bar{p}^+ \,,$$

and for $X_t = x$ and $k < x$, we have that

$$B^{\text{low}}(X_t, \delta_t) = \mathbb{E}^{\mathbb{Q}}\left(1_{\tau_k^- \leq \varsigma}\left(\gamma_k^- \, \bar{p}^-\right)|\mathcal{F}_t\right) \tag{1.54}$$

$$= \gamma_t^\top \, W^- \, \exp\left(Q^-(x - k)\right) \, \bar{p}^- \,,$$

where

$$\bar{p}^+ = \left(\bar{p}_i^+\right)_{i=1:2n} = \begin{cases} \bar{p}_i & i \in \{1, \dots, n\} \\ 0 & otherwise \end{cases}$$

and

$$\bar{p}^- = \left(\bar{p}_i^-\right)_{i=1:2n} = \begin{cases} \bar{p}_{i-n} & i \in \{n+1, \dots, 2n\} \\ 0 & otherwise \end{cases}.$$

Proof We sketch the proof of relations (1.51) and (1.53). By construction,

$$\mathbb{E}^{\mathbb{Q}}\left(1_{\tilde{\tau} \leq \varsigma}(\gamma_{\tilde{\tau}}^\top \tilde{p})|\mathcal{F}_t\right) = \mathbb{E}^{\mathbb{Q}}\left(1_{\tau_k^+ \leq \varsigma}\left(\gamma_k^+ \, \bar{p}^+\right)|\mathcal{F}_t\right).$$

Given that γ_x^+ is a Markov chain with generator Q^+, and an initial distribution W^+, we infer that

$$V^+(e_i, x) = \mathbb{E}^{\mathbb{Q}}\left(1_{\tau_k^+ \leq \varsigma}\left(\gamma_k^+ \, \bar{p}^+\right) \mid \gamma_t = e_i\,, \; X_t = x\right)$$

$$= e_i^\top \, W^+ \, \exp\left(Q^+(k - x)\right) \, \bar{p}^+ \,.$$

Using a similar reasoning as in the proof of Proposition 1.9, we can show that for $x \leq k$, $V^+(e_i, x)$ is a martingale and therefore is a solution of

$$\frac{1}{2}s_i^2 \frac{\partial^2 V^+(e_i, x)}{\partial x^2} + m_i \frac{\partial V^+(e_i, x)}{\partial x} + \sum_{j \neq i} q_{ij}^+ \left(V^+(e_i, x) - V^+(e_j, x)\right) = 0.$$

We have

$$\frac{\partial V^+(e_i, x)}{\partial x} = -e_i^\top \, W^+ Q^+ \exp\left(Q^+(k - x)\right) \, \bar{p}^+ \,,$$

$$\frac{\partial^2 V^+(e_i, x)}{\partial x^2} = e_i^\top \, W^+ \left(Q^+\right)^2 \exp\left(Q^+(k - x)\right) \, \bar{p}^+,$$

and

$$\sum_{j\neq i} q_{ij}^+ \left(V^+(e_i, x) - V^+(e_j, x)\right) = \sum_j q_{ij}^+ V^+(e_j, x).$$

Given that $\bar{p}^+$ is arbitrary, this last equation corresponds well to the system of Eqs. (1.51).   □

Solving Eqs. (1.51) and (1.52) is challenging. When the number of regimes is limited to 3 or 4, one can use a method based on eigenvalues and eigenvectors of Q^+ and Q^-. Unfortunately, for Markov chains with a high number of states, this approach becomes unstable because it requires the calculation of the determinant of a badly conditioned matrix of large dimension. However, it is possible to reduce the dimension of this problem if we recall that Σ and V are diagonal matrices. We denote by $\tilde{\mu} = \left(\tilde{\mu}_i^{\mathbb{Q}}\right)_{i=1,...,n}$ the vector of drift of X_t under $\mathbb{Q}$.

Proposition 1.11 *The matrix Q^+ may be rewritten as*

$$Q^+ = \begin{pmatrix} G_1 & G_2 \\ D^+ & -D^+ \end{pmatrix}, \tag{1.55}$$

where G_1, G_2 are $n \times n$ matrices that satisfy the following system of matrix equations:

$$O_n = \eta_{E0}^+ G_1 + \eta_{E+}^+ D^+ - D^- \eta_{E0}^+ + D^- \tag{1.56}$$

$$O_n = \eta_{E0}^+ G_2 - \eta_{E+}^+ D^+ - D^- \eta_{E+}^+$$

$$O_n = \frac{1}{2}\mathrm{diag}(\sigma^2)\left(G_1^2 + G_2 D^+\right) - \mathrm{diag}(\tilde{\mu})G_1 + B^- \eta_{E0}^+ + C$$

$$O_n = \frac{1}{2}\mathrm{diag}(\sigma^2)\left(G_1 G_2 - G_2 D^+\right) - \mathrm{diag}(\tilde{\mu})G_2 + B^- \eta_{E+}^+ + B^+$$

with $\eta^+ = \{\eta_{E0}^+, \eta_{E+}^+\}$. Here, η_{E0}^+ and η_{E+}^+ are $n \times n$ matrices.

Proof By definition of W^+, we have that

$$Q_\gamma W^+ = \begin{pmatrix} -D^- \eta_{E0}^+ + D^- & -D^- \eta_{E+}^+ \\ B^- \eta_{E0}^+ + C & B^- \eta_{E+}^+ + B^+ \\ D^+ & -D^+ \end{pmatrix}$$

and

$$MW^+ = \begin{pmatrix} -\eta_{E^0}^+ & -\eta_{E^+}^+ \\ \operatorname{diag}(\tilde{\pmb{\mu}}) & O \\ O & I_n \end{pmatrix}, \quad \Sigma^2 W^+ = \begin{pmatrix} O_n & O_n \\ \operatorname{diag}(\sigma^2) & O_n \\ O_n & O_n \end{pmatrix}.$$

If we assume that

$$Q^+ = \begin{pmatrix} G_1 & G_2 \\ G_3 & G_4 \end{pmatrix}, \quad (Q^+)^2 = \begin{pmatrix} G_1^2 + G_2 G_3 & G_1 G_2 + G_2 G_4 \\ G_1 G_3 + G_3 G_4 & G_2 G_3 + G_4^2 \end{pmatrix},$$

then

$$MW^+ Q^+ = \begin{pmatrix} -\eta_{E^0}^+ G_1 - \eta_{E^+}^+ G_3 & -\eta_{E^0}^+ G_2 - \eta_{E^+}^+ G_4 \\ \operatorname{diag}(\tilde{\pmb{\mu}}) G_1 & \operatorname{diag}(\tilde{\pmb{\mu}}) G_2 \\ G_3 & G_4 \end{pmatrix}$$

and

$$\Sigma^2 W^+ (Q^+)^2 = \begin{pmatrix} O_n & O_n \\ \operatorname{diag}(\sigma^2) \left(G_1^2 + G_2 G_3\right) & \operatorname{diag}(\sigma^2) \left(G_1 G_2 + G_2 G_4\right) \\ O_n & O_n \end{pmatrix}.$$

Injecting these expressions into Eq. (1.51) leads to the result.   □

The same result holds for Q^-:

Proposition 1.12 *The matrix Q^- may be rewritten as*

$$Q^- = \begin{pmatrix} -D^- & D^- \\ G_3 & G_4 \end{pmatrix}, \tag{1.57}$$

where G_3, G_4 are $n \times n$ matrices that satisfy the following system of matrix equations:

$$O_n = \frac{1}{2}\operatorname{diag}(\sigma^2)\left(-D^- G_3 + G_3 G_4\right) + \operatorname{diag}(\tilde{\pmb{\mu}}) G_3 + B^- + B^+ \eta_{E^-}^-$$

$$O_n = \frac{1}{2}\operatorname{diag}(\sigma^2)\left(D^- G_3 + G_4^2\right) + \operatorname{diag}(\tilde{\pmb{\mu}}) G_4 + C + B^+ \eta_{E^0}^-$$

$$O_n = -\eta_{E^-}^- D^- + \eta_{E^0}^- G_3 - D^+ \eta_{E^-}^-$$

$$O_n = \eta_{E^-}^- D^- + \eta_{E^0}^- G_4 + D^+ - D^+ \eta_{E^0}^-$$

with $\eta^- = \{\eta_{E^-}^-, \eta_{E^0}^-\}$. Here $\eta_{E^0}^-$ and $\eta_{E^-}^-$ are $n \times n$ matrices.

Table 1.9 Parameters used
for the pricing of high binary
options

σ_1	20%	p	0.5
σ_2	10%	ρ^+	200
σ_3	5%	ρ^-	200

Table 1.10 Price of high
binary options for various
strikes and initial states of the
Markov chain

Strike k	State 1	State 2	State 3
0.05	0.9503	0.949	0.9426
0.10	0.9037	0.9028	0.9013
0.15	0.8596	0.859	0.8588
0.20	0.818	0.8176	0.8177
0.25	0.7791	0.7789	0.7789
0.30	0.7429	0.7427	0.7428
0.35	0.709	0.7089	0.709
0.40	0.6773	0.6773	0.6773

To illustrate this section, we price high binary options. The matrix $Q_0^{\mathbb{Q}}$ is such that the matrix of 1 year transition probabilities under $\mathbb{Q}$ is equal to

$$
\exp\left(Q_0^{\mathbb{Q}}\right) = \begin{pmatrix} 0.6\ 0.2\ 0.2 \\ 0.2\ 0.6\ 0.2 \\ 0.2\ 0.2\ 0.6 \end{pmatrix}.
$$

The interest rate is not modulated by the chain and is set to 2%. The payoff, $\bar{p}$, is also set to one whatever the regime when X_t hits the upper boundary. The Brownian volatilities in each regime and the features of transition jumps are reported in Table 1.9.

Table 1.10 shows prices of perpetual high binary options, for different log-strikes k. Note that the matrices Q^+ and W^+ are numerically found by solving the system (1.56). Whatever the regime of δ_0, the price is inversely proportional to the strike because a higher strike postpones on average the exercise time of the binary option. The option is slightly more expensive when the Markov chain is in an initial state with a high Brownian volatility. Further numerical analysis reveals that the option price is also directly proportional to the expected size of jumps.

1.10 Further Reading

Economic cycles and their modeling by a hidden Markov chain have received a lot of attention in the econometric literature. This chapter is directly related to this line of research. For example, Guidolin and Timmermann [7] present evidence of persistent "bull" and "bear" regimes in UK stock and bond returns and consider their economic implications from the perspective of an investor's portfolio allocation. Similar results are found in Guidolin and Timmermann [9], for international stock markets. Guidolin and Timmermann [8] develop a regime switching model for asset returns with four states. From a broader perspective, regime switching models are useful to describe equity markets. Cholette et al. [5] fit skewed-t GARCH marginal distributions for international equity returns and a regime switching copula with two states. Al-Anaswah and Wilfing [1] propose a two-regime Markov switching specification to capture speculative bubbles. On the other hand, Calvet and Fisher [2, 3] show that discrete versions of multifractal processes capture thick tails and have a switching regime structure. Nakajima [21] proposes a Bayesian analysis of the stochastic volatility model with regime switching skewness. Switching regime processes are also used for investment management purposes (see Stovall [23] or Hainaut and MacGilchrist [13]). Hardy [15] and the society of actuaries (SOA), since 2004, recommend switching processes to model long-term stock returns, in actuarial applications. In Hainaut [10], multifractal diffusion is adapted for interest rate modeling. Note that the switching regime framework can be extended to manage models with multiple assets. To illustrate this, we study a bivariate model in Chap. 2. We also explore in this chapter an alternative procedure of calibration based on the Monte Carlo Markov Chain (MCMC). This is an alternative to maximum likelihood estimation based on a Bayesian learning paradigm. Finally, Hainaut and Deelstra [11, 12] show that a switching model with transition jumps can approximate both univariate and bivariate self-excited jump processes.

References

1. Al-Anaswah, N., Wilfing, B.: Identification of speculative bubbles using state-space models with Markov-switching. J. Banking Financ. **35**(5), 1073–1086 (2011)
2. Calvet, L., Fisher, A.: Forecasting multifractal volatility. J. Econ. **105**, 17–58 (2001)
3. Calvet, L., Fisher, A.: How to forecast long-run volatility: regime switching and the estimation of multifractal processes. J. Financ. Econ. **2**, 49–83 (2004)
4. Carr, P., Madan, D.: Option valuation using the fast fourier transform. J. Comput. Financ. **2**, 61–73 (1999)
5. Cholette, L., Heinen, A., Valdesogo, A.: Modelling international financial returns with a multivariate regime switching copula. J. Financ Econ. **7**(4), 437–480 (2009)
6. Engle, R.F.: Autoregressive conditional heteroskedasticity with estimates of the variance of United Kingdom inflation. Econometrica **50**, 987–10 (1982)
7. Guidolin, M., Timmermann, A.: Economic implications of bull and bear regimes in UK stock and bond returns. Econ. J. **115**, 11–143 (2005)
8. Guidolin, M., Timmermann, A.: Asset allocation under multivariate regime switching. J. Econ. Dyn. Control **31**(11), 3503–3544 (2007)
9. Guidolin, M., Timmermann, A.: International asset allocation under regime switching, skew, and kurtosis preferences. Rev. Financ. Stud. **21**(2), 889–935 (2008)
10. Hainaut, D.: A fractal version of the Hull–White interest rate model. Econ. Modell. **31**, 323–334 (2013)
11. Hainaut, D., Deelstra, G.: A self-excited switching jump diffusion (SESJD): properties, calibration and hitting time. Quant. Financ. **19**(3), 407–426 (2019)
12. Hainaut, D., Deelstra, G.: A bivariate mutually-excited switching jump diffusion (BMESJD) for asset prices. Methodol. Comput. Appl. Probab. **21**(4), 1337–1375 (2019)

13. Hainaut, D., MacGilchrist, R.: Strategic asset allocation with switching dependence. Ann. Financ. **8**(1), 75–96 (2012)
14. Hamilton, J.D.: A new approach to the economic analysis of nonstationary time series and the business cycle. Econometrica **57**(2), 357–384 (1989)
15. Hardy, M.: A regime-switching model of long-term stock returns. North Amer. Actuar. J. **5**(2), 41–53 (2001)
16. Jiang, Z., Pistorius, M.R.: On perpetual American put valuation and first-passage in a regime-switching model with jumps. Financ. Stochast. **12**(2), 331–355 (2008)
17. Jeanblanc, M., Yor, M., Chesney, M.: Mathematical Methods for Financial Markets. Springer, London (2009)
18. Kou, S.G.: A jump diffusion model for option pricing. Manag. Sci. **48**, 1086–1101 (2002)
19. Kou, S.G., Wang, H.: First passage times of a jump diffusion process. Adv. Appl. Probab. **35**, 504–531 (2003)
20. Kou, S.G., Wang, H.: Option pricing under a double exponential jump diffusion model. Manag. Sci. **50**, 1178–1192 (2004)
21. Nakajima, J.: Stochastic volatility model with regime-switching skewness in heavy-tailed errors for exchange rate returns. Stud. Nonlinear Dyn. Econ. **17**(5), 499–520 (2013)
22. Rogers, L.C.G.: Fluid models in queueing theory and Wiener–Hopf factorization of Markov chains. Ann. Appl. Probab. **4**(2), 390–413 (1994)
23. Stovall, S.: Standard & Poor's Sector Investing: How to Buy the Right Stock in the Right Industry at the Right Time. McGraw-Hill Companies, New York (1996)

Chapter 2
Estimation of Continuous Time Processes by Markov Chain Monte Carlo

From a statistical point of view, estimating parameters by log-likelihood maximization is the most reliable method. However this task is difficult to carry out when the calculation of the likelihood is computationally intensive, as for instance in the multivariate extension of switching models of Chap. 1. This chapter presents an alternative to maximum likelihood estimation based on a Bayesian learning paradigm. This estimation procedure, called Markov Chain Monte Carlo (MCMC), is a powerful framework providing information about the density of parameter estimates. In MCMC, the uncertainty about parameters is expressed and measured by probabilities. This formulation allows for a probabilistic treatment of our a priori knowledge about parameters based on simulations. The first part of this chapter reviews the main features of Markov chains. Next we review the Metropolis–Hastings (or MCMC) algorithm and apply it to fit a bivariate switching process to the S&P 500 and Nikkei indexes. This approach is combined with a particle filter in Chap. 3 and used in various contexts in subsequent chapters.

2.1 Markov Chains

Let us consider $p \in \mathbb{N}$ stock prices $S_t = \left\{ \left(S_t^k \right)_{t \geq 0} \right\}_{k=1,\ldots,p}$ defined on a probability space $(\Omega, \mathcal{F}, \mathbb{P})$ endowed with the natural filtration of this process, $\{\mathcal{F}_t\}_{t \geq 0}$. At this stage, we do not formulate any assumptions about the dynamics of stock prices. The log-return of the kth stock over a period Δ is denoted by $X_t^k = \ln \frac{S_t^k}{S_{t-\Delta}^k}$. The vector $X_t = \left(X_t^k \right)_{k=1,\ldots,p}$ is a multivariate random variable. The discrete record of T observations of p log-returns, equally spaced, with a lag Δ is a matrix $x = \{x_1, x_2, \ldots, x_T\}$, where $x_i = \left(x_{i,1}, \ldots, x_{i,p} \right)^\top \in \mathbb{R}^p$. The times of observations are $\{t_0, t_1, \ldots, t_T\}$. The probability density function (pdf) of log-returns is denoted

D. Hainaut, *Continuous Time Processes for Finance*,
Bocconi & Springer Series 12, https://doi.org/10.1007/978-3-031-06361-9_2

by $f(x_i \mid \Theta)$, where $x_t \in \mathbb{R}^p$. This pdf is defined by a vector of parameters Θ and may depend upon the history of the process.

The dimension of Θ is denoted by m. In a Bayesian setup, these parameters are realizations $\boldsymbol{\theta}$ of a multivariate random variable, with a density $\pi(\boldsymbol{\theta})$ and defined on a space of parameters $\mathcal{X} \subset \mathbb{R}^m$. The MCMC algorithm builds a discrete time Markov chain that converges in distribution toward $\pi(\boldsymbol{\theta})$.

Let us recall that a stochastic process is a sequence of random variables or vectors defined on some known state space. A Markov chain is a stochastic process in which future states are independent of past states given the present state. Think of $\mathcal{X}$ as our parameter space. "Consecutive" implies a time component, indexed by t. Consider a draw of Θ_t to be a state at iteration t. The next draw Θ_{t+1} is dependent only on the current draw Θ_t and not on any past draws.

In the rest of this chapter, Θ_t represents a random vector and we denote by $\boldsymbol{\theta}$ a realization of Θ_t. The transition between two consecutive steps of time is defined by the transition kernel.

Definition 2.1 A *transition kernel* is a function K defined on $\mathcal{X} \times \mathcal{B}(\mathcal{X})^1$ such that

$\forall \boldsymbol{\theta} \in \mathcal{X}$, $K(\boldsymbol{\theta}, \cdot)$ is a probability measure.
$\forall A \in \mathcal{B}(\mathcal{X})$, $K(\cdot, A)$ is measurable.

When $\mathcal{X}$ is discrete, the transition kernel is a matrix with elements

$$\mathbb{P}_{\boldsymbol{\theta}_t \boldsymbol{\theta}_{t-1}} = \mathbb{P}(\Theta_t = \boldsymbol{\theta}_t \mid \Theta_{t-1} = \boldsymbol{\theta}_{t-1}) \quad \boldsymbol{\theta}_{t-1}, \boldsymbol{\theta}_t \in \mathcal{X}.$$

In the continuous case, the kernel also denotes the conditional density such that $\mathbb{P}(\Theta_t \in A \mid \Theta_{t-1} = \boldsymbol{\theta}) = \int_A K(\boldsymbol{\theta}, d\boldsymbol{\theta}')$. We define a Markov chain as follows:

Definition 2.2 Given a transition kernel K, a sequence $\Theta_0, \Theta_1, \ldots, \Theta_t$ of random variables is a *Markov chain*, denoted by $(\Theta_t)_{t \geq 0}$, if for any t the conditional distribution of Θ_{t+1} given $\boldsymbol{\theta}_t, \boldsymbol{\theta}_{t-1}, \ldots, \boldsymbol{\theta}_0$ is the same as the distribution of Θ_t given $\boldsymbol{\theta}_{t-1}$:

$$\mathbb{P}(\Theta_t \in A \mid \boldsymbol{\theta}_0, \boldsymbol{\theta}_1, \ldots, \boldsymbol{\theta}_{t-1}) = \mathbb{P}(\Theta_t \in A \mid \boldsymbol{\theta}_t) = \int_A K(\boldsymbol{\theta}_t, d\boldsymbol{\theta}). \tag{2.1}$$

The chain is *time-homogeneous* if the distribution of $\Theta_{t_1}, \ldots, \Theta_{t_k}$ given Θ_{t_0} is the same as the distribution of $\Theta_{t_1-t_0}, \Theta_{t_2-t_0}, \ldots, \Theta_{t_k-t_0}$ given Θ_0 for every k and every $(k+1)$-tuple such that $t_0 \leq t_1 \leq \cdots \leq t_k$.

Example 2.3 Figure 2.1 shows an example of a Markov chain, defined on a discrete set: $\Theta_t \in \mathcal{X} = \{\boldsymbol{\theta}_L, \boldsymbol{\theta}_I, \boldsymbol{\theta}_H\}$, where $\boldsymbol{\theta}_L, \boldsymbol{\theta}_I, \boldsymbol{\theta}_H \in \mathbb{R}^m$. The arrows represent the possible transitions between two successive steps of Θ_t.

---

[1] $\mathcal{B}(\mathcal{X})$ is the sigma algebra defined on $\mathcal{X}$.

Fig. 2.1 Example of a
Markov chain, defined on a
discrete set
$\Theta_t \in X = \{\boldsymbol{\theta}_L, \boldsymbol{\theta}_I, \boldsymbol{\theta}_H\}$

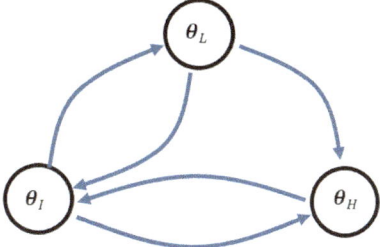

In this case, the transition kernel is given by the following matrix:

$$K(\boldsymbol{\theta}_t, \boldsymbol{\theta}_{t-1}) = \mathbb{P}(\Theta_t = \boldsymbol{\theta}_t | \Theta_{t-1} = \boldsymbol{\theta}_t) = \begin{array}{c c c c} & \mathbf{I} & \mathbf{L} & \mathbf{H} \\ \mathbf{I} & 0 & p_{IL} & p_{IH} \\ \mathbf{L} & p_{LI} & 0 & p_{LH} \\ \mathbf{H} & 1 & 0 & 0, \end{array}$$

where the p_{xy} are probabilities of transition between states.

Example 2.4 The autoregressive model AR(1) provides a simple illustration of a Markov chain on a continuous state space. Consider

$$\Theta_t = \gamma \Theta_{t-1} + \epsilon_t, \quad \gamma \in \mathbb{R}, \tag{2.2}$$

where $\epsilon_t \sim N(0, \sigma I_m)$ is a multivariate white noise of variance $\sigma^2 I_m$ and I_m is the identity matrix of dimension m. By construction, Θ_t is independent from Θ_{t-2} and Θ_{t-3} but conditional on Θ_{t-1}. The kernel function is in this case given by $K(\boldsymbol{\theta}_{t-1}, \boldsymbol{\theta}_t) \sim N(\gamma \boldsymbol{\theta}_{t-1}, \sigma I_m)$.

According to Eq. (2.1), a Markov chain has a short memory: Θ_t only depends upon the last realization of the chain, $\Theta_{t-1} = \boldsymbol{\theta}_{t-1}$. A direct consequence of this limited memory is that the expectation of any function $h(\cdot)$ of $\Theta_{t+1}, \ldots, \Theta_{t+k}$ conditionally to the information up to time t is given by

$$\mathbb{E}\left(h(\Theta_{t+1}, \ldots, \Theta_{t+k}) | \boldsymbol{\theta}_0, \ldots, \boldsymbol{\theta}_t\right) = \mathbb{E}\left(h(\Theta_{t+1}, \ldots, \Theta_{t+k}) | \boldsymbol{\theta}_t\right),$$

provided that the expectation exists.

The distribution of Θ_0, the initial state of the chain, plays an important role. In the discrete case, K is a transition matrix, and given an initial distribution $\alpha_0 = \boldsymbol{\theta}_0$, the marginal probability distribution of Θ_1 is obtained from the matrix multiplication

$$\alpha_1 = \alpha_0^\top K$$

and for Θ_t by repeated multiplication $\Theta_t \sim \alpha_t = \alpha_0^\top K^t$. If the state space of the Markov chain is continuous, the initial distribution of Θ_0 is also denoted by α_0.

Later, we will need the kernel for k transitions, which is defined as

$$K^k(\boldsymbol{\theta}, A) = \int_X K^{k-1}(\boldsymbol{\theta}', A) K(\boldsymbol{\theta}, \mathrm{d}\boldsymbol{\theta}').$$

The Chapman–Kolmogorov equation is a consequence of properties of the kernel function.

Proposition 2.5 (Chapman–Kolmogorov Equation) *For every $(m, k) \in \mathbb{N}^2$, $\boldsymbol{\theta} \in X$ and $A \in \mathcal{B}(X)$*

$$K^{m+k}(\boldsymbol{\theta}, A) = \int_X K^m(\boldsymbol{\theta}', A) K^k(\boldsymbol{\theta}, \mathrm{d}\boldsymbol{\theta}').$$

In an informal sense, the Chapman–Kolmogorov equation states that to get from $\boldsymbol{\theta}$ to A in $m + k$ steps, you pass through some $\boldsymbol{\theta}'$ on the k-th step. In the discrete case, the integral is interpreted as a product of matrices.

The property of irreducibility is a first measure of the sensitivity of the Markov chain to the initial conditions α_0. In the discrete case, the chain is irreducible if all states communicate.

Definition 2.6 Given a measure φ on $\mathbb{R}^m$, the Markov chain Θ_t with transition kernel $K(\boldsymbol{\theta}_{t-1}, \boldsymbol{\theta}_t)$ is *φ-irreducible* if for every $A \in \mathcal{B}(X)$ with $\varphi(A) > 0$, there exists a t such that $K^t(\boldsymbol{\theta}, A) > 0$ for all $\boldsymbol{\theta} \in X$. The chain is *strongly φ-irreducible* if $t = 1$ for all measurable A.

Example 2.4, Continued If $\Theta_t = \gamma \Theta_{t-1} + \epsilon_t$, where ϵ_t is a white noise on $\mathbb{R}^m$, then the chain is irreducible with respect to Lebesgue measure. But if ϵ_t is uniform on $[-1, 1]^m$ and $|\gamma| > 1$, the chain is no longer irreducible. Indeed,

$$\Theta_{t+1} - \Theta_t \geq (\gamma - 1)\Theta_{t-1} - 1 \geq 0$$

for $\Theta_t \geq \frac{1}{\theta - 1}$. The chain is thus monotonically increasing and cannot visit previous values.

Example 2.7 Figure 2.2 presents an example of an irreducible Markov chain, defined on a discrete set X. The chain is irreducible because we cannot get to A from B or C regardless of the number of steps we consider.

If X is discrete, the *period* of a state $\boldsymbol{\theta}$ of a chain is the minimal number of time steps before we could expect the chain to return to this state:[2]

$$d(\boldsymbol{\theta}) = g.c.d. \left\{ u \geq 1 ; \ K^u(\boldsymbol{\theta}, \boldsymbol{\theta}) > 0 \right\},$$

---

[2] *g.c.d.*: greatest common divisor.

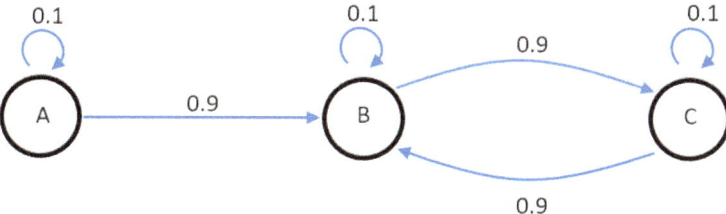

Fig. 2.2 Example of an irreducible Markov chain

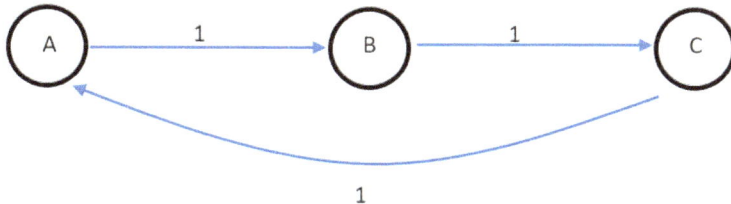

Fig. 2.3 Example of a periodic Markov chain. The period of the chain is 3

and a chain is *aperiodic* if it has period 1 for all states. Figure 2.3 shows an example of a periodic Markov chain defined on a discrete state space.

The extension to continuous Markov chains is the following:

Definition 2.8 A φ-irreducible chain (Θ_t) has a *life cycle of length d on A* if d is the g.c.d. of

$$\left\{ u \geq 1, \text{ such that } K^u(\theta, A) \geq 0 \right\}$$

$\forall \theta \in X, \forall A \in \mathcal{B}(X)$.

From an algorithmic point of view, irreducibility ensures that every set A is visited by the Markov chain, but this property is too weak to guarantee that Θ_t often enters A. Let us denote by $\eta_A = \sum_{t=1}^{\infty} I_A(\Theta_t)$ the number of visits of Ω_t to A. The following definition characterizes the frequency of visits to A:

Definition 2.9 In a finite state space X, a state $\theta \in X$ is *transient* if the average number of visits to θ, denoted by $\mathbb{E}(\eta_\theta)$, is finite and *recurrent* if $\mathbb{E}(\eta_\theta) = \infty$.

An increased level of stability is attained if the marginal distribution of Θ_t is independent from t. This is required to ensure the existence of a probability distribution π such that $\Theta_{t+1} \sim \pi$ if $\Theta_t \sim \pi$. MCMC methods are based on this requirement.

Definition 2.10 A finite measure π is *invariant* for the transition kernel $K(\cdot, \cdot)$ and the chain Θ_t if

$$\pi(B) = \int_X K(\theta, B)\pi(d\theta) \quad \forall B \in \mathcal{B}(X).$$

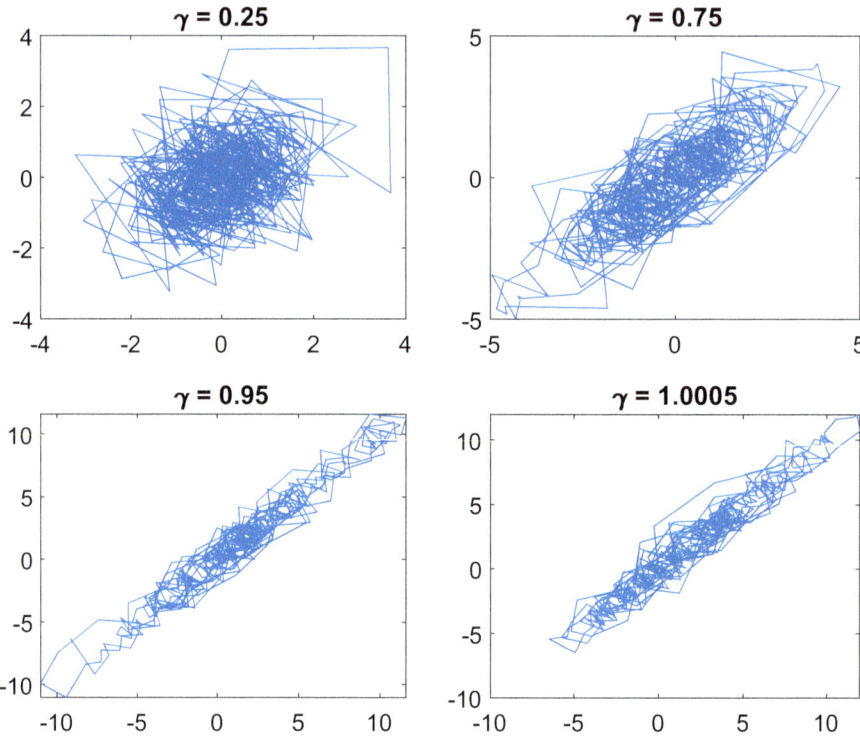

Fig. 2.4 Trajectories of four autoregressive AR(1) chains with $\sigma = 1$

When there exists an invariant probability measure for a φ-irreducible chain, the chain is called "positive." The invariant distribution is also referred to as *stationary* if π is a probability measure.

Example 2.4, Continued If $\Theta_t = \gamma \Theta_{t-1} + \epsilon_t$, the kernel is $N(\gamma \boldsymbol{\theta}_{t-1}, \sigma I_m)$, and the stationary distribution $N(\mu, \tau)$ is stationary for the AR(1) chain only if

$$\mu = \gamma \mu \quad \tau^2 = \tau^2 \gamma^2 + \sigma^2 .$$

This implies that $\mu = 0$ and

$$\tau^2 = \sigma^2 / (1 - \gamma^2) ,$$

which is possible only for $|\gamma| < 1$. Figure 2.4 exhibits the sample paths of pairs (Θ_{t-1}, Θ_t) of four univariate AR(1) chains, as defined by Eq. (2.2) and when $\sigma = 1$. We observe that all states of chains with $\gamma = 0.25$, 0.75, 0.95 are recurrent. We clearly observe that the states of the chains become transient when γ increases. When $|\gamma| \gg 1$, the chain does not show any stability and diverges.

The next result formalizes the intuition that the existence of an invariant measure prevents the probability mass from escaping to infinity:

Proposition 2.11 *If the chain* $(\Theta_t)_{t \geq 0}$ *is* positive *(invariant + φ-irreducible), then it is* recurrent.

Proof Let us recall that η_θ is the number of visits by $(\Theta_t)_{t \geq 0}$ to θ. If $(\Theta_t)_{t \geq 0}$ is transient, there exists a covering of X by uniformly transient sets A_j with bounds

$$\mathbb{E}\left(\eta_{A_j}\right) \leq M_j \quad \forall j \in \mathbb{N}.$$

From the invariance of π, we know that

$$\pi(A_j) = \int K(\theta, A_j)\pi(d\theta) = \int K^t(\theta, A_j)\pi(d\theta),$$

then for every $k \in \mathbb{N}$,

$$k\pi(A_j) = \sum_{t=0}^{k} \int K^t(\theta, A_j)\pi(d\theta) \leq \int \mathbb{E}(\eta_{A_j})\pi(d\theta) \leq M_j$$

since $\mathbb{E}(\eta_{A_j}) = \sum_{t=0}^{\infty} K^t(\theta, A_j)$. When $k \to \infty$, this shows that $\pi(A_j) = 0$, for every $j \in \mathbb{N}$. Thus, it is impossible to obtain an invariant probability. $\square$

Considering a Markov chain $(\Theta_t)_{t \geq 0}$, it is natural to establish the limit behavior of Θ_t. The existence and uniqueness of an invariant distribution π makes that distribution a natural candidate for the limiting distribution. In this case, if Θ_t converges to this invariant distribution, the chain is ergodic. More precisely, we have the following:

Definition 2.12 A chain $(\Theta_t)_{t \geq 0}$ that is invariant, irreducible, aperiodic, and positive recurrent is said to be *ergodic*. Then with probability 1, the sum $S_t(h) = \frac{1}{t}\sum_{k=1}^{t} h(\Theta_k)$, for any function $h(\cdot)$ defined on X, converges to the corresponding expectation:

$$S_t(h) \to \int_X h(\theta)\,\pi(d\theta).$$

The proof of this convergence theorem may be found in Robert and Casella [13]. The stability inherent to stationary chains can be related to another property called reversibility.

Definition 2.13 A stationary Markov chain $(\Theta_t)_{t \geq 0}$ is *reversible* if the distribution of Θ_{t+1}, conditionally on $\Theta_{t+2} = \theta$, is the same as the distribution of Θ_{t+1} conditionally on $\Theta_t = \theta$.

We will see later that this property is related to the existence of a stationary distribution.

Definition 2.14 A Markov chain with transition kernel K satisfies the *detailed balance condition* if there exists a function f such that

$$K(\theta', \theta) f(\theta') = K(\theta, \theta') f(\theta) \tag{2.3}$$

for every (θ, θ').

The balance condition is not necessary for f to be a stationary measure associated with $K(\cdot, \cdot)$, and however it does provide a sufficient condition that is used in the Metropolis–Hastings algorithm [9] as stated in the next proposition.

Proposition 2.15 *Suppose that a Markov chain with kernel K satisfies the detailed balance condition (2.3) with a function π. Then*

(1) The density π is the invariant density of the chain.
(2) The chain is reversible.

Proof (1) comes from the following relation:

$$\int_X K(\theta, B)\pi(\theta)\mathrm{d}\theta = \int_X \int_B K(\theta, \theta')\pi(\theta)\mathrm{d}\theta' \, \mathrm{d}\theta$$

$$= \int_X \int_B K(\theta', \theta)\pi(\theta')\mathrm{d}\theta' \, \mathrm{d}\theta$$

$$= \int_B \pi(\theta')\mathrm{d}\theta'$$

as $\int K(x, y)\mathrm{d}y = 1$. The proof of (2) follows from the existence of the kernel and invariant density. In this case, the detailed balance condition and reversibility are the same. $\qquad\square$

2.2 MCMC

Let us recall that the pdf of log-returns sampled at time t_i is denoted by $f(x_i \mid \Theta)$, where $x_t \in \mathbb{R}^p$. This pdf is defined by a vector of parameters Θ of dimension m. This vector is itself seen as a multivariate random variable with realizations θ and statistical distribution $\pi(\theta)$. The MCMC algorithm builds a discrete time Markov chain $(\Theta_t)_{t \geq 0}$ that converges in distribution toward $\pi(\theta)$, the stationary distribution.

Definition 2.16 A *Markov Chain Monte Carlo (MCMC) method* for the simulation of a statistical distribution $\pi(\boldsymbol{\theta})$ is any method producing an ergodic Markov chain $(\Theta_t)_{t \geq 0}$ whose stationary distribution is $\pi(\boldsymbol{\theta})$.

The main difficulty is to find valid transition kernels associated with an arbitrary stationary distribution. However, the Metropolis–Hastings algorithm is an elegant solution to this problem. Once the kernel is determined, the Markov chain is simulated, and after a burn-in period, the empirical distribution of the sample converges to π, the target density.

We select an arbitrary transition pdf that is denoted by $q(\boldsymbol{\theta}_t|\boldsymbol{\theta}_{t-1})$. In practice, we choose a distribution that is easy to simulate, admits a closed form solution, and is eventually symmetric: $q(\boldsymbol{\theta}_{t-1}|\boldsymbol{\theta}_t) = q(\boldsymbol{\theta}_t|\boldsymbol{\theta}_{t-1})$. Next we apply Algorithm 2.1.

Algorithm 2.1 Metropolis–Hastings algorithm

Main procedure:

    For $t = 0$ to maximum epoch, T

        1. Simulate $\boldsymbol{\theta}' \sim q\left(\boldsymbol{\theta}'|\boldsymbol{\theta}_t\right)$

        2. Take

$$\Theta_{t+1} = \begin{cases} \boldsymbol{\theta}' & \text{with probability } \rho(\boldsymbol{\theta}_t, \boldsymbol{\theta}') \\ \boldsymbol{\theta}_t & \text{with probability } 1 - \rho(\boldsymbol{\theta}_t, \boldsymbol{\theta}') \end{cases}$$

        where $\rho(\boldsymbol{\theta}_t, \boldsymbol{\theta}')$ is the acceptance probability

$$\rho(\boldsymbol{\theta}_t, \boldsymbol{\theta}') = \min\left\{\frac{\pi(\boldsymbol{\theta}')}{\pi(\boldsymbol{\theta}_t)} \frac{q\left(\boldsymbol{\theta}_t|\boldsymbol{\theta}'\right)}{q\left(\boldsymbol{\theta}'|\boldsymbol{\theta}_t\right)}, 1\right\}. \tag{2.4}$$

    End loop on epochs

We refer the reader to Glasserman [6] for technical details about the simulation of random samples. Figure 2.5 illustrates the Metropolis–Hastings algorithm. We consider a univariate Markov chain with a log-normal stationary distribution: $\ln \pi(\boldsymbol{\theta}) \sim N(\mu = 0.10, \sigma = 0.35)$. We assume that $q\left(\boldsymbol{\theta}'|\boldsymbol{\theta}_t\right)$ is the pdf of a uniform random variable defined on $[\boldsymbol{\theta} - \epsilon; \boldsymbol{\theta} + \epsilon]$, where $\epsilon \in \mathbb{R}^+$ is a tuning parameter set to 0.1 in the illustration. The right plot shows a sample path of Θ_t simulated with 10,000 iterations of the Metropolis–Hastings algorithm. The left plot compares the real log-normal stationary distribution to the empirical pdf of the last 5000 $\boldsymbol{\theta}_t$. We observe a clear similarity. In the next proposition, we examine the Metropolis kernel and find that it satisfies the detailed balance condition.

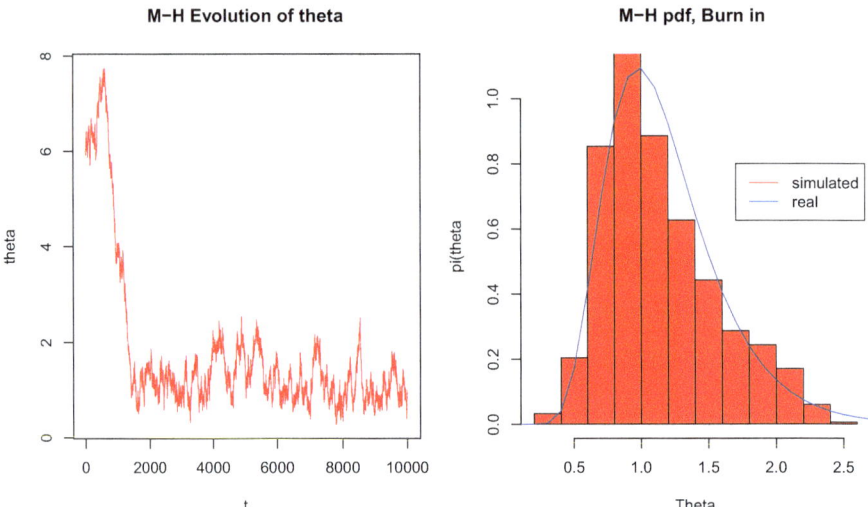

Fig. 2.5 Left plot: sample path of Θ_t simulated with the Metropolis–Hastings algorithm. Right plot: comparison of simulated and real stationary pdf of Θ_t (burn-in period: 5000 iterations)

Proposition 2.17 *Let $(\Theta_t)_{t\geq 0}$ be the chain computed by the Metropolis–Hastings algorithm. For every conditional distribution whose support includes the support X of the target distribution:*

(a) The kernel of the chain satisfies the detailed balance condition with the function π.

(b) π is a stationary distribution of the chain.

Proof The transition kernel of $(\Theta_t)_{t\geq 0}$ is the following:

$$K(\boldsymbol{\theta}_t, \boldsymbol{\theta}') = \rho(\boldsymbol{\theta}_t, \boldsymbol{\theta}')q(\boldsymbol{\theta}'|\boldsymbol{\theta}_t) + (1 - r(\boldsymbol{\theta}_t))\delta_{\{\boldsymbol{\theta}_t\}}(\boldsymbol{\theta}'),$$

where $r(\boldsymbol{\theta}_t) = \int \rho(\boldsymbol{\theta}_t, \boldsymbol{\theta}')q(\boldsymbol{\theta}'|\boldsymbol{\theta}_t)\mathrm{d}\boldsymbol{\theta}'$ and $\delta_{\{\boldsymbol{\theta}_t\}}(\cdot)$ is the Dirac mass at point $\boldsymbol{\theta}_t$. By construction, it is straightforward to check that

$$\rho(\boldsymbol{\theta}_t, \boldsymbol{\theta}')q(\boldsymbol{\theta}'|\boldsymbol{\theta}_t)\pi(\boldsymbol{\theta}_t) = \rho(\boldsymbol{\theta}', \boldsymbol{\theta}_t)q(\boldsymbol{\theta}_t|\boldsymbol{\theta}')\pi(\boldsymbol{\theta}'),$$

$$(1 - r(\boldsymbol{\theta}_t))\,\delta_{\{\boldsymbol{\theta}_t\}}(\boldsymbol{\theta}')\pi(\boldsymbol{\theta}_t) = \left(1 - r(\boldsymbol{\theta}')\right)\delta_{\{\boldsymbol{\theta}'\}}(\boldsymbol{\theta}_t)\pi(\boldsymbol{\theta}'),$$

which together establish the detailed balance condition. Statement (b) follows from Proposition 2.15.   □

Since convergence usually occurs regardless of our starting point, we can usually pick any feasible starting point (for example, picking starting draws that are in the parameter space). However, the time it takes for the chain to converge varies depending on the starting point. As a matter of practice, most people throw out a certain number of the first draws, known as the burn-in. This is to make our draws closer to the stationary distribution and less dependent on the starting point. However, it is unclear how much we should burn-in since our draws are all slightly dependent and we do not know exactly when convergence occurs.

2.3 Bayesian Inference

The discrete record of T observations of p log-returns, equally spaced, with a lag Δ is a matrix $x = \{x_1, x_2, \ldots, x_T\}$, where $x_i = \left(x_{i,1}, \ldots, x_{i,p}\right)^\top \in \mathbb{R}^p$. The times of observations are $\{t_0, t_1, \ldots, t_T\}$ and $x_{j,k} = \ln \dfrac{S_j^{(k)}}{S_{j-1}^{(k)}}$. The likelihood of the ith observation for a realization θ of $\Theta \in \mathbb{R}^m$ is denoted $f(x_i | \theta)$. The likelihood of the sample is $f(x | \theta) = \prod_{i=1}^n f(x_i | \theta)$. Furthermore, we adopt the following conventions:

- Conditionally to observations x, the m-vector of parameters θ is distributed according to $f(\theta \mid x)$, called the *posterior distribution*.
- The *prior distribution* of Θ is denoted by $f(\theta)$.

In a Bayesian framework, we aim to estimate parameters Θ given the measurement signals x. Using Bayes' rule, we determine the posterior distribution of Θ, which is equal to

$$\underbrace{f(\theta | x)}_{\text{posterior}} = \frac{f(x | \theta) \, f(\theta)}{\int_X p(x | \theta) \, p(\theta) \mathrm{d}x}$$

$$\propto \underbrace{f(x | \theta)}_{\text{Likelihood of data}} \underbrace{f(\theta)}_{\text{Prior}}.$$

If the likelihood $f(x | \theta)$ is easy to calculate, we can estimate the distribution $f(\theta | x)$ with the Metropolis–Hastings algorithm.

The transition density is denoted by $q(\theta_t | \theta_{t-1})$. If the target density $\pi(\cdot)$ of the Metropolis–Hastings algorithm is the posterior distribution of parameters $f(\theta | x)$, the acceptance probability in Eq. (2.4) is rewritten as follows:

$$\rho(\theta_t, \theta') = \min \left\{ \frac{f(\theta' | x)}{f(\theta_t | x)} \frac{q(\theta_t | \theta')}{q(\theta' | \theta_t)}, 1 \right\} \tag{2.5}$$

$$= \min \left\{ \frac{f(x | \theta') f(\theta')}{f(x | \theta_t) f(\theta_t)} \frac{q(\theta_t | \theta')}{q(\theta' | \theta_t)}, 1 \right\}.$$

Furthermore, if the condition density is chosen to be symmetric (e.g., normal distribution), then $\rho(\boldsymbol{\theta}_t, \boldsymbol{\theta}')$ becomes

$$\rho(\boldsymbol{\theta}_t, \boldsymbol{\theta}') = \min \left\{ \frac{f(\boldsymbol{x}|\boldsymbol{\theta}') f(\boldsymbol{\theta}')}{f(\boldsymbol{x}|\boldsymbol{\theta}_t) f(\boldsymbol{\theta}_t)}, \, 1 \right\}. \tag{2.6}$$

The resulting sample of parameters (after a burn-in period) $(\boldsymbol{\theta}_t)_{t=1:T}$ is next used to construct a Monte Carlo approximation of the empirical distribution of $f(\boldsymbol{\theta}|\boldsymbol{x})$. The parameter estimates are obtained by computing the following expectation:

$$\widehat{\Theta} = \mathbb{E}(\Theta|\boldsymbol{x}) = \int_{\mathcal{X}} \boldsymbol{\theta} f(\boldsymbol{\theta}\,|\,\boldsymbol{x}) \, \mathrm{d}\boldsymbol{\theta}$$

$$\approx \frac{1}{T} \sum_{t=1}^{T} \int_{\mathcal{X}} \delta_{\{\boldsymbol{\theta}_t\}}(\mathrm{d}\boldsymbol{\theta}),$$

which corresponds to a collection of Dirac atoms $\delta_{\{\boldsymbol{\theta}_t\}}(\mathrm{d}\boldsymbol{\theta})$ located at $\boldsymbol{\theta}_t$ with equal weights. This is also the sample mean of simulated $(\boldsymbol{\theta}_t)_{t=1,\ldots,T}$.

Algorithm 2.2 Metropolis–Hastings algorithm with partial update of parameters

---

Main procedure:

    For $t = 0$ to maximum epoch, T

        1. $k = t \mod n_q$. Simulate $\theta^{k'} \sim N\left(\theta_t^k, \, \sigma_\theta I_{m/n_q}\right)$ and set $\boldsymbol{\theta}' = \left(\theta_t^1, \ldots, \theta^{k'}, \ldots, \theta_t^{n_q}\right)$.

        2. Take

$$\Theta_{t+1} = \begin{cases} \boldsymbol{\theta}' & \text{with probability } \rho(\boldsymbol{\theta}_t, \boldsymbol{\theta}') \\ \boldsymbol{\theta}_t & \text{with probability } 1 - \rho(\boldsymbol{\theta}_t, \boldsymbol{\theta}') \end{cases}$$

        where $\rho(\boldsymbol{\theta}_t, \boldsymbol{\theta}')$ is the acceptance probability

$$\rho(\boldsymbol{\theta}_t, \boldsymbol{\theta}') = \min \left\{ \frac{f(\boldsymbol{x}|\boldsymbol{\theta}') f(\boldsymbol{\theta}')}{f(\boldsymbol{x}|\boldsymbol{\theta}_t) f(\boldsymbol{\theta}_t)}, \, 1 \right\}.$$

    End loop on epochs

In numerical applications, parameters are, for instance, updated with a normal transition probability

$$q(\boldsymbol{\theta}'|\boldsymbol{\theta}_t) \sim N\left(\boldsymbol{\theta}_t, \, \sigma_\theta I_m\right), \tag{2.7}$$

where I_m is the identity matrix of dimension m and σ_θ is a constant. This distribution being symmetric, the acceptance probability is given by Eq. (2.6). For models with a large number of parameters, simultaneously updating all estimates with $q(\boldsymbol{\theta}'|\boldsymbol{\theta}_t)$

leads to a high rejection rate of the proposed parameters $\boldsymbol{\theta}'$, if the standard deviation σ_θ is too high. Reducing this deviation improves the acceptance rate but can considerably slow down the convergence of the Metropolis–Hastings algorithm. An alternative is to update only a subset of parameters. Firstly, we partition the m-vector of parameters into n_q sub-vectors of size $\frac{m}{n_q}$,

$$\boldsymbol{\theta}_t = \left(\boldsymbol{\theta}_t^1, \ldots, \boldsymbol{\theta}_t^{n_q} \right).$$

During the tth iteration of the Metropolis–Hastings algorithm, we update only the kth $= t \mod n_q$ sub-vector of parameters:

$$q(\boldsymbol{\theta}^{k'}|\boldsymbol{\theta}_t^k) \sim N\left(\boldsymbol{\theta}_t^k, \sigma_\theta I_{m/n_q} \right)$$

and set the candidate vector to $\boldsymbol{\theta}' = \left(\boldsymbol{\theta}_t^1, \ldots, \boldsymbol{\theta}^{k'}, \ldots, \boldsymbol{\theta}_t^{n_q} \right)$. Algorithm 2.2 summarizes the Metropolis–Hastings framework with partial update of parameters.

Before applying the MCMC algorithm to estimate a multivariate extension of switching models, as introduced in Chap. 1, we illustrate this section with a simpler example. We consider a one-dimensional autoregressive time series,

$$x_{t+1} = \gamma x_t + \epsilon, \ \epsilon \sim N(0, \sigma),$$

where $\gamma = 0.60$ and $\sigma = 0.50$. In this case, $\boldsymbol{\theta} = (\gamma, \sigma)$ and if $\varphi(\cdot)$ is the pdf of a standard normal distribution, the likelihood of observations is given by

$$f(\boldsymbol{x}|\boldsymbol{\theta}) = \prod_{i=2}^{T} \varphi\left(\frac{x_i - \gamma x_{i-1}}{\sigma} \right).$$

As a prior distribution we choose $f(\boldsymbol{\theta}) = f_\gamma(\gamma) \times f_\sigma(\sigma)$, where

- $f_\gamma(\gamma)$ is the pdf of a uniform distribution over $[-1, 1]$.
- $f_\sigma(\sigma)$ is the pdf of a uniform distribution over $(0, 3]$.

The parameters are updated according to a bivariate normal transition probability $q(\boldsymbol{\theta}'|\boldsymbol{\theta}_t) \sim N(\boldsymbol{\theta}_t, \sigma_\theta I_2)$ with $\sigma_\theta = 0.01$. We set $\boldsymbol{\theta}_0 = (0.02, 0.01)$. The left plot of Fig. 2.6 displays the evolution of the log-likelihood $f(\boldsymbol{x}|\boldsymbol{\theta}_t)$ for the series of $\boldsymbol{\theta}_t$ sampled with the Metropolis–Hastings algorithm. The mid and right plots show simulated $\boldsymbol{\theta}_t = (\gamma_t, \sigma_t)$ with 2000 iterations of Algorithm 2.2. We observe a quick convergence to a stationary state. The parameter estimates computed with a burn-in period of 1000 iterations are $\widehat{\gamma} = 0.6051$ and $\widehat{\sigma} = 0.4845$, which are quite close to the exact values.

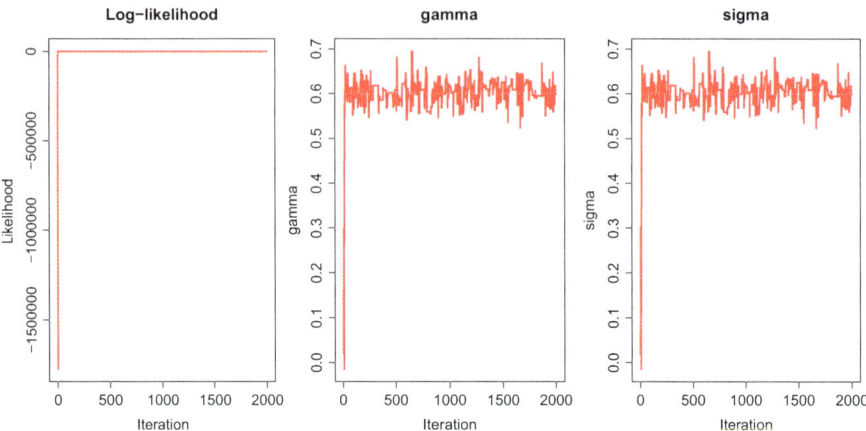

Fig. 2.6 Left plot: log-likelihood for each iteration of the Metropolis–Hastings algorithm. Mid and right plots: sampled $\boldsymbol{\theta}_t$ with 2000 iterations

2.4 Estimation of a Multivariate Switching Regime

This section illustrates how to estimate, by MCMC, a multivariate price process $S_t = \left\{ \left(S_t^k\right)_{t \geq 0} \right\}_{k=1,\dots,p}$ with switching regimes, as introduced in Chap. 1. This process is defined on a filtration $\{\mathcal{H}_t\}_{t \geq 0}$ of Ω that carries the information about asset prices. As in Chap. 1, the parameters are modulated by a hidden Markov chain, denoted by $(\delta_t)_{t \geq 0}$ and defined on the filtration $\{\mathcal{G}_t\}_{t \geq 0}$. This chain has n states and takes values from the canonical state space $E = \{e_1, \dots, e_n\}$, where $e_j = (0, \dots, 1, \dots, 0)^\top$. The set of states is denoted by $\mathcal{N} := \{1, 2, \cdots, n\}$ and the generator of δ_t is an $n \times n$ matrix $Q_0 := (q_{i,j})_{i,j \in \mathcal{N}}$.

The augmented filtration gathering information about all processes is denoted by $\{\mathcal{F}_t\}_{t \geq 0}$ and is such that $\mathcal{G}_t \cup \mathcal{H}_t \subset \mathcal{F}_t$. The instantaneous return of assets is driven by a multivariate geometric Brownian motion:

$$\frac{\mathrm{d}S_t}{S_t} = \boldsymbol{\mu}_t \mathrm{d}t + \Sigma_t \mathrm{d}\boldsymbol{W}_t \,, \tag{2.8}$$

where $\boldsymbol{W}_t$ is a vector of p Brownian motions. The vector of drift rates $\boldsymbol{\mu}_t = (\mu_i(t))_{i=1,\dots,p}$ and the $p \times p$ matrix $\Sigma_t = \left(\sigma_{i,j}(t)\right)_{i,j=1,\dots,p}$ are modulated by the Markov chain $\delta(t)$:

$$\mu_i(t) = \delta_t^\top \boldsymbol{\mu}_i \quad , \quad \sigma_{i,j}(t) = \delta_t^\top \boldsymbol{\sigma}_{i,j} \,,$$

where $\boldsymbol{\mu}_i = (\mu_{i,1}, \ldots, \mu_{i,n})^\top \in \mathbb{R}^n$ and $\boldsymbol{\sigma}_{i,j} = (\sigma_{i,j,1}, \ldots, \sigma_{i,j,n})^\top \in \mathbb{R}^n_+$. Applying the multivariate Itô lemma to the vector of log-returns $X_t = \ln \frac{S_t}{S_{t-\Delta}}$ over a time interval Δ leads to

$$dX_t = \left(\boldsymbol{\mu}_t - \frac{1}{2} \Sigma_t \Sigma_t^\top \right) dt + \Sigma_t dW_t \quad , \ t \geq t - \Delta. \qquad (2.9)$$

The discrete record of T observations of p log-returns, equally spaced, with a lag Δ is a matrix $x = \{x_1, x_2, \ldots, x_T\}$, where $x_i = (x_{i,1}, \ldots, x_{i,p})^\top \in \mathbb{R}^p$. The times of observations are $\{t_0, t_1, \ldots, t_T\}$. As the Markov chain $(\delta_t)_{t \geq 0}$ is not observable, we consider changes of regime to occur exclusively at times $\{t_0, t_1, \ldots, t_T\}$. Under this assumption and from Eq. (2.9), the probability density function (pdf) of log-returns is a multivariate normal:

$$f(x_k \mid \Theta) \sim N\left(\left(\boldsymbol{\mu}_{t_k} - \frac{1}{2} \Sigma_{t_k} \Sigma_{t_k}^\top \right) \Delta , \ \Sigma_{t_k} \Sigma_{t_k}^\top \Delta \right), \qquad (2.10)$$

where $\Theta = \left\{ (\mu_{j,k})_{j=1,\ldots,p\,,k \in N} \, , \ (\sigma_{i,j,k})_{i,j=1,\ldots,p\,,k \in N} , (q_{i,j})_{i,j \in N} \right\}$.

We adopt the notation $x_{1:k} = \{x_1, x_2, \ldots, x_k\}$. We calculate the likelihood of one observation, denoted by $f(x_k \mid \Theta, x_{1:k-1})$, with the Hamilton filter as introduced in Sect. 1.3:

$$f(x_k \mid \Theta, x_{1:k-1}) = \sum_{i=1}^{n} \sum_{j=1}^{n} p_i(t_{k-1} \mid \Theta, x_{1:k-1}) \, p_{i,j}(t_{k-1}, t_k \mid \Theta)$$

$$\times f(x_k \mid \Theta, \delta_{t_k} = e_j \, , \ \delta_{t_{k-1}} = e_i),$$

where

- $f(x_k \mid \Theta, \delta_{t_k} = e_j \, , \ \delta_{t_{k-1}} = e_i)$ is the multivariate normal density (2.10) in the ith regime.
- $p_{i,j}(t_{k-1}, t_k \mid \Theta)$ is the probability of transition from state i at time t_{k-1} to state j at time t_k for the set of parameters Θ.
- $p_i(t_{k-1} \mid \Theta, x_{1:k-1})$ is the probability of presence in state i at time t_{k-1}, conditionally to all observations up to t_{k-1}.

The probability $p_i(t_{k-1} \mid \Theta, x_{1:k-1})$ is inferred recursively from $f(x_{k-1} \mid \Theta, x_{1:k-2})$ as follows:

$$p_i(t_{k-1} \mid \Theta, x_{1:k-1}) = \frac{\sum_{j=1}^{n} p_j(t_{k-2} \mid \Theta, x_{1:k-2}) \, p_{j,i}(t_{k-2}, t_{k-1} \mid \Theta) \times f(x_{k-1} \mid \Theta, \delta_{t_{k-1}} = e_i \, , \ \delta_{t_{k-2}} = e_j)}{f(x_{k-1} \mid \Theta, x_{1:k-2})}.$$

In order to initiate the recursion, we need to determine $f(x_1 \mid \Theta)$. If the Markov chain has been running for a sufficiently long period of time, we assume that the

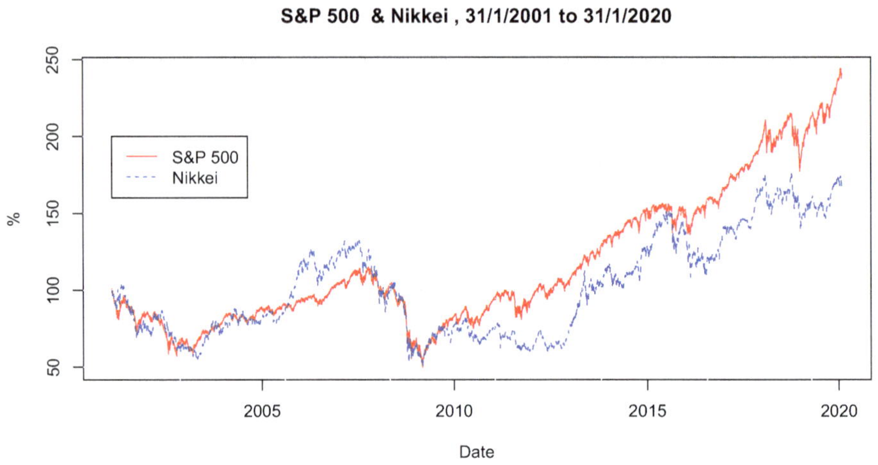

S&P 500 & Nikkei , 31/1/2001 to 31/1/2020

Fig. 2.7 Normalized Nikkei and S&P 500 daily values from 21/1/01 to 21/1/20 (100% on 31/1/2001)

probability of presence in a given state is equal to its stationary probability, denoted $\pi_i(\Theta)$ for $i = 1, \ldots, n$. The total log-likelihood used in the Metropolis–Hastings algorithm is the sum:

$$\ln f(\boldsymbol{x}|\Theta) = \ln f(\boldsymbol{x}_1|\Theta) + \ln f(\boldsymbol{x}_2|\Theta, \boldsymbol{x}_1)$$
$$+ \ln f(\boldsymbol{x}_3|\Theta, \boldsymbol{x}_{1:2}) + \cdots + \ln f(\boldsymbol{x}_T|\Theta, \boldsymbol{x}_{1:T-1}).$$

We fit a switching Brownian model (SGBM) with three regimes to the bivariate time series of the S&P 500 and Nikkei stock indexes, containing daily returns from 31/1/2001 to 31/1/2020 (4496 joint observations). Figure 2.7 compares the relative progression of these indexes. They are both scaled to 100% on 31/1/2001. In order to estimate the 21 parameters, we run 12,000 iterations of the Metropolis–Hastings algorithm with partial update of parameters ($n_q = 3$). The prior distribution is built as a product of uniform random variables that are null outside the domain of parameters. The distribution used for updating parameters is normal as in Eq. (2.7), with $\sigma_\theta = 0.01$.

Figure 2.8 shows the evolution of the log-likelihood and emphasizes the convergence to a maximum, which is around 27,750. The parameters are initialized with values close to those of the S&P 500, obtained in Chap. 1 with the Hamilton filter. We use as burn-in periods the first 8000 iterations.

Table 2.1 reports the estimates of expected returns $\mu_{1,k}$ and $\mu_{2,k}$ and standard deviations $\sqrt{\sigma_{1,1,k}^2 + \sigma_{1,2,k}^2}$, $\sqrt{\sigma_{2,2,k}^2 + \sigma_{2,1,k}^2}$ of returns in regime $k = 1, \ldots, 3$. These are computed with the 4000 last iterations. The first regime corresponds to a deep financial crisis. The average returns of the S&P 500 and Nikkei are both below -20% and the volatilities are larger than 40%. The second regime is a state of

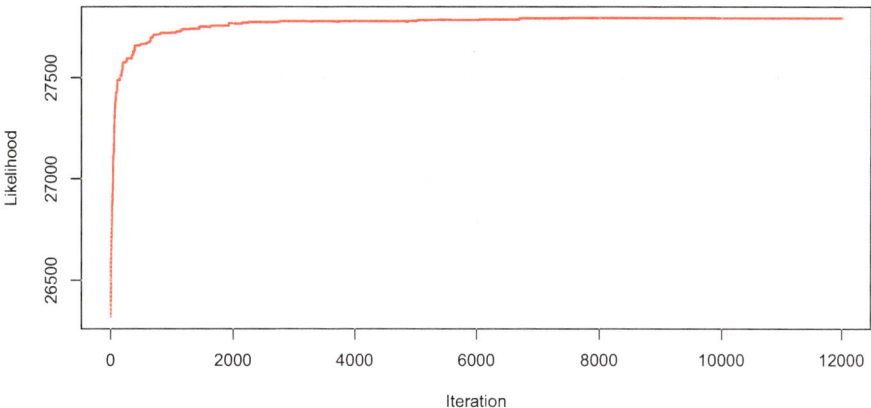

Fig. 2.8 Metropolis–Hastings algorithm: evolution of accepted log-likelihoods

Table 2.1 SGBM with 3 regimes: parameter estimates and their standard deviations (std estimates)

	Regimes	Estimates			Std estimates		
		1	2	3	1	2	3
S&P	Expected return	−0.27	0.00	0.28	0.01	0.01	0.01
	Standard deviation	0.42	0.17	0.08	0.01	0.01	0.00
Nikkei	Expected return	−0.22	0.01	0.45	0.04	0.01	0.01
	Standard deviation	0.49	0.23	0.14	0.02	0.00	0.00
	Correlation	0.24	0.15	0.22	0.01	0.01	0.01

economic stagnation, the expected returns stay close to zero, and the volatilities are around 20%. The last regime is identified with a period of strong economic growth with average returns above 25% and low volatilities. Notice that the correlation slightly increases up to 24% in periods of financial turmoil. Table 2.2 contains the transition and instantaneous probabilities of the hidden Markov chain. Notice that the daily probability of remaining in state 1 (92%) is lower than that of staying in states 2 and 3. The parameters for the S&P 500 may be compared with those of Table 1.5 of Chap. 1, which reports estimates obtained with the Hamilton filter. The standard deviations are comparable, but the expected returns exhibit larger spreads. In regimes 1 and 2, the Hamilton filter estimates returns of −33.21% and −6.65%, whereas the MCMC algorithm yields −27.12% and 0.00%.

The next step consists in filtering the probability of presence in the different regimes with the Hamilton filter. Figure 2.9 displays the results of this procedure and can be compared to Fig. 1.2 of Chap. 1 reporting the same information for the sole S&P 500. Stock markets being interconnected, we observe many similarities between these graphs. In particular, the S&P500 and the Nikkei enter into the first regime, which is assimilated to deep shocks, at the same times. In 2003, the second

Table 2.2 SGBM with 3 regimes: matrix of transition probabilities over 1 day and Q_0

	1	2	3
$p_{i,j}\,(t, t+1\,\text{day})$			
1	0.92	0.07	0.01
2	0.01	0.96	0.03
3	0.00	0.05	0.95
$q_{i,j}$			
1	−0.082	0.078	0.003
2	0.014	−0.051	0.037
3	0.001	0.049	−0.050

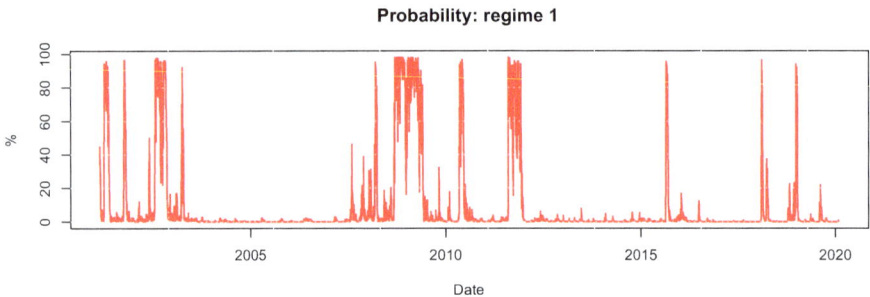

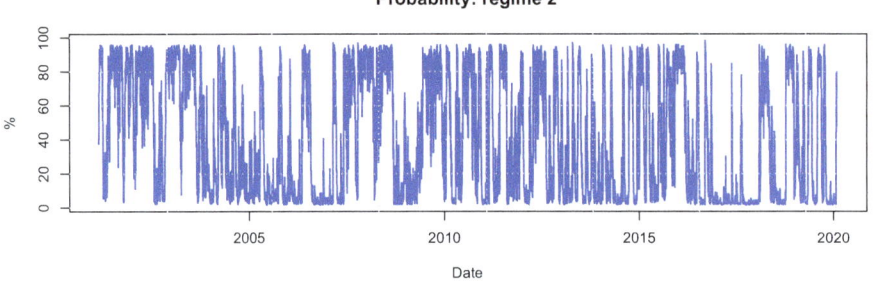

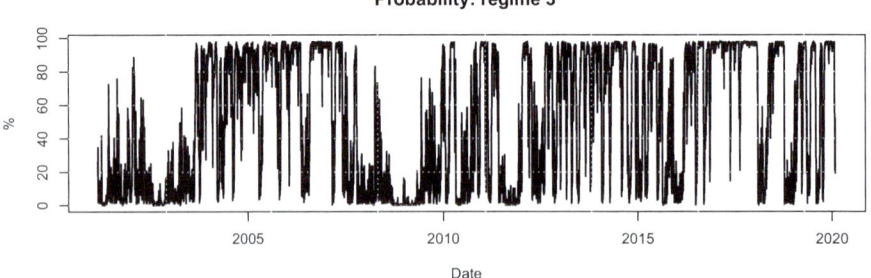

Fig. 2.9 Probabilities of presence in each regime

Gulf war drives down the S&P 500 and the Nikkei. The subprime crisis of 2008 and the double dip recession of 2011 hit both the US and Japanese markets. In 2015, the chain briefly enters into regime 1 due to turbulence on the Chinese market. The fall of oil price in 2018 due to overproduction also drives the Markov chain in the first regime.

2.5 Further Reading

The MCMC approach is a powerful tool for estimating models with latent variables. This method has been successfully applied to a wide variety of stochastic processes. We refer the reader to the seminal book of Meyn and Tweedie [10] for a detailed presentation of Markov chains in general state spaces. The paper by Roberts and Rosenthal [14] surveys various results about Markov chains on general (non-countable) state spaces. A survey of methods for Bayesian inference in continuous time asset pricing models is available in Johannes and Polson [8]. Eraker [4] uses this method for the estimation of parameters in one-factor interest rate models and a two-factor model with a latent stochastic volatility component. Chib et al. [2] study simulation-based inference in generalized models of stochastic volatility defined by heavy-tailed Student-t distributions and a jump component. Bayesian estimation and comparison of flexible, high-dimensional multivariate time series models with time varying correlations is explored in Chib et al. [3]. Griffin and Steel [7] use an MCMC method to estimate non-Gaussian Ornstein–Uhlenbeck processes for stochastic volatility. Molina et al. [11] adapt a Monte Carlo Markov Chain method to estimate the parameters of stochastic volatility models with several factors and test their approach on foreign exchange data. Li et al. [12] develop an MCMC algorithm for inferences of continuous time models with stochastic volatility and infinite-activity Lévy jumps. Zucchini et al. [15, Chapter 7] present a simulation-based "R" procedure for estimating a Poisson hidden Markov chain model. Practical aspects of the implementation of the MCMC algorithm are covered in the books by Brooks et al. [1] and Gilks et al. [5]. The MCMC algorithm is used in Chap. 3 for filtering volatility, in Chap. 4 to estimate contagion between shocks, in Chap. 10 on illiquidity, and in Chap. 9 about Volterra's equations.

References

1. Brooks, S., Gelman, A., Jones, G.L., Meng, X.L.: Handbook of Markov Chain Monte Carlo. CRC Press/Routledge Handbooks (Online), Boca Raton/Milton Park (2011)
2. Chib, S., Nardari, F., Shephard, N.: Markov chain Monte Carlo methods for stochastic volatility models. J. Econ. **108**(2), 108–281 (2002)
3. Chib, S., Nardari, F., Shephard, N.: Analysis of high dimensional multivariate stochastic volatility models. J. Econ. **134**(2), 341–371 (2006)

 4. Eraker, B.: MCMC Analysis of diffusion models with application to finance. J. Business Econ. Statist. **19**(2), 177–191 (2001)
 5. Gilks, W.R., Richardson, S., Spiegelhalter, D.J.: Markov Chain Monte Carlo in Practice. Chapman & Hall/CRC Press, Boca Raton (1996)
 6. Glasserman, P.: Monte Carlo Methods in Financial Engineering. Springer, New York (2003)
 7. Griffin, J.E., Steel, M.F.J.: Inference with non-Gaussian Ornstein–Uhlenbeck processes for stochastic volatility. J. Econ. **134**(2), 605–644 (2006)
 8. Johannes, M., Polson, N.: MCMC methods for continuous-time financial econometrics. In: Handbook of Financial Econometrics: Applications, vol. 2, pp. 1–72. Elsevier, Amsterdam (2010)
 9. Metropolis, N., Rosenbluth, A.W., Rosenbluth, M.N., Teller, A.H., Teller, E.: Equations of state calculations by fast computing machines. J. Chem. Phys. **21**, 1087–1091 (1953)
10. Meyn, S., Tweedie, R.L.: Markov Chains and Stochastic Stability, 2nd edn. Cambrdige Mathematical Library, Cambridge University Press, Cambridge (2009)
11. Molina, G., Han, C.H., Fouque, J.-P.: MCMC estimation of multiscale stochastic volatility models. In: Lee, C.F., Lee, A.C. (eds.) Handbook of Quantitative Finance and Risk Management, pp. 1109–1120. Springer, Boston (2008)
12. Li, H., Wells, M.T., Yu, C.L.: A Bayesian analysis of return dynamics with Lévy jumps. Rev. Financial Stud. **21**(5), 2345–2378 (2008)
13. Robert, C., Casella, G.: Monte Carlo Statistical Methods. Springer, New York (2004)
14. Roberts, G.O., Rosenthal, J.S.: General state space Markov chains and MCMC algorithms. Probab. Surveys **1**, 20–71 (2004)
15. Zucchini, W., MacDonald, I.L., Langrock, R.: Hidden Markov Models for Time Series: An Introduction Using R, 2nd edn. Chapman & Hall/CRC, Boca Raton (2016)

Chapter 3
Particle Filtering and Estimation

This chapter introduces an algorithm called particle filtering used for the inference of the most likely sample path of a hidden process driving stock prices in nested models. Particle filtering is a simulation-based method approximating the likelihood of observations. This approach allows us to fit processes for which the probability density function does not admit any closed form expression. The first part of the chapter introduces particle filtering applied to the Heston model for stock prices. In this dynamic, the instantaneous return is Brownian with a mean reverting variance. As the particle filter yields a non-smooth estimate of the log-likelihood function, we propose in the second part of the chapter to combine the particle filter with the Metropolis–Hastings procedure of Chap. 2 to estimate parameters. This approach, called the Particle Markov Chain Monte Carlo (PMCMC) algorithm, will be used in Chap. 10 to quantify illiquidity and in Chap. 9 to fit Volterra processes. The particle filter serves in Chap. 4 to estimate the sample path of jump intensity.

3.1 The Heston Model

To illustrate the material developed in this chapter, we consider the stochastic volatility model proposed by Heston [16]. The Heston model is a square-root diffusion model for the stochastic variance. It derives from the CIR model of Cox, Ingersoll and Ross [4] for interest rates. The stochastic variance is a process denoted by $(V_t)_{t \geq 0}$ and driven by the following dynamics:

$$\mathrm{d}V_t = \kappa \, (\gamma - V_t) \, \mathrm{d}t + \sigma \sqrt{V_t} \mathrm{d}W_t^v \, . \tag{3.1}$$

The variance reverts with a speed $\kappa > 0$ to a mean reversion level $\gamma > 0$. The volatility of the variance is a multiple σ of the square root of variance. $\left(W_t^v\right)_{t \geq 0}$ is a Brownian motion. The stock price is a stochastic process, denoted $(S_t)_{t \geq 0}$, ruled by

D. Hainaut, *Continuous Time Processes for Finance*,
Bocconi & Springer Series 12, https://doi.org/10.1007/978-3-031-06361-9_3

a geometric diffusion under the real measure $\mathbb{P}$:

$$dS_t = \mu\, S_t\, dt + S_t\sqrt{V_t}\left(\rho dW_t^v + \sqrt{1 - \rho^2}dW_t^s\right), \tag{3.2}$$

where $\left(W_t^s\right)_{t\geq 0}$ is a Brownian motion independent from $\left(W_t^v\right)_{t\geq 0}$ and $\mu \in \mathbb{R}$ is the expected instantaneous return. The parameter $\rho \in (-1, 1)$ is the correlation coefficient between the stock price and the volatility. A direct calculation leads to $\mathbb{E}\left(dW_t^v\left(\rho dW_t^v + \sqrt{1 - \rho^2}dW_t^s\right)\right) = \rho dt$, whereas the variance of $\rho dW_t^v + \sqrt{1 - \rho^2}dW_t^s$ is equal to dt. All processes are defined on a probability space Ω, endowed with their natural filtration $(\mathcal{F}_t)_{t\geq 0}$ and the probability measure, $\mathbb{P}$. From Itô's lemma for semi-martingales, any function $f(t, S_t, V_t)$ that is continuous and differentiable admits the following infinitesimal representation:

$$df = \frac{\partial f}{\partial t} + \left(\frac{\partial f}{\partial S}\mu S_t + \frac{\partial f}{\partial V}\kappa\left(\gamma - V_t\right)\right)dt \tag{3.3}$$

$$+ \frac{1}{2}S_t^2 V_t \frac{\partial^2 f}{\partial S^2}dt + \frac{1}{2}\sigma^2 V_t \frac{\partial^2 f}{\partial V^2}dt + \rho\,\sigma\, S_t\, V_t \frac{\partial^2 f}{\partial S\partial V}dt$$

$$+ \frac{\partial f}{\partial V}\sigma\sqrt{V_t}dW_t^v + \frac{\partial f}{\partial S}S_t\sqrt{V_t}\left(\rho dW_t^v + \sqrt{1 - \rho^2}dW_t^s\right).$$

The infinitesimal generator of f is the operator $\mathcal{A}f = \mathbb{E}\left(\frac{df}{dt}|\mathcal{F}_t\right)$. In this case, this is equal to

$$\mathcal{A}f(t, S, V) = \frac{\partial f}{\partial t} + \frac{\partial f}{\partial S}\mu S + \frac{\partial f}{\partial V}\kappa\left(\gamma - V\right) \tag{3.4}$$

$$+ \frac{1}{2}S^2 V\frac{\partial^2 f}{\partial S^2} + \frac{1}{2}\sigma^2 V\frac{\partial^2 f}{\partial V^2} + \rho\,\sigma SV\frac{\partial^2 f}{\partial S\partial V}.$$

Applying Itô's equation (3.3) to $d\ln S_t$ leads after integration to the following expression for the stock price:

$$S_t = S_0\exp\left(\int_0^t \mu - \frac{V_s}{2}ds + \rho\int_0^t \sqrt{V_s}\,dW_s^v\right. \tag{3.5}$$

$$\left. + \sqrt{1 - \rho^2}\int_0^t \sqrt{V_s}\,dW_s^s\right).$$

In a Black and Scholes framework, the variance is constant, and S_t is a log-normal random variable. In the Heston model, since the variance is a CIR process, the probability density function of S_t is unknown. This complicates the estimation of parameters under the real measure $\mathbb{P}$ and motivates the development of a "particle filter" in the next section.

The pricing of financial derivatives is performed under an equivalent probability measure, the risk neutral $\mathbb{Q}$, so as to exclude arbitrage opportunities. Under this measure, risky assets earn on average the risk-free rate whatever their volatility. Equivalent probability measures are constructed as follows. We consider two $\mathcal{F}_t$-adapted and square integrable processes, denoted by $(\beta_t^v)_{t\geq 0}$ and $(\beta_t^s)_{t\geq 0}$. Next we define a process $(Z_t)_{t\geq 0}$ (the Doléans-Dade exponential):

$$Z_t = \exp\left(-\frac{1}{2}\int_0^t \left(\beta_s^v\right)^2 + \left(\beta_s^s\right)^2 \, \mathrm{d}s - \int_0^t \beta_s^v \mathrm{d}W_s^v - \int_0^t \beta_s^s \mathrm{d}W_s^s\right) \quad (3.6)$$

with $Z_0 = 1$. If we apply Itô's lemma to Z_t, we immediately infer its infinitesimal dynamics:

$$\mathrm{d}Z_t = -Z_t\left(\beta_s^v \mathrm{d}W_s^v + \beta_s^s \mathrm{d}W_s^s\right).$$

Given that $\mathbb{E}\left(\mathrm{d}Z_s|\mathcal{F}_t\right) = 0$ and $\mathbb{E}\left(Z_s|\mathcal{F}_t\right) = Z_t + \int_t^s \mathbb{E}\left(\mathrm{d}Z_s|\mathcal{F}_t\right) = Z_t$, Z_t is a martingale. The process $(Z_t)_{t\geq 0}$ is also a Radon–Nikodym derivative $Z_t = \left.\frac{\mathrm{d}\mathbb{P}^\beta}{\mathrm{d}\mathbb{P}}\right|_t$ defining an equivalent probability measure, $\mathbb{P}^\beta$, to $\mathbb{P}$. We recall a well-known result in the next proposition and sketch its proof.

Proposition 3.1 *Under the measure* $\mathbb{P}^\beta$, $\mathrm{d}W_t^{s,\beta} = \mathrm{d}W_t^s + \beta_t^s \mathrm{d}t$ *and* $\mathrm{d}W_t^{v,\beta} = \mathrm{d}W_t^v + \beta_t^v \mathrm{d}t$ *are Brownian motions.*

Proof The proof consists in checking that the moment generating functions of $W_t^{s,\beta}$ and $W_t^{v,\beta}$ are those of normal random variables. We do this exercise for $W_t^{v,\beta}$. Its mgf is

$$\mathbb{E}^{\mathbb{P}^\beta}\left(e^{\omega W_t^{v,\beta}}|\mathcal{F}_0\right) = \mathbb{E}\left(Z_t\, e^{\omega W_t^v + \int_0^t \omega \beta_s^v \mathrm{d}s}|\mathcal{F}_0\right)$$

$$= e^{\frac{1}{2}\omega^2 t}\mathbb{E}\left(e^{-\frac{1}{2}\int_0^t (\beta_s^v - \omega)^2 + (\beta_s^s)^2 \mathrm{d}s - \int_0^t (\beta_s^v - \omega)\mathrm{d}W_s^v - \int_0^t \beta_s^s \mathrm{d}W_s^s}|\mathcal{F}_0\right)$$

$$= e^{\frac{1}{2}\omega^2 t}.$$

To pass from the second to the last line, we use the property that the Doléans-Dade exponential is a martingale. We recognize the mgf of a Brownian motion, which concludes the proof.   □

The equivalent measure $\mathbb{P}^\beta$ is a risk neutral one, denoted $\mathbb{Q}$, if and only if discounted prices are martingales. The next corollary states the conditions that $(\beta_t^v)_{t\geq 0}$ and $(\beta_t^s)_{t\geq 0}$ must fulfill to define $\mathbb{Q}$.

Corollary 3.2 *Let us denote the risk-free rate by $r \in \mathbb{R}$. If $\left(\beta_t^v\right)_{t \geq 0}$ and $\left(\beta_t^s\right)_{t \geq 0}$ are $\mathcal{F}_t$-adapted and square integrable processes satisfying:*

$$\frac{\mu - r}{\sqrt{V_t}} = \rho \beta_t^v + \sqrt{1 - \rho^2} \beta_t^s , \tag{3.7}$$

then the process $(Z_t)_{t \geq 0}$ in Eq. (3.6) defines a risk neutral measure.

Proof According to Proposition 3.1, the dynamics of the stock price can be rewritten as follows:

$$dS_t = \mu \, S_t \, dt - S_t \sqrt{V_t} \left(\rho \beta_t^v + \sqrt{1 - \rho^2} \beta_t^s \right) dt$$

$$+ S_t \sqrt{V_t} \left(\rho \left(dW_t^v + \beta_t^v dt \right) + \sqrt{1 - \rho^2} \left(dW_t^s + \beta_t^s dt \right) \right) .$$

The discounted stock price is a martingale under $\mathbb{P}^\beta$ if and only if the drift in this last equation is equal to the risk-free rate.   □

Condition (3.7) emphasizes that the market in the Heston model is incomplete. The risk neutral measure is indeed not unique given that there exists an infinity of solutions to Eq. (3.7). For a pair of processes (β_t^s, β_t^v) satisfying the relation (3.7), the dynamics of V_t and S_t under the risk neutral measure $\mathbb{Q}$ obey, respectively, the equations:

$$dV_t = \kappa \left(\gamma - \frac{\beta_t^v \sigma \sqrt{V_t}}{\kappa} - V_t \right) dt + \sigma \sqrt{V_t} dW_t^{v,\beta} \tag{3.8}$$

and

$$dS_t = r \, S_t \, dt + S_t \sqrt{V_t} \left(\rho dW_t^{v,\beta} + \sqrt{1 - \rho^2} dW_t^{s,\beta} \right) . \tag{3.9}$$

Equation (3.8) reveals that if $\beta_t^v \neq 0$, the variance reverts to a mean level, $\gamma_t^\beta = \gamma - \frac{\beta_t^v \sigma \sqrt{V_t}}{\kappa}$, which is itself an $\mathcal{F}_t$-process under the risk neutral measure. For a general $\beta_t^v \neq 0$, the variance is no longer a CIR process, and the moment generating function of the log-return of S_t is unknown. Since this prevents us from pricing options, we exclusively consider processes β_t^v of the form:

$$\beta_t^v = \beta_v \sqrt{V_t} , \tag{3.10}$$

where $\beta_v \in \mathbb{R}$ is constant. With this assumption, we have that $\beta_t^s = \frac{\mu - r - \rho \beta_v V_t}{\sqrt{1 - \rho^2} \sqrt{V_t}}$, and the variance dynamics is form-invariant under $\mathbb{Q}$:

$$\mathrm{d}V_t = \tilde{\kappa} \left(\tilde{\gamma} - V_t \right) \mathrm{d}t + \sigma \sqrt{V_t} \mathrm{d}W_t^{v,\beta}, \tag{3.11}$$

where $\tilde{\kappa}$ and $\tilde{\gamma}$ are the modified speed and level of mean reversion:

$$\tilde{\kappa} = \kappa + \beta_v \sigma, \qquad \tilde{\gamma} = \frac{\kappa \gamma}{\kappa + \beta_v \sigma}.$$

Notice that $\beta_v > -\frac{\kappa}{\sigma}$; otherwise, the mean reversion level $\tilde{\gamma}$ is negative under $\mathbb{Q}$.

The pricing of European options is performed under this measure by a Discrete Fourier Transform, in a similar manner to Sect. 1.6 of Chap. 1. This method requires the moment generating function of $(X_t)_{t \geq 0}$, the log-return of S_t. This log-return under the risk neutral measure is led by the SDE:

$$\mathrm{d}X_t = \mathrm{d}\ln\left(\frac{S_t}{S_0}\right) \tag{3.12}$$

$$= \left(r - \frac{V_t}{2}\right) \mathrm{d}t + \rho \sqrt{V_t} \, \mathrm{d}W_t^{v,\beta} + \sqrt{1 - \rho^2} \sqrt{V_t} \, \mathrm{d}W_t^{s\beta},$$

and the next proposition presents its mgf.

Proposition 3.3 *The mgf of X_s under the risk neutral $\mathbb{Q}$, for $s \geq t$ with $\omega \in \mathbb{C}_-$, is given by the following expression:*

$$\mathbb{E}^{\mathbb{Q}}\left(e^{\omega X_s} \mid \mathcal{F}_t\right) = \left(\frac{S_t}{S_0}\right)^{\omega} \exp\left(A(\omega, t, s) + B(\omega, t, s) V_t\right), \tag{3.13}$$

where $A(\omega, t, s)$ and $B(\omega, t, s)$ are such that

$$\begin{cases} \frac{\partial A(\omega,t,s)}{\partial t} &= -\omega r - B(\omega, t, s) \, \tilde{\kappa} \, \tilde{\gamma} \\ \frac{\partial B(\omega,t,s)}{\partial t} &= \frac{\omega}{2} - \frac{1}{2}\omega^2 + B(\omega, t, s) \left(\tilde{\kappa} - \rho \sigma \omega\right) \\ & \quad - \frac{1}{2}\sigma^2 B(\omega, t, s)^2 \end{cases} \tag{3.14}$$

with the terminal boundary conditions:

$$A(\omega, s, s) = 0 \quad B(\omega, s, s) = 0.$$

Proof Let us define $f(t, X_t, V_t) = \mathbb{E}^{\mathbb{Q}}\left(e^{\omega X_s} \mid \mathcal{F}_t\right)$. The infinitesimal generator of this function is the operator defined by

$$\mathscr{A}f = \lim_{h \to 0} \frac{\mathbb{E}^{\mathbb{Q}}\left(f\left(t + h, X_{t+h}, V_{t+h}\right) - f(t, X_t, V_t) \mid \mathcal{F}_t\right)}{h}.$$

This limit is null by definition of f. Therefore, we infer that if $X_t = X$ and $V_t = V$, the function f is a solution of the equation:

$$0 = \frac{\partial f}{\partial t} + \frac{\partial f}{\partial X}\left(r - \frac{V}{2}\right) + \frac{\partial f}{\partial V}\tilde{\kappa}\,(\tilde{\gamma} - V) \qquad (3.15)$$

$$+ \frac{1}{2}V\frac{\partial^2 f}{\partial X^2} + \frac{1}{2}\sigma^2 V\frac{\partial^2 f}{\partial V^2} + \rho\sigma V\frac{\partial^2 f}{\partial X \partial V}.$$

Let us further assume that f is an exponential affine function of X_t and V_t:

$$f(t, X, V) = \exp\left(A(\omega, t, s) + B(\omega, t, s)V + C(\omega, t, s)X\right),$$

where $A(\omega, t, s)$, $B(\omega, t, s)$, and $C(\omega, t, s)$ are the time-dependent functions with terminal conditions $A(\omega, s, s) = 0$, $B(\omega, s, s) = 0$, and $C(\omega, s, s) = \omega$. The partial derivatives of f with respect to the state variables are

$$\frac{\partial f}{\partial t} = \left(\frac{\partial}{\partial t}A(\omega, t, s) + \frac{\partial}{\partial t}B(\omega, t, s)V + \frac{\partial}{\partial t}C(\omega, t, s)X\right)f,$$

$$\frac{\partial f}{\partial V} = B(\omega, t, s)f, \qquad \frac{\partial^2 f}{\partial V^2} = B(\omega, t, s)^2 f.$$

$$\frac{\partial f}{\partial X} = C(\omega, t, s)f, \qquad \frac{\partial^2 f}{\partial X^2} = C(\omega, t, s)^2 f,$$

and

$$\frac{\partial^2 f}{\partial X \partial V} = = B(\omega, t, s)C(\omega, t, s)f.$$

If we insert these partial derivatives into Eq. (3.15), we obtain the equality:

$$0 = \left(\frac{\partial A}{\partial t} + \frac{\partial B}{\partial t}V + \frac{\partial C}{\partial t}X\right) + C\left(r - \frac{V}{2}\right) + B\tilde{\kappa}\,(\tilde{\gamma} - V)$$

$$+ \frac{1}{2}VC^2 + \frac{1}{2}\sigma^2 VB^2 + \rho\sigma V BC.$$

Since this equation is satisfied for all values of X, we infer that $\frac{\partial C}{\partial t} = 0$ and $C = \omega$. Grouping other terms leads to the following system of ordinary differential equations:

$$\begin{cases} \frac{\partial A(\omega, t, s)}{\partial t} = -\omega r - B(\omega, t, s)\tilde{\kappa}\,\tilde{\gamma} \\ \frac{\partial B(\omega, t, s)}{\partial t} = \frac{\omega}{2} - \frac{1}{2}\omega^2 + B(\omega, t, s)\,(\tilde{\kappa} - \rho\sigma\omega) \\ \qquad\qquad\quad - \frac{1}{2}\sigma^2 B(\omega, t, s)^2. \end{cases} \qquad (3.16)$$

□

The system of ODEs (3.14) admits an analytical solution provided in the following corollary.

Corollary 3.4 *Let us define the following constants:*

$$d = \sqrt{(\rho\sigma\omega - \tilde{\kappa})^2 + \sigma^2 \left(\omega - \omega^2\right)}$$

and

$$g = \frac{\tilde{\kappa} - \rho\sigma\omega + d}{\tilde{\kappa} - \rho\sigma\omega - d}.$$

The functions $A(\omega, t, s)$ and $B(\omega, t, s)$ solving Eq. (3.14) are given by

$$A(\omega, t, s) = r\,\omega\,(s - t)$$

$$+ \frac{\tilde{\kappa}\tilde{\gamma}}{\sigma^2}\left((\tilde{\kappa} - \rho\sigma\omega + d)\,(s - t) - 2\ln\left(\frac{1 - g e^{d(s-t)}}{1 - g}\right)\right) \tag{3.17}$$

and

$$B(\omega, t, s) = \frac{\tilde{\kappa} - \rho\sigma\omega + d}{\sigma^2}\,\frac{1 - e^{d\,(s-t)}}{1 - g\,e^{d\,(s-t)}}. \tag{3.18}$$

One can prove this corollary by checking that the derivatives of Eqs. (3.17) and (3.18) fulfill the ODEs (3.14).

We can easily compute by Fast Fourier Transform the pdf of the log-return in the Heston model, both under the real or risk neutral measure. Using Eq. 3.13 of Proposition 3.3 allows us to calculate $\Upsilon_{t,T}(\omega) = \mathbb{E}^{\mathbb{Q}}\left(e^{\omega X_T} \mid \mathcal{F}_t\right)$ in Proposition 1.8 of Chap. 1. Figure 3.1 illustrates this and compares the pdfs of the two-year log-return computed by FFT and simulation. The parameters are $\tilde{\kappa} = 1.8$, $\tilde{\gamma} = 0.15^2$, $\sigma = 0.1$, $\rho = -0.8$, and $r = 0.01$. We observe a clear similarity and a light left tail asymmetry.

3.2 Filtering of Stochastic Volatility

This section introduces the particle filtering, also called sequential Monte Carlo (SMC), which is used to determine the most likely evolution of hidden processes such as the volatility in the Heston model. There is a considerable literature on the development of simulation-based methods to perform filtering of nonlinear Gaussian state space models. Leading contributions to the literature are reviewed by Doucet et al. [8]. Below, we consider a sample of discrete observations of log-returns $X_t = \ln\frac{S_{t+\Delta}}{S_t}$. The sample is denoted by $x = \{x_1, x_2, \ldots, x_T\}$. The interval

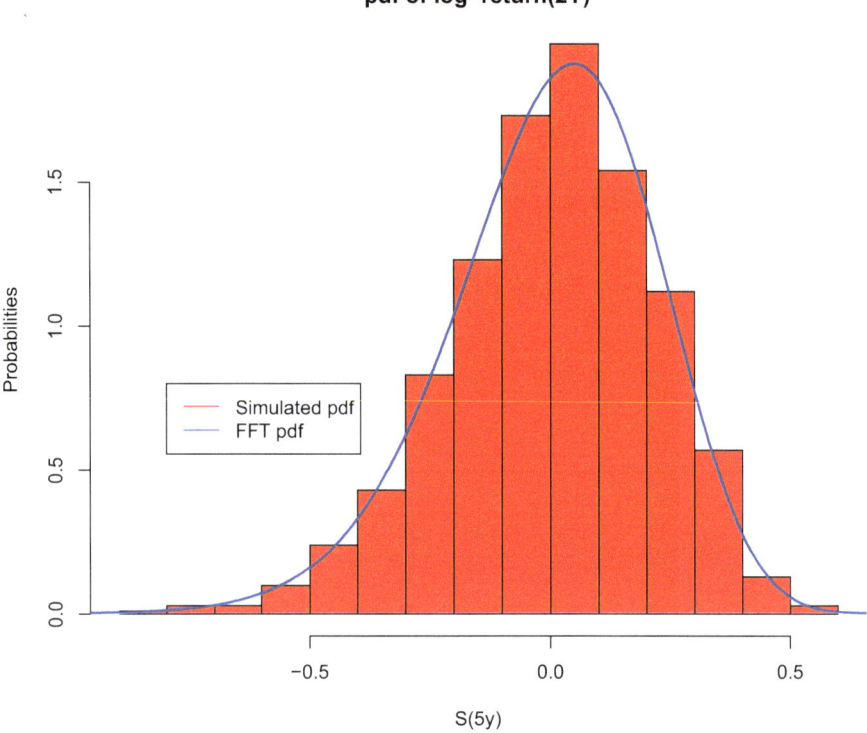

pdf of log-return(2Y)

Fig. 3.1 Blue line: pdf of the 2-year log-return, $X_{2\,years}$, computed by FFT with $M = 2^6$ discretization steps and $x_{\max} = 3$. Red histogram: empirical distribution of simulated 2-year log-return

of time between two consecutive observations is Δ, and the sampling times are $\{t_1, \ldots, t_T\}$. We adopt the notations $V_j = V_{t_j}$ and $X_j = X_{t_j}$, whereas v_j and x_j are the realizations of V_j, X_j. The dynamics of log-returns is approached by discretizing Equation (3.5):

$$X_{j+1} = \left(\mu - \frac{V_j}{2}\right) \Delta + \rho \sqrt{V_j}\, \Delta W^v_{j+1} + \sqrt{1 - \rho^2} \sqrt{V_j}\, \Delta W^s_{j+1},$$

where $\Delta W^v_{j+1} \sim N\left(0, \sqrt{\Delta}\right)$ and $\Delta W^s_{j+1} \sim N\left(0, \sqrt{\Delta}\right)$. The dynamics of V_t is approached in the same manner by

$$V_{j+1} = V_j + \kappa \left(\gamma - V_j\right) \Delta + \sigma \sqrt{V_j} \Delta W^v_{j+1}. \tag{3.19}$$

Notice that the parameters $\theta = \{\mu, \kappa, \gamma, \sigma, \rho\}$ are assumed to be known at this stage. The purpose of the particle filter is to estimate the most likely sample path of the variance process, based on observations of log-returns. An estimation method is proposed in the next section.

We denote by Δw_j^v the realizations of ΔW_j^v. The information $V_j = v_j$, and $\Delta W_j^v = \Delta w_j^v$ is stored in a "particle": a vector denoted $\boldsymbol{u}_j = \left(v_j, \Delta w_j^v \right)$. The process $\left(\boldsymbol{u}_j \right)_{j=1,\dots,T}$ is a Markov chain. The visible information up to time t_j is contained in a vector $\boldsymbol{x}_{1:j} = \{x_1, \dots, x_j\}$, and the density of the "measurement" $f(x_j | \boldsymbol{u}_j)$, conditionally to the information in the particle, is a normal probability density function. If we denote by $\phi(\cdot)$ the pdf of an $N(0, 1)$, we have that

$$f(x_j|\boldsymbol{u}_j) = \frac{1}{\sqrt{1-\rho^2}\sqrt{v_{j-1}\Delta}}\phi\left(\frac{x_j - \left(\mu - \frac{1}{2}v_{j-1} \right)\Delta - \rho\sqrt{v_{j-1}}\,\Delta w_j^v}{\sqrt{1-\rho^2}\sqrt{v_{j-1}\Delta}} \right).$$

On the other hand, we can simulate the transition density $f(\boldsymbol{u}_{j+1} | \boldsymbol{u}_j)$ with Eq. (3.19). The pdf of $\boldsymbol{u}_0$ is $f(\boldsymbol{u}_0)$, and the posterior distribution of $\boldsymbol{u}_j$ is denoted by $f(\boldsymbol{u}_j | \boldsymbol{x}_{1:j})$. As $P(A|B) = \frac{P(A \cup B)}{P(B)}$, this posterior distribution can be rewritten as follows:

$$f(\boldsymbol{u}_j | \boldsymbol{x}_{1:j}) = \frac{f(\boldsymbol{x}_{1:j}, \boldsymbol{u}_j)}{f(\boldsymbol{x}_{1:j})}, \tag{3.20}$$

and according to Bayes' rule, the denominator satisfies the equality:

$$f(\boldsymbol{x}_{1:j}) = f(\boldsymbol{x}_{1:j-1}, x_j) = f(x_j | \boldsymbol{x}_{1:j-1})f(\boldsymbol{x}_{1:j-1}).$$

Since $f(x_j | \boldsymbol{x}_{1:j-1}, \boldsymbol{u}_j) = f(x_j | \boldsymbol{u}_j)$, the numerator of Eq. (3.20) is also equal to

$$f(\boldsymbol{x}_{1:j}, \boldsymbol{u}_j) = f(x_j | \boldsymbol{x}_{1:j-1}, \boldsymbol{u}_j)f(\boldsymbol{x}_{1:j-1}, \boldsymbol{u}_j)$$
$$= f(x_j | \boldsymbol{u}_j)f(\boldsymbol{u}_j | \boldsymbol{x}_{1:j-1})f(\boldsymbol{x}_{1:j-1}).$$

The posterior distribution of $\boldsymbol{u}_j$ conditionally to the available information at t_j is then rewritten as follows:

$$f(\boldsymbol{u}_j | \boldsymbol{x}_{1:j}) = \frac{f(x_j | \boldsymbol{u}_j)}{\int f(x_j | \boldsymbol{u}_j)f(\boldsymbol{u}_j | \boldsymbol{x}_{1:j-1})\mathrm{d}v_j}f(\boldsymbol{u}_j | \boldsymbol{x}_{1:j-1}), \tag{3.21}$$

where

$$f(\boldsymbol{u}_j | \boldsymbol{x}_{1:j-1}) = \int f(\boldsymbol{u}_j | \boldsymbol{u}_{j-1})f(\boldsymbol{u}_{j-1} | \boldsymbol{x}_{1:j-1})\mathrm{d}\boldsymbol{u}_{j-1}. \tag{3.22}$$

Algorithm 3.1 Particle filtering of the volatility process

Initial step:

   draw N values of $v_0^{(i)}$ for $i = 1, \ldots, N$, from an initial distribution $f(v_0)$.

Main procedure:

   For $j = 1$ to maximum epoch, T

   Prediction step:

   Draw a sample $\Delta w_j^{v\,(i)}$ from an $N(0, \sqrt{\Delta}), i = 1, \ldots N$.

   Update $v_j^{(i)}$ using the relation

$$v_j^{(i)} = v_{j-1}^{(i)} + \kappa \left(\gamma - v_{j-1}^{(i)} \right) \Delta + \sigma \sqrt{v_{j-1}^{(i)}} \Delta w_j^{v(i)} .$$

   Correction step:

   The "particle" $\boldsymbol{u}_j^{(i)} = \left(v_j^{(i)}, \Delta w_j^{v\,(i)} \right)$ has a probability of occurrence equal to

$$p_j^{(i)} = \frac{f(x_j \mid \boldsymbol{u}_j^{(i)})}{\sum_{i=1:N} f(x_j \mid \boldsymbol{u}_j^{(i)})} ,$$

   where

$$f(x_j \mid \boldsymbol{u}_j^{(i)}) = \frac{1}{\sqrt{1 - \rho^2} \sqrt{v_{j-1}^{(i)} \Delta}}$$

$$\times \phi \left(\frac{x_j - \left(\mu - \frac{1}{2} v_{j-1}^{(i)} \right) \Delta - \rho \sqrt{v_{j-1}^{(i)}} \Delta w_j^{v(i)}}{\sqrt{1 - \rho^2} \sqrt{v_{j-1}^{(i)} \Delta}} \right)$$

   Resampling step:

   Resample with replacement N particles according to the importance weights $p_j^{(i)}$.

   The new importance weights are set to $p_j^{(i)} = \frac{1}{N}$.

   End loop on epochs

The calculation of $f(\boldsymbol{u}_j \mid \boldsymbol{x}_{1:j})$ can then be performed in three steps. The first is a prediction step in which we approach $f(\boldsymbol{u}_j \mid \boldsymbol{x}_{1:j-1})$ by simulations, based on relation (3.22). In the correction step, we next calculate the probabilities $f(\boldsymbol{u}_j \mid \boldsymbol{x}_{1:j})$ using Eq. (3.21). In practice, the integral in the prediction step is replaced by a Monte Carlo simulation of N "particles," denoted $\boldsymbol{u}_j^{(i)}$ for $i = 1, \ldots, N$. In the third step, we perform a resampling in order to keep track of the most likely sample paths of the volatility. The particle filter is presented in Algorithm 3.1.

Finally, the filtered variance for the period j is computed as the sum of particles weighted by their probability of occurrence:

$$\widehat{V}_j = \mathbb{E}\left(V_j \mid \boldsymbol{x}_{1:j} \right) = \sum_{i=1:N} v_j^{(i)} p_j^{(i)} ,$$

whereas the log-likelihood of the whole sample is approached before resampling by the sum:

$$\log f(\boldsymbol{x}|\boldsymbol{\theta}) = \sum_{j=1}^{T} \log f\left(x_j \mid x_{j-1}\right) \tag{3.23}$$

$$= \sum_{j=1}^{T} \log \left(\int f(x_j \mid \boldsymbol{u}_j) f(\boldsymbol{u}_j | x_{j-1}) \mathrm{d}\boldsymbol{u}_j \right)$$

$$= \sum_{j=1}^{T} \log \left(\sum_{i=1}^{N} p_j^{(i)} f(x_j \mid \boldsymbol{u}_j^{(i)}) \right).$$

Notice that there exist smoothing procedures such as the forward–backward smoother (FBS) or maximum a posteriori smoother (MAP). However, these procedures are time consuming, and the gain of accuracy is sometimes limited. We refer the interested reader to Chopin and Papaspiliopoulos [6, Chapter 12] for a detailed presentation of smoothing methods.

In order to illustrate this section, we simulate a sample path of the Heston model with parameters provided in Table 3.1. The time horizon of the simulation is set to 10 years, and we consider 255 time steps per year. Figure 3.2 compares simulated to filtered stock volatilities ($\sqrt{V_t}$). We run the filter with 500 particles. We clearly observe that the filtered sample path of volatilities is close to the real one.

Table 3.1 Parameters used for simulating a sample path of the Heston model

Heston model: parameters			
μ	0.05	ρ	-0.8
κ	1.80	σ	0.10
γ	$(0.15)^2$	V_0	$(0.15)^2$

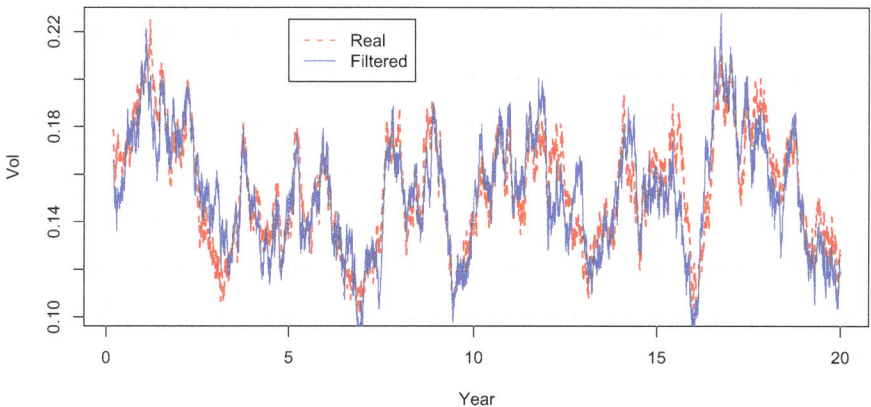

Fig. 3.2 Real and filtered sample paths of volatilities over 10 years

3.3 Estimation with a Rolling Window

In the next section, we combine a particle filter with the Metropolis–Hastings algorithm (see Chap. 2 for details) to fit the Heston model. Although this procedure converges in theory to parameter estimates, we often experience numerical difficulties when the algorithm is initialized with random values. It is advisable to obtain an approximate estimate of the parameters and to initialize the MCMC algorithm with them. One way to obtain good parameter approximation is to approach the time series $(V_t)_{t \geq 0}$ by the observed variance calculated over a rolling window of time. Under this assumption, it is possible to find maximum likelihood estimators of κ, γ, and σ.

We assume that the log-return $X_t = \ln\left(\frac{S_{t+\Delta}}{S_t}\right)$ is sampled at times

$$\{t_{-h} \ldots, t_0, t_1, \ldots, t_T\},$$

where $h \in \mathbb{N}$ is the length of the rolling window on which the variance is computed. Sampling times are equispaced by the time interval Δ. The realizations of the log-return are denoted by $\{x_{-(h-1)} \ldots, x_0, x_1, \ldots, x_T\}$. The sample of variances at times $\{t_1, \ldots, t_T\}$ is approached by $\boldsymbol{v} = \{v_1, v_2, \ldots, v_T\}$:

$$v_j = \frac{1}{\Delta h} \sum_{k=0}^{h} \left(x_{j-k} - \bar{x}_{j-h:j}\right)^2, \quad j = 1, \ldots, T, \tag{3.24}$$

where $\bar{x}_{j-h:j} = \frac{1}{h+1} \sum_{k=0}^{h} x_{j-k}$. According to Eq. (3.1), variations of the variance $\Delta v_j = v_{j+1} - v_j$ are approximately normal:

$$\Delta v_j \sim N\left(\kappa\left(\gamma - v_j\right)\Delta, \ \sigma\sqrt{v_j \Delta}\right).$$

Therefore,

$$(\widehat{\kappa}, \widehat{\gamma}, \widehat{\sigma}) = \arg\max_{\kappa, \gamma, \sigma} \sum_{j=1}^{T-1} \ln\left(\frac{1}{\sigma\sqrt{v_j\Delta}}\phi\left(\frac{\Delta v_j - \kappa\left(\gamma - v_j\right)\Delta}{\sigma\sqrt{v_j\Delta}}\right)\right).$$

In order to illustrate the reliability of this approximation, we simulate a daily sample path of the Heston model with the parameters of Table 3.1. Figure 3.3 compares the variance to its estimate (3.24) calculated with a rolling window of $h = 50$ days. This graph reveals that the rolling variance overestimates the real variances but captures its trend well. In a similar manner, we infer from Eq. (3.2) that the log-return is also Gaussian:

$$x_j \sim N\left(\left(\mu - \frac{v_j}{2}\right)\Delta, \ \sqrt{v_j\Delta}\right).$$

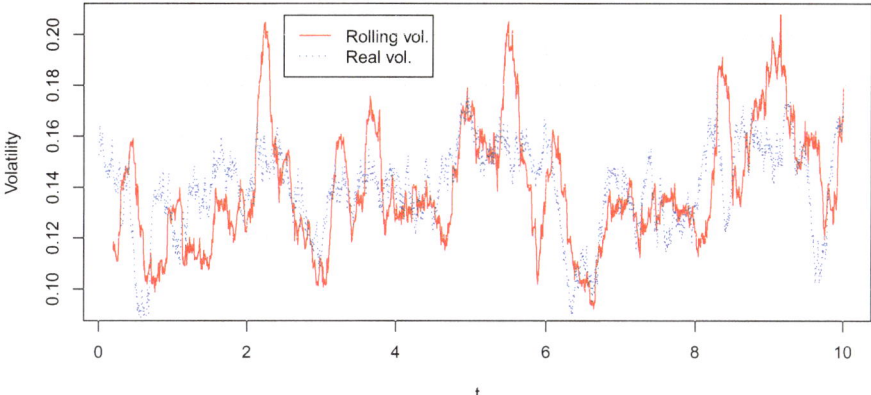

Fig. 3.3 Comparison of real and rolling volatilities over 10 years

Table 3.2 Comparison of parameters with their estimates. The rolling variance is computed on a window of 50 days

Estimators		Real values	
$\widehat{\mu}$	0.0946	μ	0.05
$\widehat{\kappa}$	1.8412	κ	1.80
$\widehat{\gamma}$	0.0209	γ	0.0225
$\widehat{\sigma}$	0.0932	σ	0.10
$\widehat{\rho}$	−0.1277	ρ	−0.80

Therefore, the drift μ is estimated by log-likelihood maximization.

$$\widehat{\mu} = \arg\max_{\mu} \sum_{j=1}^{T} \ln\left(\frac{1}{\sqrt{v_j \Delta}}\phi\left(\frac{x_j - \left(\mu - \frac{v_j}{2}\right)\Delta}{\sqrt{v_j \Delta}}\right)\right).$$

We might consider approaching the correlation by its empirical equivalent between the time series x_j and Δv_j, eventually normalized. However, this method is inaccurate. Indeed, approaching the variance with a rolling window introduces a lag between the realization of V_{t_j} and its estimate v_j, as illustrated in Fig. 3.3. In numerical applications, we estimate the correlation by the empirical equivalent between the rolling variance v_j and the rolling mean $\bar{x}_{j-h:j}$. Table 3.2 compares real and estimated parameters. We will use these estimates as the starting point of the MCMC algorithm in the next section.

3.4 Bayesian Estimation

An inherent problem of particle filters is that the estimate of the likelihood is not a smooth function of the parameters. From a practical viewpoint, the calibration of a model by maximizing the log-likelihood (e.g., with a gradient descent) is then uncertain. An alternative that is explored by Chen and Poon [5] consists in implementing an expectation–maximization (EM) algorithm. In this section, we opt for a variant of the Markov Chain Monte Carlo (MCMC) approach.

MCMC methods have become one of the standard tools of the statistician's apparatus, and such methods now deal with high dimensionality and complex patterns of dependence in statistical models. Here we review the Particle Markov Chain Monte Carlo (PMCMC) procedure (Andrieu et al. [2]) used in numerical applications to estimate the Heston model. The set of unknown parameters is denoted by $\Theta = \{\mu, \kappa, \gamma, \sigma, \rho\}$ in the sequel and is used to index the probability distribution function. We adopt a Bayesian approach, and Θ is a multivariate random variable with realizations $\boldsymbol{\theta}$. The domain of Θ is $\mathcal{X}$. The parameter's posterior distribution, based on the sample $\boldsymbol{x} = \{x_1, \ldots, x_T\}$, is

$$f(\boldsymbol{\theta} \mid \boldsymbol{x}) = \frac{f(\boldsymbol{\theta}) f(\boldsymbol{x} \mid \boldsymbol{\theta})}{\int_{\mathcal{X}} f(\boldsymbol{\theta}') f(\boldsymbol{x} \mid \boldsymbol{\theta}') \mathrm{d}\boldsymbol{\theta}'}, \tag{3.25}$$

where $f(\boldsymbol{\theta})$ and $f(\boldsymbol{x} \mid \boldsymbol{\theta})$ denote, respectively, the parameter's prior distribution and the likelihood of the data. The density $f(\boldsymbol{\theta} \mid \boldsymbol{x})$ is built by the PMCMC method. It generates a sample from $f(\boldsymbol{\theta} \mid \boldsymbol{x})$ by creating a Markov chain with the same stationary distribution as the parameter's posterior distribution. Once the Markov chain has reached stationarity after a transient phase, called the "burn-in" period, samples from the posterior distribution are simulated. The standard MCMC algorithm requires a point-wise estimate of $f(\boldsymbol{x} \mid \boldsymbol{\theta})$, which is not available in our model. Instead, $f(\boldsymbol{x} \mid \boldsymbol{\theta})$ is approached by its estimate (3.23), obtained by the particle filter introduced in the previous section.

The construction of the Markov chain consists of two steps, repeated iteratively. At the beginning of the $(k + 1)$th iteration, we propose a candidate parameter $\boldsymbol{\theta}'$ from a proposal distribution $q(\boldsymbol{\theta}' \mid \boldsymbol{\theta}_k)$ given the previous state of the Markov chain, $\boldsymbol{\theta}_k$. The proposal distribution has a support that covers the target distribution.

In the second step, we determine if the state should be updated by $\boldsymbol{\theta}'$. For this purpose, the acceptance probability of the Metropolis–Hastings Algorithm 2.2 is computed as follows:

$$\rho(\boldsymbol{\theta}', \boldsymbol{\theta}_k) = \min \left\{ \frac{f(\boldsymbol{\theta}' \mid \boldsymbol{x})}{f(\boldsymbol{\theta}_k \mid \boldsymbol{x})} \frac{q(\boldsymbol{\theta}_k \mid \boldsymbol{\theta}')}{q(\boldsymbol{\theta}' \mid \boldsymbol{\theta}_k)}, 1 \right\}. \tag{3.26}$$

This determines the probability that the candidate parameter will be assigned as the next state of the Markov chain, $\boldsymbol{\theta}' \to \boldsymbol{\theta}_{k+1}$. Intuitively, if we disregard the influence of the proposal distribution $q(\cdot, \cdot)$, a candidate is accepted if it increases the posterior likelihood $f(\boldsymbol{\theta}' \mid \boldsymbol{x}) > f(\boldsymbol{\theta}_k \mid \boldsymbol{x})$. The presence of $q(\cdot)$ in Eq. (3.26) allows a small decrease in the posterior likelihood, so as to explore the space of parameters.

In numerical applications, the transition distribution $q(\boldsymbol{\theta}' \mid \boldsymbol{\theta}_k)$ is assumed to be normal, $\boldsymbol{\theta}' \sim N(\boldsymbol{\theta}_k, \sigma_\theta I_m)$, where I_m is the identity matrix. As this distribution is symmetric, $q\left(\boldsymbol{\theta}_k|\boldsymbol{\theta}'\right) = q\left(\boldsymbol{\theta}'|\boldsymbol{\theta}_k\right)$, the acceptance probability simplifies to

$$\rho(\boldsymbol{\theta}', \boldsymbol{\theta}_k) = \min\left\{\frac{f(\boldsymbol{x}\mid\boldsymbol{\theta}')f(\boldsymbol{\theta}')}{f(\boldsymbol{x}\mid\boldsymbol{\theta}_k)f(\boldsymbol{\theta}_k)}, 1\right\}.$$

The resulting samples $\boldsymbol{\theta}_{1:n}$ (n iterations after the burn-in period) now serve to build the empirical distribution of $f(\boldsymbol{\theta}\mid\boldsymbol{x})$, which is defined by

$$f(\boldsymbol{\theta}\mid\boldsymbol{x}) = \frac{1}{n}\sum_{k=1}^{n}\delta_{\boldsymbol{\theta}_k}(\mathrm{d}\boldsymbol{\theta}),$$

where $\delta_{\boldsymbol{\theta}_k}(\mathrm{d}\boldsymbol{\theta})$ are the Dirac atoms located at $\boldsymbol{\theta} = \boldsymbol{\theta}_k$, with equal weights. The estimate of parameters with respect to the posterior distribution is then

$$\widehat{\Theta} = \mathbb{E}\left(\Theta|\boldsymbol{x}\right) \approx \frac{1}{n}\sum_{k=1}^{n}\boldsymbol{\theta}_k.$$

In order to benchmark the goodness of fit of the PMCMC algorithm, we test it with simulated data. We simulate a sample path of the Heston model with the parameters of Table 3.1 over a period of 10 years. The variance is computed on a rolling window of $h = 50$ days, and the parameters of the stochastic variance are estimated by log-likelihood maximization. These estimates are reported in the first column of Table 3.3. Next we run 2000 iterations of the PMCMC algorithm with 600 particles. As the initial parameters of the Markov chain are not too far from the optimal ones, the algorithm converges relatively quickly. We consider a burn-in period of 1800 iterations. Average values of the chain over the last 200 iterations are reported in the second column of Table 3.3. We see that the parameter estimates are not far from their real value, except for the expected return, which is very sensitive to the dataset. Particular care must be taken regarding the reliability of this estimate.

Figure 3.4 compares the sample path of real and filtered volatilities, with 600 particles and the parameter estimates of Table 3.3. Globally, the filter successfully detects the trend followed by the hidden volatility. Nevertheless, we observe that it under- or overestimates the volatility when the real volatility deviates briefly and widely from the mean reversion level. In a second test, we fit the Heston model to the time series of the S&P 500 stock index from 31/1/2001 to 31/1/2020

Table 3.3 Comparison of parameters with their estimates. The rolling variance is computed on a window of 50 days

Estimators rolling variance		Estimators PMCMC			Real values	
$\widehat{\mu}$	0.0946	$\widehat{\mu}$	0.1317	μ	0.05
$\widehat{\kappa}$	1.8412	$\widehat{\kappa}$	1.6439	κ	1.80
$\widehat{\gamma}$	0.0209	$\widehat{\gamma}$	0.0198	γ	0.0225
$\widehat{\sigma}$	0.0932	$\widehat{\sigma}$	0.0918	σ	0.10
$\widehat{\rho}$	-0.1277	$\widehat{\rho}$	-0.7543	ρ	-0.80

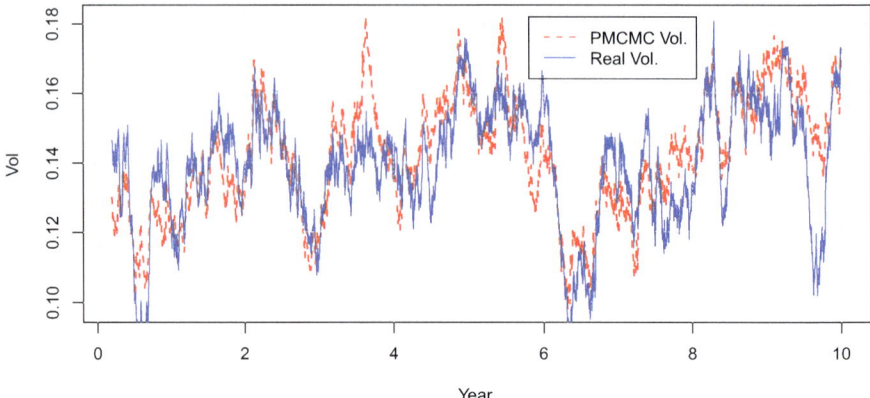

Fig. 3.4 Comparison of real and filtered volatilities with the PMCMC over 10 years

Table 3.4 Parameter
estimates for the S&P 500

	Estimators rolling variance		Estimators PMCMC	MCMC std. err.
$\widehat{\mu}$	0.1047	$\widehat{\mu}$	0.1232	0.0102
$\widehat{\kappa}$	0.4002	$\widehat{\kappa}$	0.7171	0.1016
$\widehat{\gamma}$	0.0267	$\widehat{\gamma}$	0.1016	0.0039
$\widehat{\sigma}$	0.1610	$\widehat{\sigma}$	0.4234	0.0049
$\widehat{\rho}$	−0.5390	$\widehat{\rho}$	−0.6359	0.0448

(4779 observations). Again we approach the variance by computing it with a rolling window of 50 days. The parameter estimates obtained by log-likelihood maximization are shown in the first column of Table 3.4. The volatility reverts to $\sqrt{\widehat{\gamma}} = 16.34\%$, and the correlation between log-return and variance is significantly negative. The average stock return is around 10%. Next we run the PMCMC algorithm with 3000 iterations. The log-likelihood converges after 2000 iterations. The second and third columns of Table 3.4 report the averages and standard deviations of the chain over the last 1000 iterations. The speed of mean reversion rises from 0.40 to 0.71, and the volatility reverts to a higher level $\sqrt{\widehat{\gamma}} = 31.87\%$. The expected return climbs to 12.32%. Figure 1.4 compares the filtered volatility with the MCMC parameters of Table 3.4 to its estimate with a rolling window. The filtered sample path is more erratic than the rolling volatility. We also observe peaks of volatility during major financial crises: in 2003, during the second Gulf war, during the subprime crisis of 2008, and in the double dip recession of 2011. Lower peaks are also observed in 2015–2016 due to turbulence on the Chinese market and in 2018 when oil price fell due to overproduction. This graph may be compared to Fig. 1.4, which shows the filtered volatility of a multifractal switching model. Both models agree on periods during which the volatility reached its highest values (Fig. 3.5).

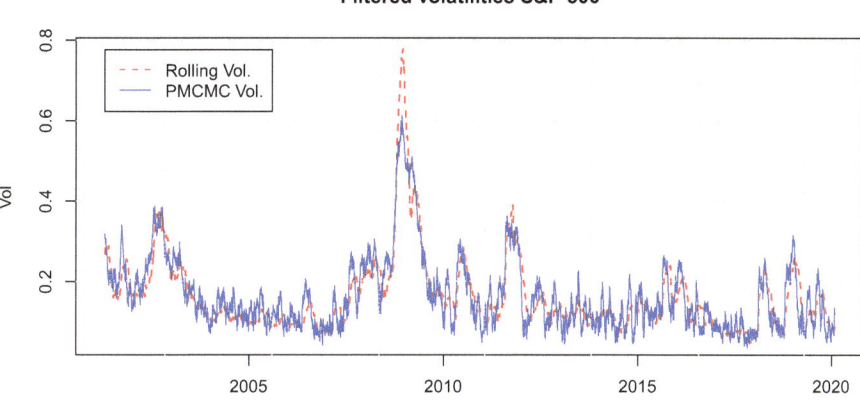

Fig. 3.5 S&P 500, Comparison of filtered volatilities obtained by MCMC to rolling volatilities (a window of 50 days)

3.5   Further Reading

When dealing with a non-Gaussian and nonlinear specification, simulation-based methods offer strong advantages over the alternative approaches. Among the most popular in the financial literature, we find the method of simulated moments (Duffie and Singleton [7]), indirect inference methods (Gourieroux, Monfort and Renault [13]), and the efficient method of moments (EMM) (Gallant and Tauchen [12]). A last approach consists in approximating the log-likelihood function as done by Atiya and Wall [3] for the Heston model. To estimate a jump-diffusion process from equity returns, Andersen, Benzoni and Lund [1] use an EMM approach. Eraker et al. [9] develop a likelihood-based estimation approach for estimating multivariate jump-diffusion using Markov Chain Monte Carlo (MCMC) methods. Chen and Poon [5], Fulop and Duan [10], and Fulop et al. [11] use particle filtering to estimate models with stochastic volatility and self-excited jump processes. Hainaut and Moraux [15] study a switching self-excited jump-diffusion and estimate its parameters by PMCMC. Hainaut and Goutte [14] use a similar approach to fit a microstructure model for stock prices. More recently, Kouritzin and Mackay [17] used sequential Monte Carlo within a simulation for path-dependent option pricing. We will retrieve the PMCMC algorithm in Chap. 4 to estimate a model with self-exciting jumps, in Chap. 10 to quantify illiquidity and in Chap. 9 to fit Volterra processes.

References

1. Andersen, T., Benzoni, L., Lund, J.: An empirical investigation of continuous-time equity return models. J. Finance **57**, 1239–1284 (2002)
2. Andrieu, C., Doucet, A., Holenstein, R.: Particle Markov chain Monte Carlo methods. J. R. Statist. Soc. B. **72**(3), 269–342 (2010)
3. Atiya, A.F., Wall, S.: An analytic approximation of the likelihood function for the Heston model volatility estimation problem. Quant. Finance **9**(3), 289–296 (2009)
4. Cox, J.C., Ingersoll, J.E., Ross S.A.: A theory of the term structure of interest rates. Econometrica **53**(2), 385–408 (1985)
5. Chen, K., Poon, S.-H.: Variance swap premium under stochastic volatility and self-exciting jumps, Working Paper SSRN-id2200172 (2013)
6. Chopin, N., Papaspiliopoulos, O.: An Introduction to Sequential Monte Carlo. Springer Series in Statistics. Springer Nature Switzerland AG, Cham (2020)
7. Duffie, D., Singleton, K.: Simulated moments estimation of Markov models of asset prices. Econometrica **61**, 929–952 (1993)
8. Doucet, A., De Freitas, J.F.G., Gordon, N.: Sequential Monte Carlo Methods in Practice. Cambridge University Press, Cambridge (2000)
9. Eraker, B., Johannes, M., Polson, N.: The impact of jumps in volatility and returns. J. Finance **58**, 1269–1300 (2003)
10. Fulop, A., Duan, J.-C.: Density-tempered marginalized sequential Monte Carlo samplers. J. Bus. Econ. Stat. **33**, 192–202 (2015)
11. Fulop, A., Li, J., Yu, J.: Self-exciting jumps, learning, and asset pricing implications. Rev. Financial Stud. **28**, 876–912 (2015)
12. Gallant, A.R., Tauchen, G.: Which moments to match? Econ. Theory **12**, 657–681 (1996)
13. Gourieroux, C., Monfort, A., Renault, E.: Indirect inference. J. Appl. Econ. **8**, S85–S118 (1993)
14. Hainaut, D., Goutte, S.: A switching microstructure model for stock prices. Math. Financial Econ. **13**(3), 459–490 (2019)
15. Hainaut, D., Moraux, F.: A switching self-exciting jump diffusion process for stock prices. Ann. Finance **15**, 267–306 (2019)
16. Heston, S.L.: A closed-form solution for options with stochastic volatility with applications to bond and currency options. Rev. Financial Stud. **6**(2), 327–343 (1993)
17. Kouritzin, M., Mackay, A.: Branching particle pricers with Heston examples. Int. J. Theoret. Appl. Finance **23**(1), 2050003 (2020)

Chapter 4
Modeling of Spillover Effects in Stock Markets

The switching regime processes of Chap. 1 and the Heston model of Chap. 3 cannot duplicate the clustering of jumps in stock markets caused, for example, by an onslaught of bad news. Self-excited processes offer an interesting way to introduce such spillover effects into the market dynamics. In this approach, the occurrence of a shock depends on the recent history of prices. This dynamic was initially introduced by Hawkes [19, 20] and Hawkes and Oakes [22] to model earthquake aftershocks. In the most common and simplest specification, the intensity of jumps, akin to the instantaneous probability of a shock, increases as soon as a jump in price occurs. The influence of this jump on the intensity then decays exponentially over time. This chapter reviews the features of self-exciting jump-diffusions and provides the theoretical background to read Chap. 5 about non-Markov extensions of such processes.

4.1   The Self-exciting Jump-Diffusion (SEJD)

Shocks in financial markets are induced by a point process denoted $(L_t)_{t \geq 0}$. We define it as the sum of N_t jumps of size $e^{J_j} - 1$:

$$L_t = \sum_{j=1}^{N_t} \left(e^{J_j} - 1 \right) . \tag{4.1}$$

The counting process $(N_t)_{t \geq 0}$ has an intensity $(\lambda_t)_{t \geq 0}$. Each random jump (J_j) is an independent copy of a random variable J.

   In our numerical applications, we consider a double exponential distribution of J. Double exponential jumps (DEJ) are defined by three parameters (p, ρ^+, ρ^-): $\rho^+ \in \mathbb{R}^+$, $\rho^- \in \mathbb{R}^-$, and $p \in (0, 1)$. The features of DEJ were detailed in Sect. 1.2.

D. Hainaut, *Continuous Time Processes for Finance*,
Bocconi & Springer Series 12, https://doi.org/10.1007/978-3-031-06361-9_4

We recall that the pdf of J is a function $\nu(z)$ equal to

$$\nu\left(z\right) = p\rho^{+}e^{-\rho^{+}z}\mathbb{1}_{\{z\geq 0\}} - (1-p)\,\rho^{-}e^{-\rho^{-}z}\mathbb{1}_{\{z<0\}}. \tag{4.2}$$

The intensity of the jump process N_t is akin to the instantaneous probability of observing a new shock. To introduce memory into the equity dynamics, the jump arrival intensity, denoted by $\lambda = (\lambda_t)_t$, depends on past jumps, as follows:

$$\mathrm{d}\lambda_t = \alpha\,(\theta - \lambda_t)\,\mathrm{d}t + \eta\mathrm{d}N_t\,. \tag{4.3}$$

With such a specification, the intensity reverts with speed $\alpha \in \mathbb{R}^+$ to a long run mean level $\theta \in \mathbb{R}^+$. When a jump of market price occurs, the intensity instantaneously increases by an amount η. This mechanism allows for contagion between jumps and explains why jumps can arrive in groups during an economic crisis. We assume that the price process of a stock, denoted by $S = (S_t)_t$, is ruled by a diffusion and a jump process according to the following stochastic differential equation (SDE):

$$\frac{\mathrm{d}S_t}{S_{t^-}} = \mu\mathrm{d}t + \sigma\mathrm{d}W_t + \mathrm{d}L_t - \lambda_t\mathbb{E}\left(e^J - 1\right)\mathrm{d}t \tag{4.4}$$

$$= \mu\mathrm{d}t + \sigma\mathrm{d}W_t + \left(e^J - 1\right)\mathrm{d}N_t - \lambda_t\mathbb{E}\left(e^J - 1\right)\mathrm{d}t\,,$$

where μ is a constant drift term, $\mathrm{d}W_t$ denotes the increment of the Brownian motion $W = (W_t)_{t\geq 0}$, and σ is a constant diffusion coefficient ($\sigma \in \mathbb{R}^+$). The third term of the above equation represents the influence of shocks to the price that can occur in the interval $(t^-, t + \mathrm{d}t)$. The last term is the jump compensator: its presence ensures that the average stock return is equal to $\mu\mathrm{d}t$. The second equality arises naturally because the probability of observing more than one jump in an infinitesimal period is negligible.

According to Itô's lemma for jump processes, any function $f\,(t, S_t, N_t, \lambda_t)$ that is respectively C^1 and C^2 with respect to time, intensity, and stock price admits the following representation:

$$\mathrm{d}f = \left(\frac{\partial f}{\partial t} + \mu S_t\frac{\partial f}{\partial S} + \frac{1}{2}\sigma^2 S_t^2\frac{\partial^2 f}{\partial S^2}\right)\mathrm{d}t + \sigma S_t\frac{\partial f}{\partial S}\,\mathrm{d}W_t \tag{4.5}$$

$$- \lambda_t\mathbb{E}\left(e^J - 1\right)S_t\frac{\partial f}{\partial S}\mathrm{d}t + \alpha(\theta - \lambda_t)\frac{\partial f}{\partial \lambda}\mathrm{d}t$$

$$+ \left(f\left(t, S_t e^J, N_t + 1, \lambda_t + \eta\right) - f\right)\mathrm{d}N_t\,.$$

The infinitesimal generator of f is the operator $\mathcal{A}f = \mathbb{E}\left(\frac{df}{dt} \mid \mathcal{F}_t\right)$ and is therefore equal to

$$\mathcal{A}f(t, S, n, \lambda) = \frac{\partial f}{\partial t} + \mu S \frac{\partial f}{\partial S} + \frac{1}{2}\sigma^2 S^2 \frac{\partial^2 f}{\partial S^2} \tag{4.6}$$

$$- \lambda \mathbb{E}\left(e^J - 1\right) S \frac{\partial f}{\partial S} + \alpha(\theta - \lambda) \frac{\partial f}{\partial \lambda}$$

$$+ \lambda \int_{-\infty}^{-\infty} f\left(t, Se^z, n+1, \lambda+\eta\right) - f \, \nu(dz).$$

Using Eq. (4.5), we can check by direct differentiation that the solution to Eq. (4.3) is the following process:

$$\lambda_t = \theta + (\lambda_0 - \theta)\,e^{-\alpha t} + \int_0^t \eta e^{-\alpha(t-s)}dN_s. \tag{4.7}$$

The integrand in the last term of Eq. (4.7) is called the *kernel function* and defines the memory of the jump process. In this setting, the kernel is a decreasing exponential function, reflecting that the influence of past jumps on the current intensity decays exponentially. Figure 4.1 shows the simulated paths of λ_t and N_t in the single-factor model, illustrating that past jumps are forgotten at an exponential rate.

Finally, applying Itô's lemma (1.6) to $\ln S_t$ allows us to infer that the stock value at time t is the following exponential:

$$S_t = S_0 \exp\left(\left(\mu - \frac{\sigma^2}{2}\right)t - \mathbb{E}\left(e^J - 1\right)\int_0^t \lambda_s ds + \sigma W_t + \sum_{j=1}^{N_t} J_j\right).$$

The log-return $X_t := \ln \frac{S_t}{S_0}$ is therefore equal to

$$X_t = \left(\mu - \frac{\sigma^2}{2}\right)t - \mathbb{E}\left(e^J - 1\right)\int_0^t \lambda_s ds + \sigma W_t + \sum_{i=1}^{N_t} J_j. \tag{4.8}$$

The infinitesimal generator of a smooth function $f(t, X_t, N_t, \lambda_t)$ is equal to

$$\mathcal{A}f(t, X, n, \lambda) = \frac{\partial f}{\partial t} + \left(\mu - \frac{\sigma^2}{2}\right)\frac{\partial f}{\partial X} + \frac{1}{2}\sigma^2 \frac{\partial^2 f}{\partial X^2} \tag{4.9}$$

$$- \lambda \mathbb{E}\left(e^J - 1\right)\frac{\partial f}{\partial X} + \alpha(\theta - \lambda)\frac{\partial f}{\partial \lambda}$$

$$+ \lambda \int_{-\infty}^{-\infty} f\left(t, X+z, n+1, \lambda+\eta\right) - f(\cdot) \, \nu(dz).$$

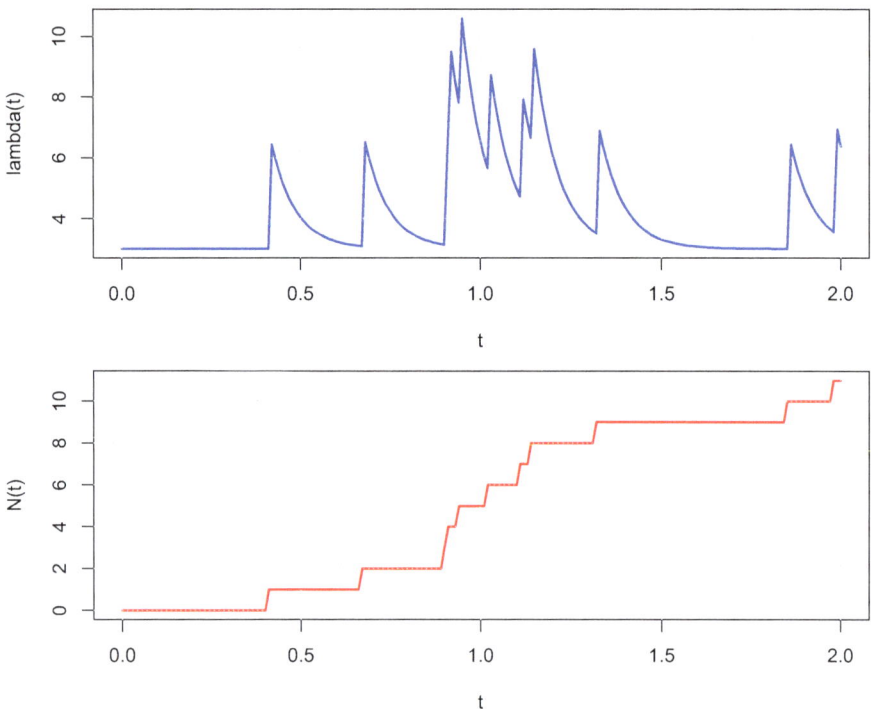

Fig. 4.1 Simulation of a sample path of a self-excited process: $\theta = 3$, $\alpha = 15$, and $\eta = 2$

Before studying the properties of S_t and X_t, we present the statistical features of time intervals between successive jumps.

4.2   Likelihood of Inter-Arrival Times

We assume that the jump process is observable and focus on the distribution of times at which a shock is observed. These jump times are denoted by $(T_k)_{k \in N_t}$. Imagine that up to now, $k - 1$ jumps have occurred. If the last arrival time is $T_{k-1} = u$, we denote by $f_k(t \mid u)$ the density (pdf) of the next jump arrival time: $T_k \mid T_{k-1} = u$. The cumulative distribution function (cdf) of this random variable is equal to

$$F_k(t \mid u) = \mathbb{P}\,(T_k \leq t \mid T_{k-1} = u)$$

$$= \int_u^t f_k(s \mid u)\mathrm{d}s\,.$$

According to Eq. (4.7), the intensity is a deterministic function of time between two successive jumps. To lighten our notation, we denote this deterministic function by $\lambda_k^*(t) = (\lambda_t | T_{k-1} = u)$:

$$\lambda_k^*(t) = \theta + e^{-\alpha(t-u)} (\lambda_u - \theta) \; , \; t \geq u = T_{k-1}. \tag{4.10}$$

The survival function is the probability:

$$H_k(t \mid u) = 1 - F_k(t \mid u) = \mathbb{P}\left(T_k \geq t \mid T_{k-1} = u\right) .$$

The *intensity*, also called the *hazard rate* in statistics, is by definition equal to the instantaneous probability of observing a jump, that is

$$\lambda_k^*(t) = \lim_{\Delta \to 0} \frac{\mathbb{P}\left(t \leq T_k \leq t + \Delta \mid T_{k-1} = u\right)}{H(t \mid u) \, \Delta}$$

$$= \lim_{\Delta \to 0} \frac{\mathbb{P}\left(t \leq T_k \leq t + \Delta \mid T_{k-1} = u \, , \, T_k \geq t\right)}{\Delta} .$$

Given that $\mathbb{P}\left(t \leq T_k \leq t + \Delta \mid T_{k-1} = u\right) = \int_t^{t+\Delta} f_k(s \mid u) ds$, the previous limit becomes

$$\lambda_k^*(t) = \frac{f_k(t \mid u)}{1 - F_k(t \mid u)}$$

$$= -\frac{d}{dt} \ln \left(1 - F_k(t \mid u)\right) ,$$

for all $T_k > t \geq u$. By direct integration, we infer that the hazard rate is linked to the cdf of jump arrival times by the relation:

$$F_k(t \mid u) = 1 - \exp\left(-\int_u^t \lambda_k^*(s) ds\right) . \tag{4.11}$$

Bayes' rule allows us to infer a simple expression for the conditional probability of survival:

$$\mathbb{P}\left(T_k \geq t \mid T_{k-1} = u \, , \, T_k \geq s\right) = \frac{\mathbb{P}\left(T_k \geq t \, , \, T_k > s \mid T_{k-1} = u\right)}{\mathbb{P}\left(T_k \geq s \mid T_{k-1} = u \, , \, \right)}$$

$$= \frac{\mathbb{P}\left(T_k \geq t \mid T_{k-1} = u\right)}{\mathbb{P}\left(T_k \geq s \mid T_{k-1} = u \, , \, \right)}$$

$$= \exp\left(-\int_s^t \lambda_k^*(v) dv\right) ,$$

for $u \leq s \leq t$. The next proposition is helpful for estimating parameters of the jump process.

Proposition 4.1 *Let us denote by $\Theta = \{\alpha, \theta, \eta\}$ the triplet of parameters involved in the dynamics of λ_t. The log-likelihood of a time series of n jump times $(t_k)_{k=1,\ldots,n}$ is denoted by $\ln \mathcal{L}(\Theta \mid t_n, \ldots, t_1)$ and is equal to*

$$\mathcal{L}(\Theta \mid t_n, \ldots, t_1) = \sum_{k=1}^{n} \ln f(t_k \mid t_{k-1}) , \qquad (4.12)$$

where $f(t_k \mid t_{k-1})$ is the pdf of inter-arrival times:

$$f(t_k \mid t_{k-1}) = \lambda(t_k-) \qquad (4.13)$$

$$\times \exp\left(-\theta(t_k - t_{k-1}) - (\lambda(t_{k-1}) - \theta)\frac{\left(1 - e^{-\alpha(t_k - t_{k-1})}\right)}{\alpha}\right) .$$

The sample path of the intensity, $\lambda(\cdot)$, is such that

$$\begin{cases} \lambda(s) & = \theta + e^{-\alpha(s-t_{k-1})}(\lambda(t_{k-1}) - \theta) \quad s \in [t_{k-1}, t_k), \\ \lambda(t_k) & = \lambda(t_k-) + \eta . \end{cases} \qquad (4.14)$$

Proof Equation (4.14) results from the definition of the intensity. Differentiating Eq. (4.11) leads to the pdf of inter-arrival times:

$$f(t_k \mid t_{k-1}) = \lambda(t_k-) \exp\left(-\int_{t_{k-1}}^{t_k-} \lambda(s)\mathrm{d}s\right) .$$

Integrating (4.10) leads directly to Eq. (4.13). The likelihood of jump times is

$$\mathcal{L} = \mathbb{P}(T_n \in [t_n, t_n + \mathrm{d}t] \ldots, T_k \in [t_k, t_k + \mathrm{d}t], \ldots, T_1 \in [t_1, t_1 + \mathrm{d}t])$$

$$= \prod_{k=1}^{n} \mathbb{P}(T_k \in [t_k, t_k + \mathrm{d}t] \mid T_{k-1} \in [t_{k-1}, t_{k-1} + \mathrm{d}t])$$

$$= \prod_{k=1}^{n} f(t_k \mid t_{k-1})$$

with $t_0 = 0$.   □

If the times of jumps are visible, we can easily estimate the parameters by log-likelihood maximization. The parameter estimates are denoted by $\widehat{\Theta}$:

$$\widehat{\Theta} = \underset{\Theta}{\arg\max}(\mathcal{L}(\Theta \mid t_n, \ldots, t_1)) .$$

4.3 Jump Detection with the "Peak Over Threshold" Method

Estimating the parameters of a jump-diffusion process using a time series is challenging and requires advanced econometric techniques. Our model involves one latent process $\lambda = (\lambda_t)_{t \geq 0}$ that we need to back out using a filtering technique. There is a considerable body of literature on how to perform filtering of non-Gaussian and nonlinear state-space models. In this section, we employ a peak over threshold approach, similar to that in Embrechts et al. [11]. This approach is robust and easy to implement. Furthermore, parameter estimates found in this way may be used for initializing the MCMC algorithm, as detailed in Sect. 4.5.

The discrete record of n observations of log-returns, equally spaced with a lag Δ, is $\{x_1, x_1, x_2, \ldots, x_n\}$. The times of observation are denoted by $\{s_0, s_1, \ldots, s_n\}$. A jump is believed to occur when this return is above or below certain thresholds. These thresholds, denoted by $g(\alpha_1)$ and $g(\alpha_2)$, depend on the lag between observations and the confidence levels α_1 and α_2. To define these, we fit a pure Gaussian process: $x_k \sim \mu_g \Delta + \sigma_g W_\Delta$. The unbiased estimators of μ_g and σ_g are

$$\widehat{\mu}_g = \frac{1}{n\Delta} \sum_{j=1}^{n} x_j \qquad \widehat{\sigma}_g^2 = \frac{1}{(n-1)\Delta} \sum_{j=1}^{n} \left(x_j - \widehat{\mu}_g \right)^2 .$$

If $\Phi(\cdot)$ denotes the cdf of the standard normal distribution, $g(\alpha_1)$, $g(\alpha_2)$ are set to the α_1 and α_2 percentiles of the Brownian term: $g(\alpha_i) = \widehat{\mu}_g \Delta + \widehat{\sigma}_g \sqrt{\Delta} \Phi^{-1}(\alpha_i)$ for $i = 1, 2$. The time of the kth jump is therefore

$$t_k = \min\{s_j \in \{s_1, \ldots, s_n\} \mid x_j \geq g(\alpha_1) \text{ or } x_j \leq g(\alpha_2) , \ j \geq k\},$$

and the sample path of $(N_t)_{t \geq 0}$ is approached by the following time series:

$$N(s_j) = \max\{k \in \mathbb{N} \mid t_k \leq s_j\} .$$

The levels of confidence, α_1 and α_2, are optimized such that the skewness and the kurtosis of x_i for periods without jumps are close to those of a normal distribution.

We have applied this procedure to the daily log-return of the S&P 500 from 31/1/01 to 31/1/20. Over this period and for $\Delta = 1/252$, we find $\widehat{\mu}_g = 4.6258\%$ and $\widehat{\sigma}_g = 18.6405\%$. The optimal confidence levels α_1 and α_2 are respectively equal to 0.9623 and 0.0665. The skewness and kurtosis of log-returns for days without detected jumps are equal to -0.0010 and 3.0008. We assume that jumps occur when the log-return breaches the upper and lower thresholds $g(\alpha_1) = 2.1065\%$ and $g(\alpha_2) = -1.7459\%$. Figure 4.2 shows the time series of filtered jumps. We detect 144 positive and 262 negative jumps. Grouped jumps in returns are clearly visible from 2001 to 2003 (Internet bubble burst and second Gulf war), from September 2008 to the end of 2009 (the U.S. credit crunch period), and from September 2011 to February 2012 (the second period of the double dip recession) and around the end of 2018. Shocks during these periods do not display any clear trend: negative

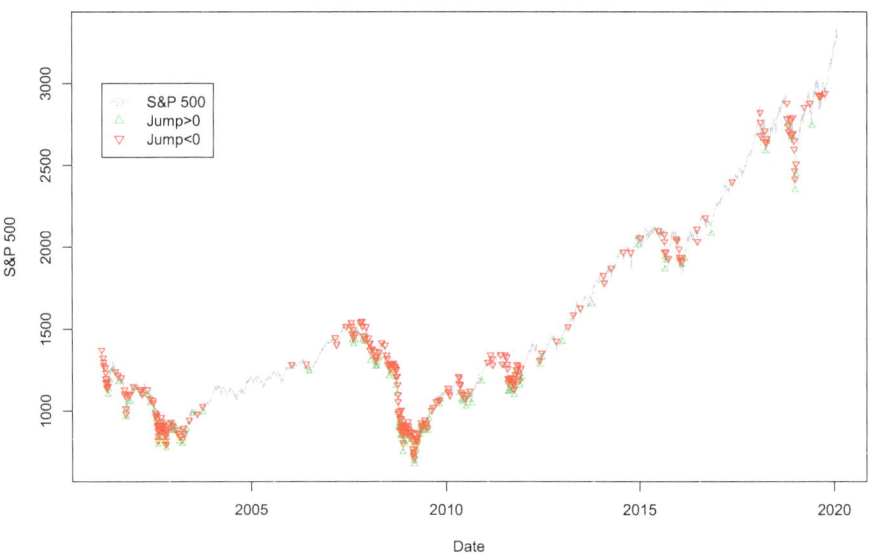

Fig. 4.2 Positive and negative shocks, S&P 500 from 31/1/2001 to 31/2/2020

movements alternate regularly with large positive movements. Once jumps are
detected, for parameter estimates $\{\widehat{\alpha}, \widehat{\theta}, \widehat{\eta}\}$, the sample path of λ_t is approached by

$$\begin{cases} \lambda(s_j) & = \widehat{\theta} + e^{-\widehat{\alpha}(s_j - t_{k-1})}\left(\lambda(t_{k-1}) - \widehat{\theta}\right) & s_j \in [t_{k-1}, t_k), \\ \lambda(s_j) & = \lambda(s_j-) + \widehat{\eta} & s_j = t_k. \end{cases} \qquad (4.15)$$

The estimators are found by maximizing the log-likelihood of inter-arrival times,
Eq. (4.1) in Proposition 4.1. Applying this procedure to the S&P 500 provides the
following parameter estimates: $\widehat{\alpha} = 14.55$, $\widehat{\theta} = 3.53$ and $\widehat{\eta} = 12.23$. Figure 4.3
shows the sample path of λ_t constructed with Eq. (4.15). The periods during which
the intensity reaches its highest values correspond to major recent economic crises.

The estimation of parameters defining the dynamics of stock prices is also
performed by log-likelihood maximization. In the absence of jumps, the log-return
has a normal distribution. If a jump occurs, we know from Chap. 1, Proposition 1.3,
that the pdf of the sum $\sigma W_\Delta + J$ is

$$h(z|\sigma, p, \rho^+, \rho^-) = p\rho^+ \exp\left(\frac{1}{2}\left(\rho^+\right)^2 \sigma^2 \Delta - \rho^+ z\right) \Phi\left(\frac{z - \rho^+ \sigma^2 \Delta}{\sqrt{\Delta}\sigma}\right)$$

$$- (1-p)\,\rho^- \exp\left(\frac{1}{2}\left(\rho^-\right)^2 \sigma^2 \Delta - \rho^- z\right)\left(1 - \Phi\left(\frac{z - \rho^- \sigma^2 \Delta}{\sqrt{\Delta}\sigma}\right)\right).$$

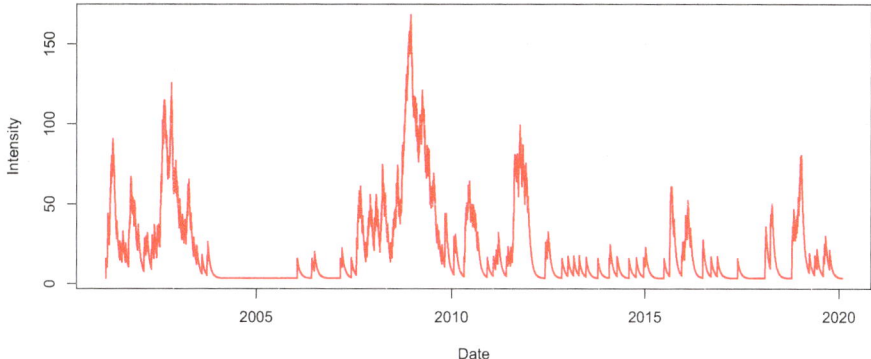

Fig. 4.3 Filtered sample path of λ_t for the S&P 500 from 31/1/2001 to 31/2/2020

According to Eq. (4.8), for parameter estimates $\{\widehat{\mu}, \widehat{\sigma}, \widehat{p}, \widehat{\rho^+}, \widehat{\rho^-}\}$, the sample path of the log-return drift is approached by

$$\mu_X(s_j)\Delta = \left(\widehat{\mu} - \frac{\widehat{\sigma}^2}{2}\right)\Delta - \mathbb{E}\,\widehat{(e^J - 1)}\lambda(s_j)\Delta \quad j = 1, \ldots, n\,,$$

where $\mathbb{E}\,\widehat{(e^J - 1)} = \widehat{p}\frac{1}{\rho^+ - 1} + (1 - \widehat{p})\left(\frac{1}{\rho^- - 1}\right)$. If $\phi(\cdot)$ is the pdf of an $N(0, 1)$, the parameter estimates are obtained by log-likelihood maximization:

$$\{\widehat{\mu}, \widehat{\sigma}, \widehat{p}, \widehat{\rho^+}, \widehat{\rho^-}\} = \arg\max \sum_{j=1}^{n} \ln\left(\mathbf{1}_{\{s_j \neq (t_k)_k\}} \frac{1}{\widehat{\sigma}\sqrt{\Delta}}\phi\left(\frac{x_j - \mu_X(s_j)\Delta}{\widehat{\sigma}\sqrt{\Delta}}\right)\right.$$

$$\left. + \mathbf{1}_{\{s_j = (t_k)_k\}}h\left(x_j - \mu_X(s_j)\Delta\,|\widehat{\sigma}, \widehat{p}, \widehat{\rho^+}, \widehat{\rho^-}\right)\right)\,.$$

Table 4.1 presents the results of the calibration to the S&P dataset. When a jump occurs, the probability of observing a positive shock is around 35%. The average size of positive jumps ($+3.04\%$) is larger in absolute value than that of negative shocks (-2.61%). The Brownian volatility is equal to 11.71%. Since the total volatility of log-returns is 18.64%, we infer that jumps are responsible for 37.17% of the global volatility.

Table 4.1 Parameter estimates for the SEJD fitted to the S&P 500

SEJD parameters			
$\widehat{\mu}$	0.1074	$\widehat{\sigma}$	0.1171
$\widehat{\rho^+}$	32.8562	$\widehat{\rho^-}$	-38.3253
$\widehat{p}$	0.3547	$\widehat{\alpha}$	14.5470
$\widehat{\theta}$	3.5326	$\widehat{\eta}$	12.2271

Estimation by an advanced method such as the particle MCMC requires the accurate simulation of sample paths. Due to the self-excitation, particular care must be taken when simulating jumps. The next section presents a sampling method based on "thinning."

4.4 Sampling of Self-excited Processes

Sampling a Hawkes process by thinning is similar to simulating an inhomogeneous Poisson process by the same method. For this reason, we review some properties of Poisson processes.

The inter-arrival time, τ, for a homogeneous Poisson process with parameter λ^*, has an exponential distribution with pdf $f_\tau(t) = \lambda^* e^{-\lambda^* t}$ and cdf $F_\tau(t) = 1 - e^{-\lambda^* t}$. Because both $F_\tau(t)$ and $F_\tau^{-1}(t)$ have a closed form expression, we can use the inverse transform technique to sample waiting times. We know that $F_\tau^{-1}(\tau)$ is a uniform random variable $\mathcal{U}_{[0,1]}$. Therefore, sampling a waiting interval τ for a Poisson process is done by

$$\text{sampling } U \sim \mathcal{U}_{[0,1]} \text{ and setting } s = -\frac{1}{\lambda^*} \ln U . \tag{4.16}$$

On the other hand, the thinning property of Poisson processes states that a Poisson process with an intensity λ can be split into two independent processes with intensities λ_1 and λ_2, so that $\lambda = \lambda_1 + \lambda_2$. A jump is caused by the first or the second process with respective probabilities $\frac{\lambda_1}{\lambda}$ and $\frac{\lambda_2}{\lambda}$. From this property, we can see that we can simulate a non-homogeneous Poisson process with intensity function $\lambda(t)$ by thinning a homogeneous Poisson process with intensity $\lambda^* \geq \lambda(t)$ for all $t \geq 0$.

The thinning procedure of Ogata [23] to sample Hawkes processes is presented in Algorithm 4.1. For any bounded $\lambda(t)$, we can find a constant λ^* so that $\lambda(t) \leq \lambda^*$ in a given time interval. From Sect. 4.2, we know that between two consecutive event times $[t_k, t_{k+1})$:

$$\lambda(t) = \theta + e^{-\alpha(t-t_k)} \left(\lambda(t_k) - \theta\right) , \tag{4.17}$$

and $\lambda(t_k) = \lambda(t_k-) + \eta$. Hence, $\lambda(t_k)$ is an upper bound of event intensity over this interval. We explain the sampling of event time t_{k+1}, after having already sampled $t_1, \ldots, t_k$. We start our time counter $T = t_k$. We sample an inter-arrival time τ, using Eq. (4.16) with $\lambda^* = \lambda(t_k)$.

We update the time counter $T = T + \tau$. We accept or reject this inter-arrival time according to the ratio of the true event rate to the thinning rate λ^* (step 3 of the Algorithm). If accepted, we record the $(k+1)$th event time as $t_{k+1} = T$. Otherwise, we repeat the sampling of an inter-arrival time until one is accepted. Note that, even if an inter-arrival time is rejected, the time counter T is updated, i.e., the principle of thinning a homogeneous Poisson process with a higher intensity value.

Algorithm 4.1 Sampling algorithm of N jumps of a self-excited process

Set current time $T = 0$ and jump counter $k = 1$

While $k \leq N$:

1. Set the upper bound of Poisson intensity $\lambda^* = \lambda(T)$ with Eq. (4.17).
2. Sample $U \sim \mathcal{U}_{[0,1]}$, and set $\tau = -\frac{1}{\lambda^*} \ln U$.
3. Update current time: $T = T + \tau$.
4. Sample $R \sim \mathcal{U}_{[0,1]}$.
5. If $R \leq \frac{\lambda(T-)}{\lambda^*}$, then $t_k = T$, $k = k + 1$.
6. Otherwise reject the sample and return to step 1.

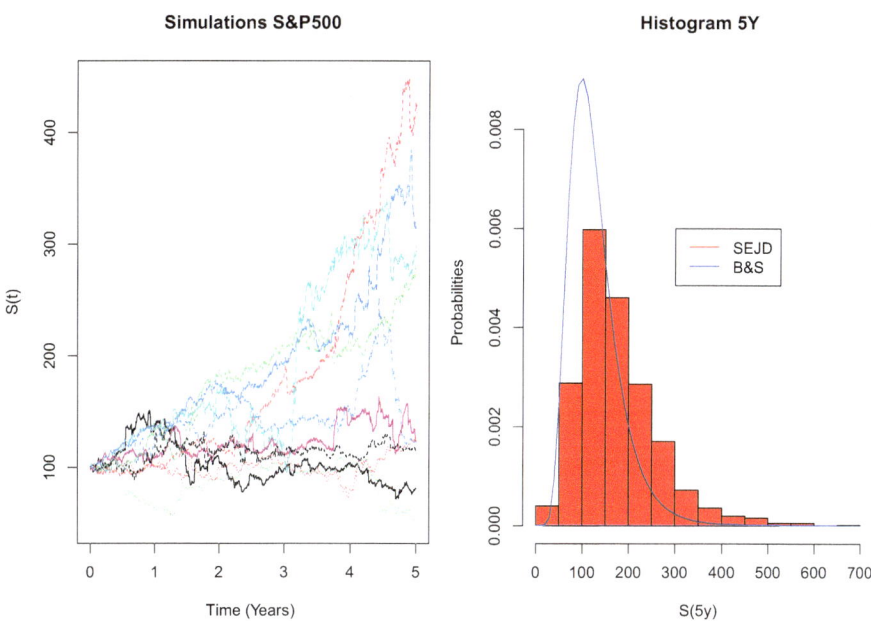

Fig. 4.4 Left plot: simulated sample paths of S_t with S&P 500 parameter estimates (SEJD with $S_0 = 100$). Right plot: comparison of the pdf of $S_{t=5\,years}$ to the pdf in the Black & Scholes setting (B&S)

The left plot of Fig. 4.4 shows simulated sample paths of a self-exciting jump-diffusion with the parameters of Table 4.1. Let us recall that these parameters fit the model for the time series of the S&P 500 from 31/1/01 to 31/1/20. The right plot compares the empirical pdfs of $S_{t=5\,years}$ in the SEJD and the Black and Scholes (B&S) model. In this last setting, the log-likelihood estimates of the mean and volatility are 4.633% and 18.64%. We see that the SEJD has a fatter right tail than the B&S model.

This is confirmed by Table 4.2, which reports the means, volatilities, 5% and 95% percentiles of $S_{t=5\,years}$ for the SEJD and B&S models fitted to the S&P 500. Despite

Table 4.2 Statistics on the
5-year value of S_t ($S_0 = 100$)
in the SEJD and B&S models
fitted to the S&P 500 (31/1/01
to 31/1/20)

	SEJD	B&S
$\mathbb{E}\left(S_{5\,y}\right)$	174.88	126.05
$\sqrt{\mathbb{V}\left(S_{5\,y}\right)}$	87.19	54.90
5% percentile	68.28	58.22
95% percentile	333.00	229.38

a higher volatility, the 5% and 95% percentiles of the SEJD are higher than those
computed with a log-normal distribution. Surprisingly, the SEJD better captures the
upside potential of financial markets and is not riskier than the B&S model.

4.5 Particle Filtering of the Hawkes Intensity and MCMC Estimation

The SEJD model depends upon an unobservable state variable, λ_t, which drives the
jump dynamic. We describe a sequential Monte Carlo method, also called a particle
filter method, to guess the state variables when the parameters of the SEJD are
known. This approach is adapted from the particle filter of Sect. 3.2.

The filter is based on a time-discretization of Eq. (4.8) that defines the model. We
denote by Δ the length of the time interval. The continuously compounded return
(over the period Δ) at time $t_j = j\Delta$, defined by $X_j = \ln \frac{S_{j\Delta}}{S_{(j-1)\Delta}}$, satisfies the
following equation in discrete time:

$$X_j = \left(\mu - \frac{\sigma^2}{2} - \lambda_j \mathbb{E}\left(e^J - 1\right)\right)\Delta + \sigma\,\Delta W_j + \Delta L'_j, \tag{4.18}$$

where ΔW_j stands for a standard normal random variable and $\Delta L'_j =$
$\sum_{k=N_{(j-1)\Delta}}^{N_{j\Delta}} J_k$, whereas λ_j is the intensity at time t_j. We recall that we assume
that the vector parameters, denoted by $\theta = \{\alpha, \theta, \eta, \mu, \sigma, p, \rho^+, \rho^-\}$, are known.

We denote by $u_j = (\lambda_j, \Delta L'_j)$ the "particle" that puts together information about
the intensity at time $t = j\Delta$ and the realized jump over the interval $[(j-1)\Delta, \Delta]$.
The model admits a useful state-space representation where Eq. (4.18) provides a
measurement equation or system (the "space") that defines the relationship between
the (possibly observed) return and the hidden state variables (the intensity and jump
process). The vector $u_j = (\lambda_j, \Delta L_j)$ can help us to find the transition system (the
"state") that describes the dynamics of the jump process.

As usual, we denote by $x = \{x_1, x_2, \ldots, x_T\}$ the sample of observed continuously compounded returns. Conditionally to information contained in u_j, the return

density $f(x_j | u_j)$ is Gaussian

$$f(x_j | u_j) = N\left(\left(\mu - \frac{\sigma^2}{2} - \lambda_j \mathbb{E}\left(e^J - 1\right)\right)\Delta + \Delta L'_j, \ \sigma\sqrt{\Delta}\right).$$

Alternatively, it is possible to simulate the transition density $f(u_{j+1} | u_j)$ with Ogata's Algorithm 4.1. The density of u_0 is $f(u_0)$, and the posterior distribution of u_j is denoted by $f(u_j | x_{1:j})$. Using standard Bayesian arguments, the expression for the posterior distribution is given by

$$f(u_j | x_{1:j}) = \frac{f(x_j | u_j)}{\int f(x_j | u_j) f(v_j | x_{1:j-1}) du_j} f(u_j | x_{1:j-1}), \qquad (4.19)$$

where

$$f(u_j | x_{1:j-1}) = \int f(u_j | u_{j-1}) f(u_{j-1} | x_{1:j-1}) du_{j-1}. \qquad (4.20)$$

The calculation of $f(u_j | x_{1:j})$ is done in two steps. In the first prediction step, we estimate $f(u_j | x_{1:j-1})$ by the relation (4.20). In the correction step, we calculate the probabilities $f(u_j | x_{1:j})$ by a Monte Carlo simulation of N particles, denoted $u_j^{(i)} = (\lambda_j^{(i)}, \Delta L'^{(i)}_j)$, with importance weights $p_j^{(i)}$ for $i = 1, \ldots, N$, as detailed in Algorithm 4.2. Finally, the filtered intensity for the period j is computed as the sum of particles, weighted by their probabilities of occurrence:

$$\widehat{\lambda}_j = \mathbb{E}\left(\lambda_j | x_{1:j}\right) = \sum_{i=1:N} \lambda_j^{(i)} p_j^{(i)},$$

and the log-likelihood of the sample is approached by

$$\log f(x | \theta) = \sum_{j=1}^{T} \log f\left(x_j | x_{j-1}\right) \qquad (4.21)$$

$$= \sum_{j=1}^{T} \log \int f(x_j | u_j) f(u_j | x_{j-1}) du_j$$

$$= \sum_{j=1}^{T} \log\left(p_j^{(i)} \sum_{i=1}^{N} f(x_j | u_j^{(i)})\right).$$

To illustrate this section, we simulate one sample path of the SEJD with daily steps over a horizon of 10 years. We use the parameters of Table 4.1 that fits the S&P 500 and run the filter with 100 particles. Figure 4.5 compares simulated and filtered

---

Algorithm 4.2 Particle filtering of the Hawkes intensity

Initial step:

Draw N values of $\lambda_0^{(i)}$ for $i = 1, \ldots, N$, from an initial distribution $f(\lambda_0)$.

Main procedure:

For $j = 1$ to maximum epoch, T

Prediction step:

Starting from $\lambda_{j-1}^{(i)}$ and $L_{j-1}^{'(i)}$ sample $\lambda_j^{(i)}$ and $L_j^{'(i)}$
with Ogata's algorithm, $i = 1, \ldots N$.

Correction step:

The "particle" $\boldsymbol{u}_j^{(i)} = \left(\lambda_j^{(i)}, L_j^{'(i)}\right)$ has a probability of occurrence equal to

$$p_j^{(i)} = \frac{f(x_j \mid \boldsymbol{u}_j^{(i)})}{\sum_{i=1:N} f(x_j \mid \boldsymbol{u}_j^{(i)})},$$

where

$$f(x_j \mid \boldsymbol{u}_j^{(i)}) = \frac{1}{\sqrt{\sigma^2 \Delta}}$$

$$\times \phi \left(\frac{x_j - \left(\mu - \frac{\sigma^2}{2} - \lambda_j^{(i)} \mathbb{E}\left(e^J - 1\right)\right) \Delta - \boldsymbol{\Delta} L_j^{'(i)}}{\sqrt{\sigma}\,\Delta} \right).$$

Resampling step:

Resample with replacement N particles according to
the importance weights $p_j^{(i)}$.

The new importance weights are set to $p_j^{(i)} = \frac{1}{N}$.

End loop on epochs

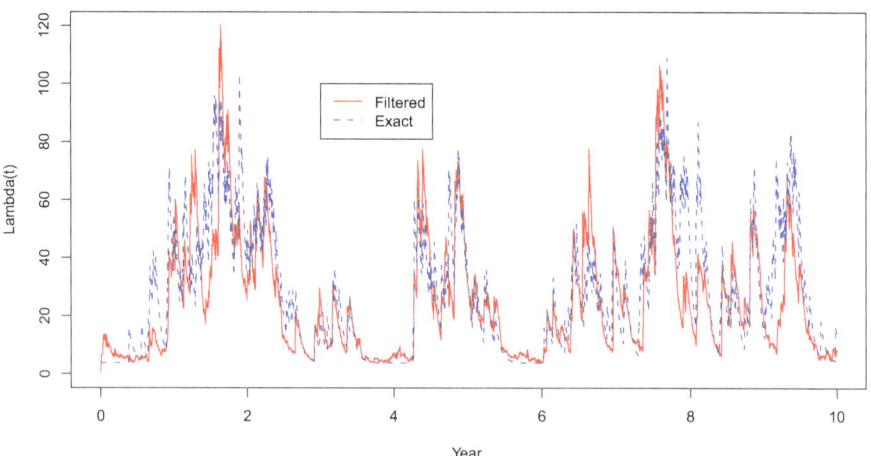

Fig. 4.5 Blue dotted curve: simulated sample path of $(\lambda_t)_{t \geq 0}$. Red curve: filtered sample path

$(\lambda_t)_{t \geq 0}$. The similarity between these sample paths is remarkable and confirms the ability of the particle filter to guess the evolution of the hidden intensity.

In a similar manner to Sect. 3.4, we can estimate the SEJD with a particle Markov Chain Monte Carlo algorithm. The unknown parameters are distributed according to a multivariate random variable denoted Θ with realizations $\boldsymbol{\theta}$. The domain of Θ is $\mathcal{X}$. The parameter's posterior distribution, based on the sample $\boldsymbol{x} = \{x_1, \ldots, x_T\}$, is

$$f(\boldsymbol{\theta} \mid \boldsymbol{x}) = \frac{f(\boldsymbol{\theta}) f(\boldsymbol{x} \mid \boldsymbol{\theta})}{\int_{\mathcal{X}} f(\boldsymbol{\theta}') f(\boldsymbol{x} \mid \boldsymbol{\theta}') \mathrm{d}\boldsymbol{\theta}'}, \qquad (4.22)$$

where $f(\boldsymbol{\theta})$ and $f(\boldsymbol{x} \mid \boldsymbol{\theta})$ are the parameter's prior distribution (e.g., uniform distribution on $\mathcal{X}$) and the likelihood of the data. The density $f(\boldsymbol{\theta} \mid \boldsymbol{x})$ is built by the PMCMC method. It generates a sample from $f(\boldsymbol{\theta} \mid \boldsymbol{x})$ by creating a Markov chain with the same stationary distribution as the parameter's posterior distribution. As a closed form estimate of $f(\boldsymbol{x} \mid \boldsymbol{\theta})$ is not available in our model, we approach it by Eq. (4.21).

The construction of the Markov chain consists of two steps, repeated iteratively. During the $(k + 1)$th iteration, we draw a candidate parameter $\boldsymbol{\theta}'$ from a proposal distribution $q(\boldsymbol{\theta}' | \boldsymbol{\theta}_k)$. The proposal distribution has a support that covers the target distribution. For instance, we may assume that

$$q(\boldsymbol{\theta}' | \boldsymbol{\theta}_k) \sim N(\boldsymbol{\theta}_k, \sigma_\theta I_m),$$

where $\sigma_\theta \in \mathbb{R}^+$ and I_m is the identity matrix. In the second step, we determine if we need to update the state by $\boldsymbol{\theta}'$. For this purpose, the acceptance probability of the Metropolis–Hastings algorithm 2.2 is computed as follows:

$$\rho(\boldsymbol{\theta}', \boldsymbol{\theta}_k) = \min \left\{ \frac{f(\boldsymbol{x} \mid \boldsymbol{\theta}') f(\boldsymbol{\theta}')}{f(\boldsymbol{x} \mid \boldsymbol{\theta}_k) f(\boldsymbol{\theta}_k)}, 1 \right\}.$$

The resulting samples $\boldsymbol{\theta}_{1:n}$ (n iterations after the burn-in period) serve to build the empirical distribution of $f(\boldsymbol{\theta} \mid \boldsymbol{x})$, which is defined by

$$f(\boldsymbol{\theta} \mid \boldsymbol{x}) = \frac{1}{n} \sum_{k=1}^{n} \delta_{\boldsymbol{\theta}_k}(\mathrm{d}\boldsymbol{\theta}),$$

where $\delta_{\boldsymbol{\theta}_k}(\mathrm{d}\boldsymbol{\theta})$ are the Dirac atoms located at $\boldsymbol{\theta} = \boldsymbol{\theta}_k$, with equal weights. The estimate of parameters with respect to the posterior distribution is then

$$\widehat{\Theta} = \mathbb{E}(\Theta | \boldsymbol{x}) \approx \frac{1}{n} \sum_{k=1}^{n} \boldsymbol{\theta}_k.$$

Table 4.3 Parameter
estimates for the SEJD fitted
to the S&P 500 by MCMC
and POT

SEJD parameters			
POT		MCMC	
$\widehat{\mu}$	0.1074	$\widehat{\mu}$	0.1171
$\widehat{\rho^+}$	32.8562	$\widehat{\rho^+}$	32.3693
$\widehat{p}$	0.3547	$\widehat{p}$	0.3287
$\widehat{\theta}$	3.5326	$\widehat{\theta}$	3.5258
$\widehat{\sigma}$	0.1171	$\widehat{\sigma}$	0.1171
$\widehat{\rho^-}$	-38.3253	$\widehat{\rho^-}$	-38.3247
$\widehat{\alpha}$	14.5470	$\widehat{\alpha}$	14.5171
$\widehat{\eta}$	12.2271	$\widehat{\eta}$	12.2131
$\log f(x\mid\theta)$	16,451.40	$\log f(x\mid\theta)$	16,576.46

In theory, the PMCMC algorithm converges to the optimal parameter estimates. Nevertheless, numerical experiments show that this method fails to retrieve parameters from a simulated sample path. The main reason is a lack of identifiability of the SEJD model and in particular of the variance. The variance comes both from the Brownian motion and self-exciting jumps. Therefore, for a given observed variance, there exist multiple combinations of SEJD parameters replicating it. Our numerical analysis reveals that the MCMC algorithm will select parameter estimates with the lowest possible Brownian volatility, while the parameters of the jump process are tuned in such a way to replicate at best the time series.

Nevertheless, the MCMC algorithm can be used to fine-tune parameters found by the POT method other than the Brownian volatility, σ. To illustrate this, we fit the SEJD model to the time series of the S&P 500. We run 1000 MCMC iterations and use as an initial set of parameters those found by the POT method (see Table 4.1). The volatility is considered as an exogenous factor and is fixed at $\sigma = 11.71\%$. Table 4.3 compares POT and MCMC estimates, averaged over the last 300 simulations. The log-likelihood is clearly improved by the MCMC method.

Figure 4.6 shows the most likely sample paths of jump intensities computed with a particle filter and POT and PMCMC parameter estimates. Both curves are very close, except in 2008, during which the filtered intensity with POT parameters is clearly higher.

4.6 Properties of Jump Intensity and Log-Return in the SEJD

By construction, the counting process $(N_t)_{t\geq0}$ is not a Markov process, but the pair $Y_t = (\lambda_t, N_t)_{t\in\mathbb{R}^+}$ is Markov in the state space $D = \mathbb{R}_+ \times \mathbb{N}$. The natural filtration of this bivariate process is denoted by $\{\mathcal{G}_t\}_{t\geq0}$. Therefore, the infinitesimal generator

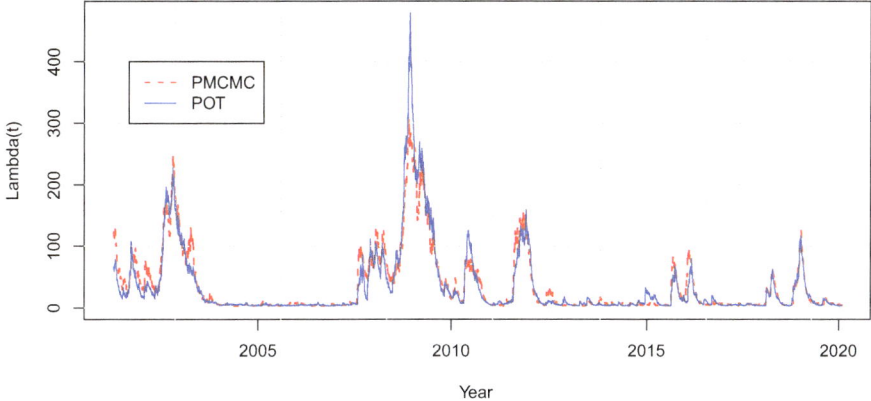

Fig. 4.6 Blue curve: sample path of $(\lambda_t)_{t \geq 0}$ by particle filtering for the S&P 500 (POT estimates). Red dotted curve: sample path of $(\lambda_t)_{t \geq 0}$ by particle filtering for the S&P 500 (MCMC estimates)

of Y_t is the operator acting on a sufficiently regular function $f \: : \: D \to \mathbb{R}$ such that

$$\mathcal{A}f(y) = \lim_{h \to 0} \frac{\mathbb{E}\left(f\left(Y_{t+h}\right) - f(y) \mid \mathcal{G}_t\right)}{h} \: ,$$

where $Y_t = \mathbf{y} = (\lambda, n)$. This generator is given by

$$\mathcal{A}f(y) = \alpha \left(\theta - \lambda\right) \frac{\partial f}{\partial \lambda}(y) + \lambda \left(f(\lambda + \eta \, , \, n + 1) - f(\mathbf{y})\right) . \qquad (4.23)$$

On the other hand, the compensated process

$$M_t = f(Y_t) - f(Y_0) - \int_0^t \mathcal{A}f(Y_u)du$$

is a martingale relative to its natural filtration. Thus, for $s > t$, we have that

$$\mathbb{E}\left(f(Y_s) - \int_0^s \mathcal{A}f(Y_u)du \mid \mathcal{G}_0\right) = f(Y_t) - \int_0^t \mathcal{A}f(Y_u)du$$

by the martingale property. This leads to the Dynkin formula:

$$\mathbb{E}\left(f\left(Y_s\right) \mid \mathcal{G}_t\right) = f(Y_t) + \mathbb{E}\left(\int_t^s \mathcal{A}f(Y_u)du \mid \mathcal{G}_t\right) . \qquad (4.24)$$

This last formula allows us to calculate the first two moments of the intensity.

Proposition 4.2 *The first and second moments of λ_t are respectively given by*

$$\mathbb{E}\left(\lambda_t \mid \mathcal{G}_0\right) = e^{(\eta-\alpha)t}\lambda_0 + \frac{\alpha\theta}{\eta - \alpha}\left(e^{(\eta-\alpha)t} - 1\right). \qquad (4.25)$$

If we define $\rho_1 = \eta^2\lambda_0$ and $\rho_2 = \frac{\eta^2\alpha\theta}{\eta-\alpha}$, then

$$\mathbb{V}\left(\lambda_t \mid \mathcal{G}_t\right) = \frac{\rho_1 + \rho_2}{\eta - \alpha}\left(e^{2(\eta-\alpha)t} - e^{(\eta-\alpha)t}\right) + \frac{\rho_2}{2\left(\eta - \alpha\right)}\left(1 - e^{2(\eta-\alpha)t}\right). \quad (4.26)$$

Proof To lighten the notation, we momentarily denote $\mathbb{E}\left(.\mid\mathcal{G}_0\right)$ by $\mathbb{E}_0(\cdot)$. From Eq. (4.7) and given that

$$\mathbb{E}_0(\mathrm{d}N_s) = \mathbb{E}_0\left(\mathbb{E}_s\left(\mathrm{d}N_s\right)\right) = \mathbb{E}_0(\lambda_s),$$

we infer that

$$\mathbb{E}_0\left(\lambda_t\right) = \theta + \left(\lambda_0 - \theta\right)e^{-\alpha t} + \eta\int_0^t e^{-\alpha(t-s)}\mathbb{E}_0\left(\lambda_s\right)\mathrm{d}s.$$

Differentiating this last equation with respect to time, we obtain that $\mathbb{E}_0\left(\lambda_t\right)$ is the solution of an ordinary differential equation (ODE):

$$\frac{\mathrm{d}\mathbb{E}_0\left(\lambda_t\right)}{\mathrm{d}t} = -\alpha e^{-\alpha t}\left(\lambda_0 - \theta\right) + \eta\mathbb{E}_0\left(\lambda_t\right) - \alpha\eta\int_0^t e^{-\alpha(t-s)}\mathbb{E}_0\left(\lambda_s\right)\mathrm{d}s$$

$$= \eta\mathbb{E}_0\left(\lambda_t\right) + \alpha\left(\theta - \mathbb{E}_0\left(\lambda_t\right)\right). \qquad (4.27)$$

The solution of this ODE is given by Eq. (4.25). On the other hand, using the Dynkin formula (4.24) leads to the following expression for the second moment of the intensity:

$$\mathbb{E}_0\left(\lambda_t^2\right) = \lambda_0^2 + \mathbb{E}_0\left(\int_0^t 2\alpha\left(\theta - \lambda_s\right)\lambda_s + \lambda_s\left(2\eta\lambda_s + \eta^2\right)\mathrm{d}s\right).$$

Differentiating this expression with respect to time leads to the following ODE for the second moment:

$$\frac{\mathrm{d}\mathbb{E}_0\left(\lambda_t^2\right)}{\mathrm{d}t} = \left(2\alpha\theta + \eta^2\right)\mathbb{E}_0\left(\lambda_t\right) + 2\left(\eta - \alpha\right)\mathbb{E}_0\left(\lambda_t^2\right).$$

On the other hand, from Eq. (4.27), we infer that:

$$\frac{\mathrm{d}\mathbb{E}_0\left(\lambda_t\right)^2}{\mathrm{d}t} = 2\left(\eta - \alpha\right)\mathbb{E}_0\left(\lambda_t\right)^2 + 2\alpha\theta\mathbb{E}_0\left(\lambda_t\right).$$

From these two last equations, the variance is a solution of an ordinary differential equation:

$$\frac{d\mathbb{V}_0(\lambda_t)}{dt} = \eta^2 \mathbb{E}_0(\lambda_t) + 2(\eta - \alpha)\mathbb{V}_0(\lambda_t)$$

$$= \eta^2 \lambda_0 e^{(\eta - \alpha)t} + \frac{\eta^2 \alpha \theta}{\eta - \alpha}\left(e^{(\eta - \alpha)t} - 1\right)$$

$$+ 2(\eta - \alpha)\mathbb{V}_0[\lambda_t].$$

Using the notations $\rho_1 = \eta^2 \lambda_0$ and $\rho_2 = \frac{\eta^2 \alpha \theta}{\eta - \alpha}$ allows us to rewrite this ODE as follows:

$$\frac{d\mathbb{V}_0(\lambda_t)}{dt} - 2(\eta - \alpha)\mathbb{V}_0(\lambda_t) = (\rho_1 + \rho_2)e^{(\eta - \alpha)t} - \rho_2,$$

which admits as solution:

$$\mathbb{V}_0(\lambda_t) = e^{2(\eta - \alpha)t}\left[(\rho_1 + \rho_2)\int_0^t e^{-(\eta - \alpha)s}ds - \rho_2 \int_0^t e^{-2(\eta - \alpha)s}ds\right]. \qquad \square$$

We draw an important conclusion from Eqs. (4.25) and (4.26): the expectation and variance are well defined when $t \to \infty$ if and only if $\eta - \alpha < 0$. If this condition of stability is not fulfilled, the process "explodes" to infinity. In contrast, when $\eta - \alpha < 0$, the expected value of the jump arrival intensity tends to a constant as t becomes large, and the long-term value of λ_t is equal to $-\frac{\alpha \theta}{\eta - \alpha}$. The long-term value of the variance is $\frac{\eta^2 \alpha \theta}{2(\eta - \alpha)^2}$. The parameter estimates for the S&P 500 in Table 4.1 satisfy this relation of stability. Furthermore, the asymptotic number of jumps per year is $\lambda_\infty = 21.41$, which agrees with our observations: 406 detected jumps on 19 years indeed correspond to 21.36 jumps per year. The next proposition establishes the expression of the autocovariance of the jump intensity.

Proposition 4.3 *The covariance between λ_s and λ_t for $t \leq s$ is proportional to the variance*

$$\mathbb{C}[\lambda_t \lambda_s \mid \mathcal{G}_0] = e^{(\eta - \alpha)(s - t)}\mathbb{V}[\lambda_t \mid \mathcal{G}_0]. \tag{4.28}$$

Proof To lighten the notation, we momentarily denote $\mathbb{E}(. \mid \mathcal{G}_t)$ by $\mathbb{E}_t(\cdot)$. Using nested expectations, we have that $\mathbb{E}_0[\lambda_t \lambda_s] = \mathbb{E}_0[\lambda_t \mathbb{E}_t[\lambda_s]]$, and from Proposition 4.2, we know that

$$\mathbb{E}_t[\lambda_s] = e^{(\eta - \alpha)(s - t)}\lambda_t + \frac{\alpha \theta}{\eta - \alpha}\left(e^{(\eta - \alpha)(s - t)} - 1\right)$$

and

$$\mathbb{E}_0 \left[\lambda_t \lambda_s \right] = e^{(\eta - \alpha)(s-t)} \mathbb{E}_0 \left[\lambda_t^2 \right] + \frac{\alpha \theta}{\eta - \alpha} \left(e^{(\eta - \alpha)(s-t)} - 1 \right) \mathbb{E}_0 \left[\lambda_t \right] . \qquad (4.29)$$

On the other hand, we can formulate $\mathbb{E}_0 \left[\lambda_s \right]$ as a function of $\mathbb{E}_0 \left[\lambda_t \right]$:

$$\mathbb{E}_0 \left[\lambda_s \right] = e^{(\eta - \alpha)(s-t)} \mathbb{E}_0 \left[\lambda_t \right] + \frac{\alpha \theta}{\eta - \alpha} \left(e^{(\eta - \alpha)(s-t)} - 1 \right) .$$

Therefore, the product of moments is equal to

$$\mathbb{E}_0 \left[\lambda_s \right] \mathbb{E}_0 \left[\lambda_t \right] = e^{(\eta - \alpha)(s-t)} \mathbb{E}_0 \left[\lambda_t \right]^2 \qquad (4.30)$$
$$+ \frac{\alpha \theta}{\eta - \alpha} \left(e^{(\eta - \alpha)(s-t)} - 1 \right) \mathbb{E}_0 \left[\lambda_t \right] .$$

Subtracting Eq. (4.29) from (4.30) allows us to reformulate the covariance as a function of the variance of λ_t.   □

The next proposition presents the joint moment generating function (mgf) of λ_t and of the log-return, $X_t := \ln \frac{S_t}{S_0}$, as defined by Eq. (4.8). The log-return is defined on a filtration $\{\mathcal{H}_t\}_{t \geq 0}$ of Ω that carries the information about the asset price. The information about the jump process and its intensity is contained in a filtration $\{\mathcal{G}_t\}_{t \geq 0}$. The augmented filtration that gathers information about all processes is denoted by $\{\mathcal{F}_t\}_{t \geq 0}$ and is such that $\mathcal{G}_t \cup \mathcal{H}_t \subset \mathcal{F}_t$. The next result involves the moment generating function of stock price jumps. We denote the latter by $\psi(\omega)$ and in the case of a double exponential distribution:

$$\psi(\omega) = \mathbb{E} \left(e^{\omega J} \right) = p \frac{\rho^+}{\rho^+ - \omega} + (1 - p) \frac{\rho^-}{\rho^- - \omega}.$$

Proposition 4.4 *The mgf of X_s for $s \geq t$ and $\omega \in \mathbb{R}$ is the exponential of an affine function of the intensity:*

$$\mathbb{E} \left(e^{\omega_1 X_s + \omega_2 \lambda_s} \mid \mathcal{F}_t \right) = \left(\frac{S_t}{S_0} \right)^{\omega_1} \exp \left(A(t, s) + B(t, s) \lambda_t \right) , \qquad (4.31)$$

where $A(t, s)$ and $B(t, s)$ are solutions of the system of ODEs:

$$\begin{cases} \frac{\partial A(t,s)}{\partial t} = -\omega_1 \left(\mu - \frac{\sigma^2}{2} \right) - \omega_1^2 \frac{\sigma^2}{2} - \alpha \theta B(t, s) \\ \frac{\partial B(t,s)}{\partial t} = \alpha B(t, s) + \omega_1 (\psi(1) - 1) - \left[e^{B(t,s) \eta} \psi(\omega_1) - 1 \right] \end{cases} \qquad (4.32)$$

with the terminal conditions: $A(s, s) = 0$ and $B(s, s) = \omega_2$.

Proof Let us define $f(t, X_t, N_t, \lambda_t) = \mathbb{E}\left(e^{\omega_1 X_s + \omega_2 \lambda_s} \mid \mathcal{F}_t\right)$ with $t \leq s$. The infinitesimal of this function is provided by Eq. (4.9), and by definition of f, we have that $\mathscr{A}f = 0$:

$$0 = \frac{\partial f}{\partial t} + \left(\mu - \frac{\sigma^2}{2}\right)\frac{\partial f}{\partial X} + \frac{1}{2}\sigma^2 \frac{\partial^2 f}{\partial X^2} + \alpha(\theta - \lambda)\frac{\partial f}{\partial \lambda} \tag{4.33}$$

$$-\lambda \mathbb{E}\left(e^J - 1\right)\frac{\partial f}{\partial X} + \lambda \int_{-\infty}^{-\infty} f\left(t, X + z, n + 1, \lambda + \eta\right) - f(\cdot)\, \nu(\mathrm{d}z).$$

We postulate that $f(\cdot)$ is an exponential affine function of λ_t and X_t:

$$f = \exp\left(A(t, s) + B(t, s)\lambda + C(t, s)X\right),$$

where $A(t, s)$, $B(t, s)$, $C(t, s)$ are time-dependent functions. Under this assumption, the partial derivatives of f are given by

$$\frac{\partial f}{\partial t} = \left(\frac{\partial A(t, s)}{\partial t} + \frac{\partial B(t, s)}{\partial t}\lambda + \frac{\partial C(t, s)}{\partial t}X\right) f,$$

$$\frac{\partial f}{\partial X} = C(t, s)f \quad \frac{\partial^2 f}{\partial X^2} = C(t, s)^2 f \quad \frac{\partial f}{\partial \lambda} = B(t, s)f.$$

The integrand in Eq. (4.33) is rewritten as follows:

$$\int_0^{+\infty} f\left(t, X + z, \lambda + \eta\right) - f(\cdot)\, \mathrm{d}\nu(z) = f\left[e^{B(t,s)\eta}\psi\left(C(t, s)\right) - 1\right].$$

Injecting these expressions into Eq. (4.33) leads to the following relation:

$$0 = \left(\frac{\partial A(t, s)}{\partial t} + \frac{\partial B(t, s)}{\partial t}\lambda + \frac{\partial C(t, s)}{\partial t}X\right) \tag{4.34}$$

$$+ C(t, s)\left(\mu - \frac{\sigma^2}{2}\right) + C(t, s)^2 \frac{\sigma^2}{2} - \lambda\mathbb{E}\left(e^J - 1\right)C$$

$$+ \alpha(\theta - \lambda)B(t, s) + \lambda\left[e^{B(t,s)\eta}\psi\left(C(t, s)\right) - 1\right].$$

This last relation is being fulfilled for all values of λ and X, we infer that $\frac{\partial C(t,s)}{\partial t} = 0$, and therefore $C(t, s) = \omega_1$. Since the multiplier of λ in Eq. (4.34) must be null, we infer that $B(t, s)$ is the solution of an ODE:

$$\frac{\partial B(t, s)}{\partial t} - \omega_1 \mathbb{E}\left(e^J - 1\right) - \alpha B(t, s) + \left[e^{B(t,s)\eta}\psi\left(\omega_1\right) - 1\right] = 0.$$

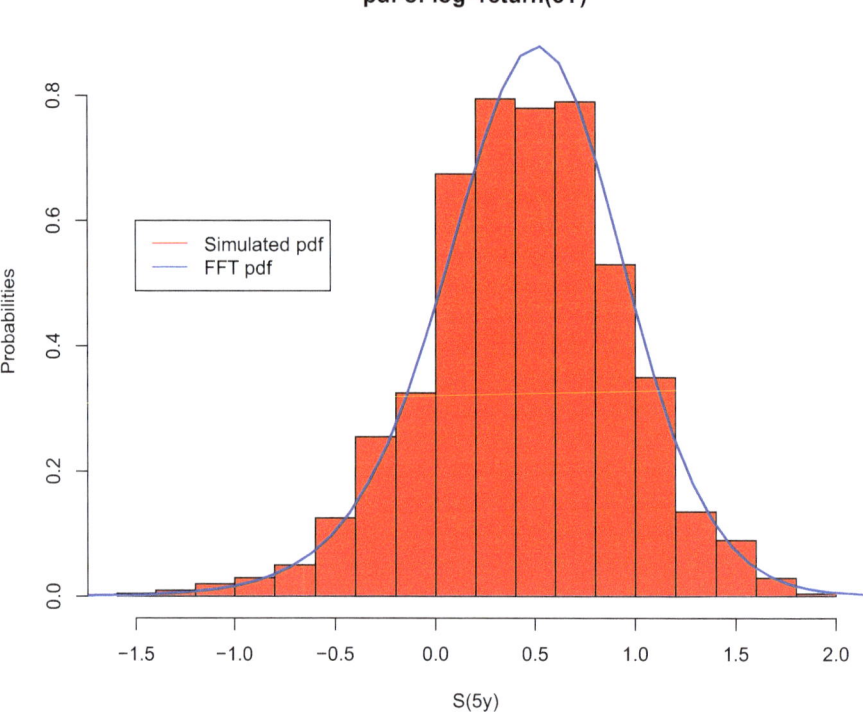

Fig. 4.7 Blue line: pdf of the 5-year log-return, $X_{5\,\text{years}}$, computed by FFT with $M = 2^6$ discretization steps and $x_{\text{max}} = 3$. Red histogram: empirical distribution of the simulated 5-year log-return

The sum of terms that do not depend upon X and λ in Eq. (4.34) is also null. This provides us with the ODE defining $A(t, s)$:

$$0 = \frac{\partial A(t, s)}{\partial t} + \omega_1 \left(\mu - \frac{\sigma^2}{2} \right) + \omega_1^2 \frac{\sigma^2}{2} + \alpha\,\theta\,B(t, s)\,. \qquad\qquad \Box$$

This last result may be used to compute the probability density function of the log-return by Discrete Fourier Transform. Using Eq. 4.31 of Proposition 4.4 allows us to calculate $\Upsilon_{t,T}(\omega) = \mathbb{E}\left(e^{\omega X_T} \mid \mathcal{F}_t \right)$ in Proposition 1.8 of Chap. 1. Figure 4.7 illustrates this and compares the pdfs of the 5-year log-return computed by FFT and by simulation. We observe a clear similarity and a light left tail asymmetry.

4.7 Change of Measure

We continue to explore the properties of the SEJD with a discussion about the choice of a risk neutral measure. The market, such as modeled, is incomplete. A consequence of this incompleteness is the existence of several equivalent measures that are all potential candidates for the definition of a risk neutral measure. In this section, we focus on a family of changes of measure that are induced by exponential martingales of the form:

$$M_t(\xi, \varphi) := \exp\left(\kappa_1(\xi)\lambda_t + \xi \sum_{j=1}^{N_t} J_j + \kappa_2(\xi)t\right) \tag{4.35}$$

$$\times \exp\left(-\frac{1}{2}\int_0^t \varphi(s)^2 \mathrm{d}s - \int_0^t \varphi(s)\mathrm{d}W_s\right),$$

where $\varphi(s)$ is an $\mathcal{F}_s$-adapted process and ξ is constant. $\kappa_1(\xi)$ and $\kappa_2(\xi)$ are the functions of ξ that correspond to the price of jump risk. In our framework, jumps are random, and the affine change of measure modifies both frequencies and distribution. We focus on processes $\varphi(s)$ that are linear functions of the intensity:

$$\varphi(s) = \varphi_1 + \varphi_2\lambda_s.$$

The next proposition details the conditions under which the process $M = (M_t)_t$ is a local martingale:

Proposition 4.5 *If for any parameter ξ there exist solutions $\kappa_1(\cdot)$ and $\kappa_2(\cdot)$ of the system*

$$\kappa_1(\xi)\alpha - \left(e^{\kappa_1(\xi)\eta}\psi(\xi) - 1\right) = 0, \tag{4.36}$$

$$\kappa_2(\xi) + \kappa_1(\xi)\alpha\theta = 0,$$

then $(M_t(\xi))_t$ is a local martingale.

Proof If we denote by m_t the logarithm of M_t, then

$$m_t = \kappa_1(\xi)\lambda_t + \xi \sum_{j=1}^{N_t} J_j + \kappa_2(\xi)t \tag{4.37}$$

$$-\frac{1}{2}\int_0^t \varphi(s)^2 \mathrm{d}s - \int_0^t \varphi(s)\mathrm{d}W_s,$$

and its infinitesimal dynamics is given by

$$dm_t = \kappa_1(\xi)\,\alpha\,(\theta - \lambda_t)\,dt + \kappa_2(\xi)\,dt + (\kappa_1(\xi)\,\eta + \xi\,J)\,dN_t$$
$$-\frac{1}{2}\varphi(t)^2 dt - \varphi(t) dW_t\,.$$

The random measure of J is denoted by $\Xi(\cdot)$ and is such that $J = \int_{-\infty}^{\infty} \Xi(dz)$. We denote by $[m_t, m_t]^c$ the quadratic variation of the continuous part of m_t. Applying Itô's lemma for semi-martingales to M_t gives us the following relation:

$$dM_t = M_t\,dm_t + \frac{1}{2}M_t d\,[m_t, m_t]^c$$
$$+M_t \int_{-\infty}^{\infty} \left(e^{\kappa_1(\xi)\eta + \xi z} - 1 - (\kappa_1(\xi)\,\eta + \xi z)\right) \Xi(dz)dN_t\,,$$

which may be expanded as follows:

$$dM_t = M_t \left(\begin{array}{c} \kappa_1(\xi)\,\alpha\,(\theta - \lambda_t)\,dt + \kappa_2(\xi)\,dt + (\kappa_1(\xi)\,\eta + \xi\,J)\,dN_t \\ -\frac{1}{2}\varphi(t)^2 dt - \varphi(t) dW_t \end{array}\right)$$
$$+\frac{1}{2}M_t d\,[m_t, m_t]^c + M_t \int_{-\infty}^{\infty} \left(e^{(\kappa_1\eta + \xi z)} - 1 - (\kappa_1(\xi)\,\eta + \xi z)\right) \Xi(dz)dN_t\,,$$

and after the introduction of the compensator of the jump process,

$$dM_t = \kappa_1(\xi)\,\alpha\theta\,M_t dt + \kappa_2(\xi)\,M_t dt - \varphi(t)M_t dW_t$$
$$-M_t\lambda_t \left(\kappa_1(\xi)\,\alpha - \int_{-\infty}^{\infty} \left(e^{(\kappa_1(\xi)\eta + \xi z)} - 1\right) \nu(dz)\right) dt$$
$$+M_t \int_{-\infty}^{\infty} \left(e^{(\kappa_1(\xi)\eta + \xi z)} - 1\right) (\Xi(dz)dN_t - \lambda_t\nu(dz)dt)$$

because $\frac{1}{2}M_t d\,[m_t, m_t]^c = \frac{1}{2}\varphi(t)^2 M_t dt$. Since the integral with respect to $\Xi(dz)dN_t - \lambda_t\nu(dz)dt$ is a local martingale, M is also a local martingale if and only if the relations (4.36) hold.   □

An equivalent measure $\mathbb{Q}^{\xi,\varphi}$ can then be defined by the ratio:

$$\left.\frac{d\mathbb{Q}^{\xi,\varphi}}{d\mathbb{P}}\right|_{\mathcal{F}_t} = \frac{M_t(\xi,\varphi)}{M_0(\xi,\varphi)}\,. \tag{4.38}$$

In the sequel, we simply write $\kappa_1(\xi) = \kappa_1$ and $\kappa_2(\xi) = \kappa_2$, but we should not forget the dependence of these parameters on ξ. This new measure is particularly interesting because it preserves the structure of the jump process, as demonstrated by the next propositions.

Proposition 4.6 *Let us denote by* $N^Q = \left(N_t^Q\right)_t$ *the counting process with the following intensity:*

$$\lambda_t^Q = e^{\kappa_1 \eta} \psi(\xi) \lambda_t$$

under $\mathbb{Q}^{\xi,\varphi}$*. Then the dynamics of* $\left(\lambda_t^Q\right)_t$ *under* $\mathbb{Q}^{\xi,\varphi}$ *is ruled by the following SDE:*

$$d\lambda_t^Q = \alpha \left(\theta^Q - \lambda_t^Q\right) dt + \eta^Q d N_t^Q \,,$$

where

$$\theta^Q = \theta e^{\kappa_1 \eta} \psi(\xi) \,,$$
$$\eta^Q = \eta e^{\kappa_1 \eta} \psi(\xi) \,.$$

Proof If m_t is the logarithm of M_t, as defined by Eq. (4.37), the mgf of λ_T under $\mathbb{Q}^{\xi,\varphi}$ is equal to

$$\mathbb{E}^Q \left(e^{\omega \lambda_T^Q} | \mathcal{F}_t\right) = \mathbb{E}\left(e^{m_T - m_t + \omega e^{\kappa_1 \eta} \psi(\xi) \lambda_T} | \mathcal{F}_t\right)$$
$$= e^{-m_t} \mathbb{E}\left(e^{m_T + \omega e^{\kappa_1 \eta} \psi(\xi) \lambda_T} | \mathcal{F}_t\right) .$$

If $f(t, \lambda_t, m_t)$ denotes $\mathbb{E}\left(e^{m_T + \omega e^{\kappa_1 \eta} \psi(\xi) \lambda_T} | \mathcal{F}_t\right)$, then according to Itô's lemma, it solves the following stochastic differential equation:

$$0 = \frac{\partial f}{\partial t} + (\kappa_1 \alpha (\theta - \lambda) + \kappa_2) \frac{\partial f}{\partial m} + \alpha (\theta - \lambda) \frac{\partial f}{\partial \lambda} \qquad (4.39)$$
$$- \frac{1}{2} \varphi(t)^2 \frac{\partial f}{\partial m} + \frac{1}{2} \varphi(t)^2 \frac{\partial^2 f}{\partial m^2}$$
$$+ \lambda \int_{-\infty}^{+\infty} [f(t, \lambda + \eta, m + (\kappa_1 \eta + \xi z)) - f(.)] \, d\nu(z) \,.$$

We conjecture that $f(\cdot)$ is an exponential affine function of state variables:

$$f = \exp\left(A(t, T) + e^{\kappa_1 \eta} \psi(\xi) B(t, T) \lambda_t + C(t, T) m_t\right) ,$$

with the terminal conditions $A(T, T) = 0$, $B(T, T) = \omega$, and $C(T, T) = 1$. If we temporarily write $\psi^b = e^{\kappa_1 \eta} \psi(\xi)$, then the partial derivatives of $f(\cdot)$ are

$$\frac{\partial f}{\partial t} = \left(\frac{\partial A}{\partial t} + \psi^b \lambda \frac{\partial B}{\partial t} + \frac{\partial C}{\partial t} m \right) f,$$

$$\frac{\partial f}{\partial m} = C f \qquad \frac{\partial^2 f}{\partial m^2} = C^2 f \qquad \frac{\partial f}{\partial \lambda} = B \psi^b f.$$

Inserting these expressions into Eq. (4.39) leads to the following relation (after grouping terms)

$$0 = \frac{\partial A}{\partial t} + \left(\alpha \theta \psi^b B + \alpha \theta \kappa_1 C + \kappa_2 C \right) - \frac{1}{2} \varphi(t)^2 C + \frac{1}{2} \varphi(t)^2 C^2$$

$$+ \lambda \left(\psi^b \frac{\partial B}{\partial t} - \kappa_1 \alpha C - \alpha \psi^b B + \int_{-\infty}^{+\infty} \left[e^{B \psi^b \eta + C (\kappa_1 \eta + \xi z)} - 1 \right] d\nu(z) \right)$$

$$+ m \left(\frac{\partial}{\partial t} C \right).$$

As this equality is satisfied whatever the values of m_t, we infer that $C(t, s) = 1$ as $\frac{\partial}{\partial t} C(t, s) = 0$, and we get that

$$0 = \frac{\partial A}{\partial t} + \alpha \theta \psi^b B + \alpha \theta \kappa_1 + \kappa_2,$$

$$0 = \psi^b \frac{\partial}{\partial t} B - \kappa_1 \alpha - \alpha \psi^b B + \int_{-\infty}^{+\infty} \left[e^{B \psi^b \eta + (\kappa_1 \eta + \xi z)} - 1 \right] d\nu(z).$$

Using condition (4.36), this system is simplified as follows:

$$\frac{\partial A}{\partial t} = -\alpha \left[\theta \psi^b \right] B,$$

$$\frac{\partial B}{\partial t} = \alpha B - \left[e^{B \psi^b \eta} - 1 \right],$$

and by comparison with the results of Proposition 4.4, the proof is complete.   □

It remains to determine $\varphi(t)$, the $\mathcal{F}_t$-adapted process involved in the definition of $\frac{d\mathbb{Q}^{\xi, \varphi}}{d\mathbb{P}}$ such that the discounted asset price is a martingale under the risk neutral measure.

Proposition 4.7 *If the $\mathcal{F}_t$-adapted process defining the martingale (4.35) is equal to*

$$\varphi(t) = \frac{\mu + \lambda_t \left[e^{\kappa_1 \eta} \psi(\xi) \left(\psi^Q(1) - 1 \right) - (\psi(1) - 1) \right] - r}{\sigma}, \qquad (4.40)$$

then the equivalent measure $\mathbb{Q}^{\xi,\varphi}$ is risk neutral. The log-return, $X_t := \ln\frac{S_t}{S_0}$, is driven by the following dynamics under this measure $\mathbb{Q}^{\xi,\varphi}$:

$$dX_t = \left(r - \frac{1}{2}\sigma^2 - \lambda_t^Q \mathbb{E}^Q\left(e^{J^Q} - 1\right)\right) dt + \sigma\, dW_t^Q + J^Q dN_t,$$

where J^Q is a random variable with an mgf equal to

$$\psi^Q(\omega) = \mathbb{E}\left(e^{\omega J^Q}\right) = \frac{\psi(\omega + \xi)}{\psi(\xi)}.$$

The asset price is in this case ruled by the following SDE:

$$dS_t = r S_t dt + \sigma\, S_t\, dW_t^Q \tag{4.41}$$
$$+ S_t\left[\left(e^{J_t^Q} - 1\right) dN_t^Q - \lambda_t^Q \mathbb{E}^Q\left(e^{J^Q} - 1\right) dt\right].$$

Proof If m_t is the logarithm of M_t, as defined by Eq. (4.37), the mgf of X_T under $\mathbb{Q}^{\xi,\varphi}$ is equal to

$$\mathbb{E}^Q\left(e^{\omega X_T}|\mathcal{F}_t\right) = \mathbb{E}\left(e^{m_T - m_t + \omega X_T}|\mathcal{F}_t\right)$$
$$= e^{-m_t}\mathbb{E}\left(e^{m_T + \omega X_T}|\mathcal{F}_t\right).$$

If $f(\cdot)$ denotes $\mathbb{E}\left(e^{m_T + \omega X_T}|\mathcal{F}_t\right)$, then, according to Itô's lemma, it solves the following stochastic differential equation:

$$0 = \frac{\partial f}{\partial t} + (\kappa_1\alpha(\theta - \lambda) + \kappa_2)\frac{\partial f}{\partial m} + \alpha(\theta - \lambda)\frac{\partial f}{\partial \lambda} - \frac{1}{2}\varphi(t)^2\frac{\partial f}{\partial m} \tag{4.42}$$
$$+ \frac{1}{2}\varphi(t)^2\frac{\partial^2 f}{\partial m^2} + \frac{\partial f}{\partial X}\left(\mu - \frac{\sigma^2}{2} - \lambda\mathbb{E}\left(e^J - 1\right)\right) + \frac{\partial^2 f}{\partial X^2}\frac{\sigma^2}{2} - \frac{\partial^2 f}{\partial X\partial m}\varphi\sigma$$
$$+ \lambda\int_{-\infty}^{+\infty} f(t, X + z, \lambda + \eta, m_t + (\kappa_1\eta + \xi z)) - f\, d\nu(z).$$

We assume that $f(\cdot)$ is an exponential affine function of state variables:

$$f = \exp\left(A(t, T) + B(t, T)e^{\kappa_1\eta}\psi(\xi)\lambda_t + C(t, T)X_t + D(t, T)m_t\right),$$

with the terminal conditions $A(T, T) = 0$, $B(T, T) = 0$, $C(T, T) = \omega$, and $D(T, T) = 1$. If we write $\psi^b = e^{\kappa_1\eta}\psi(\xi)$, the partial derivatives of $f(\cdot)$ are given by

$$\frac{\partial f}{\partial t} = \left(\frac{\partial A}{\partial t} + \psi^b \lambda \frac{\partial B}{\partial t} + \frac{\partial C}{\partial t} X_t + \frac{\partial D}{\partial t} m \right) f ,$$

$$\frac{\partial f}{\partial X} = C f, \qquad \frac{\partial^2 f}{\partial X^2} = C^2 f,$$

$$\frac{\partial f}{\partial m} = D f, \qquad \frac{\partial^2 f}{\partial m^2} = D^2 f,$$

$$\frac{\partial^2 f}{\partial X \partial m} = C D f, \qquad \frac{\partial f}{\partial \lambda} = B \psi^b f.$$

Inserting these expressions into Eq. (4.42) leads to the following relation (after grouping terms):

$$0 = \frac{\partial A}{\partial t} + \left(\alpha \theta \psi^b B + \left(\mu - \frac{\sigma^2}{2} \right) C + \frac{\sigma^2}{2} C^2 + \alpha \theta \kappa_1 D + \kappa_2 D \right)$$

$$- C D \varphi \sigma + \lambda \left(\psi^b \frac{\partial B}{\partial t} - \kappa_1 \alpha D - \alpha \psi^b B - C \mathbb{E} \left(e^J - 1 \right) \right)$$

$$+ \lambda \int_{-\infty}^{+\infty} \left[e^{B \psi^b \eta + C z + D(\kappa_1 \eta + \xi z)} - 1 \right] d\nu(z) + X_t \left(\frac{\partial C}{\partial t} \right) + m_t \left(\frac{\partial D}{\partial t} \right),$$

and we infer that $D(t, s) = 1$, $C(t, s) = \omega$ as $\frac{\partial}{\partial t} D(t, s) = 0$ and $\frac{\partial}{\partial t} C(t, s) = 0$. On the other hand, as $\kappa_2 = -\alpha \theta \kappa_1$, this last equation becomes

$$0 = \frac{\partial A}{\partial t} + \left(\alpha \theta \psi^b B + \left(\mu - \frac{\sigma^2}{2} \right) \omega + \frac{\sigma^2}{2} \omega^2 \right)$$

$$- \omega \varphi \sigma + \lambda \left(\psi^b \frac{\partial B}{\partial t} - \kappa_1 \alpha - \alpha \psi^b B - \omega \mathbb{E} \left(e^J - 1 \right) \right)$$

$$+ \lambda \int_{-\infty}^{+\infty} \left[e^{B \psi^b \eta + \kappa_1 \eta + (\omega + \xi) z} - 1 \right] d\nu(z).$$

If we remember that $\varphi = \varphi_1 + \varphi_2 \lambda_t$, we obtain the following expression:

$$0 = \frac{\partial A}{\partial t} + \left(\alpha \theta \psi^b B + \left(\mu - \frac{\sigma^2}{2} - \varphi_1 \sigma \right) \omega + \frac{\sigma^2}{2} \omega^2 \right)$$

$$+ \lambda \left(\psi^b \frac{\partial B}{\partial t} - \kappa_1 \alpha - \alpha \psi^b B - \omega \left(\mathbb{E} \left(e^J - 1 \right) + \varphi_2 \sigma \right) \right)$$

$$+ \lambda \int_{-\infty}^{+\infty} \left[e^{B \psi^b \eta + \kappa_1 \eta + (\omega + \xi) z} - 1 \right] d\nu(z).$$

Finally, as $\kappa_1\alpha = \psi^b - 1$, we obtain

$$0 = \frac{\partial A}{\partial t} + \alpha\theta\psi^b B + \left(\mu - \frac{\sigma^2}{2} - \varphi_1\sigma\right)\omega + \frac{\sigma^2}{2}\omega^2$$

$$0 = \frac{\partial}{\partial t}B - \alpha B - \omega\frac{(\psi(1) - 1 + \varphi_2\sigma)}{\psi^b} + \left[\frac{e^{B\psi^b\eta}e^{\kappa_1\eta}\psi(\omega + \xi)}{\psi^b} - 1\right].$$

Choosing $\varphi_1 = \frac{\mu - r}{\sigma}$ and $\varphi_2 = \frac{1}{\sigma}\left[\psi^b\left(\psi^Q(1) - 1\right) - (\psi(1) - 1)\right]$ allows us to rewrite this system as follows:

$$\frac{\partial}{\partial t}A = -\omega\left(r - \frac{\sigma^2}{2}\right) - \omega^2\frac{\sigma^2}{2} - \alpha\left(\theta\psi^b\right)B,$$

$$\frac{\partial}{\partial t}B = \alpha B + \omega\left(\psi^Q(1) - 1\right) - \left[e^{B\eta^Q}\psi^Q(\omega) - 1\right].$$

We infer the dynamics of X_t under the measure $\mathbb{Q}^{\xi,\varphi}$ by comparing with the results of Proposition 4.4. As $dS_t = d(e^{X_t})$, Eq. (4.41) is proven by applying Itô's lemma.

$\square$

Note that the function φ can be split into two parts to highlight that the risk premium is the sum of two components:

$$\varphi(t) = \frac{\mu - r}{\sigma} + \lambda_t\frac{e^{\kappa_1\eta}\psi(\xi)\left(\psi^Q(1) - 1\right) - (\psi(1) - 1)}{\sigma}.$$

The terms of this sum are respectively the risk premiums for the Brownian motion and for the jump risk. On the other hand, we show that under the risk neutral measure, jumps still have a double exponential distribution, as stated in the following result:

Proposition 4.8 *Under $\mathbb{Q}^{\xi,\varphi}$, jumps J_i^Q are double exponential random variables with a density equal to*

$$\nu^Q(z) = p^Q\rho^{+Q}e^{-\rho^{+Q}z}1_{\{z\geq 0\}} - (1 - p^Q)\rho^{-Q}e^{-\rho^{-Q}z}1_{\{z<0\}}, \qquad (4.43)$$

and where the parameters are adjusted as follows:

$$\rho^{+Q} = \rho^+ - \xi,$$

$$\rho^{-Q} = \rho^- - \xi,$$

$$p^Q = \frac{p\rho^+\rho^{-Q}}{\left(p\rho^+\rho^{-Q} + (1 - p)\rho^-\rho^{+Q}\right)}.$$

Proof By construction, the moment generating function for jumps under the risk neutral measure is the ratio

$$\psi^Q(\omega) = \frac{\psi(\omega+\xi)}{\psi(\xi)},$$

where

$$\psi(\omega+\xi) = p\frac{\rho^+}{\rho^+ - \omega - \xi} + (1-p)\frac{\rho^-}{\rho^- - \omega - \xi}$$

$$= \frac{p\rho^+\left(\rho^{-Q} - \omega\right) + (1-p)\rho^-\left(\rho^{+Q} - \omega\right)}{\left(\rho^{+Q} - \omega\right)\left(\rho^{-Q} - \omega\right)},$$

and

$$\psi(\xi) = \frac{p\rho^+\rho^{-Q} + (1-p)\rho^-\rho^{+Q}}{\rho^{+Q}\rho^{-Q}}.$$

Since

$$\frac{\psi(\omega+\xi)}{\psi(\xi)} = \frac{\frac{p\rho^+\rho^{-Q}}{p\rho^+\rho^{-Q}+(1-p)\rho^-\rho^{+Q}}\rho^{+Q}\left(\rho^{-Q} - \omega\right)}{\left(\rho^{+Q} - \omega\right)\left(\rho^{-Q} - \omega\right)}$$

$$+ \frac{\frac{(1-p)\rho^-\rho^{+Q}}{p\rho^+\rho^{-Q}+(1-p)\rho^-\rho^{+Q}}\rho^{-Q}\left(\rho^{+Q} - \omega\right)}{\left(\rho^{+Q} - \omega\right)\left(\rho^{-Q} - \omega\right)}$$

one can appropriately rearrange this equation to complete the proof. □

The constant ξ that serves to define the new measure is the cost of the risk for the jump component in the price process. Note that ξ must be such that $\alpha > \eta e^{\kappa_1\eta}\psi(\xi) = \eta^Q$; otherwise, the jump process is no longer stable. As illustrated in Table 4.4, ξ significantly influences the parameters defining the price process under

Table 4.4 This table illustrates the relationship between ξ and the jump process parameters, under $Q^{\xi,\varphi}$. The other parameters used to produce this table are those fitted by the POT procedure and reported in Table 4.1. Notice that when $\xi = 0$, the dynamics under $\mathbb{P}$ and $Q^{\xi,\varphi}$ are exactly the same

ξ	θ^Q	η^Q	p^Q	ρ^{+Q}	ρ^{-Q}	λ_∞^Q
-2	4.2	17.05	0.33	34.86	-36.33	72.22
-1.5	3.99	16.18	0.34	34.36	-36.83	37.53
-1	3.86	15.66	0.34	33.86	-37.33	28.59
-0.5	3.77	15.28	0.35	33.36	-37.83	24.14
0	3.69	14.98	0.35	32.86	-38.33	21.43
0.5	3.64	14.75	0.36	32.36	-38.83	19.61
1	3.59	14.56	0.37	31.86	-39.33	18.33
1.5	3.55	14.41	0.37	31.36	-39.83	17.4
2	3.52	14.29	0.38	30.86	-40.33	16.73

the risk neutral measure and the asymptotic level of the intensity. The numerical analysis reveals that for negative values of ξ, the jump frequency is higher than that under the real measure. For positive values of ξ, we observe on average fewer jumps under $\mathbb{Q}$ than under $\mathbb{P}$.

4.8 Further Reading

Hawkes processes combined with diffusions are used extensively to explain the daily dynamics of financial assets. For example, Aït-Sahalia et al. [1, 2] develop a multivariate setting with self- and mutual excitations to examine whether some jumps result from past shocks in stocks and CDS markets. Hainaut and Moraux [16] use a self-excited process to model the clustering of jumps and price and hedge options in this framework. Bacry et al. [4] and Hawkes [21] provide detailed literature reviews on applications of Hawkes processes in finance. Carr and Wu [6] and Fulop et al. [13] develop mono-asset settings, where the memory of negative jumps is the key to evaluating asset risk. Hainaut [14, 15] proposes two models for interest rates, based on Hawkes processes. Self-exciting processes are also commonly used to model "high-frequency" data. Engle and Russell [12] point out the fundamental role of Hawkes self-exciting processes in modeling the duration between two transactions. Readers interested in high-frequency applications may consult Bowsher [5], Chavez-Demoulin and McGill [7], Bacry et al. [3], and Da Fonseca and Zaatour [8] for more recent contributions. Dassios and Zhao [9] investigate the properties of a model with dynamic contagion. Dassios and Zhao [10] proposed a more efficient sampling algorithm for Hawkes processes based on a decomposition into several sub-processes. Hawkes processes may be extended in several directions. For example, parameters of the Hawkes process can be modulated by a Hidden Markov chain, as done in Chap. 1 for Brownian diffusions. This approach is explored in Hainaut and Moraux [18] for stock prices. On the other hand, a Hawkes process can be approached by a switching regime model with jump at transitions as shown in Hainaut and Deelstra [17]. The next chapter exclusively focuses on self-exciting jumps with a non-exponential memory.

References

1. Aït-Sahalia, Y., Laeven, R.J.A., Pelizzon, L.: Mutual excitation in eurozone sovereign CDS. J. Economet. **183**(2), 151–167 (2014)
2. Aït-Sahalia, Y., Cacho-Diaz, J., Laeven, R.J.A.: Modeling financial contagion using mutually exciting jump processes. J. Financ. Econ. **117**, 585–606 (2015)
3. Bacry, E., Delattre, S., Hoffmann, M., Muzy, J.F.: Modelling microstructure noise with mutually exciting point processes. Quant. Finan. **13**(1), 65–67 (2013)
4. Bacry, E., Mastromatteo, I., Muzy, J.F.: Hawkes processes in finance. Market Microstruct. Liquidity **1**(1), 1550005 (2015)

5. Bowsher, C.G.: Modelling security markets in continuous time: Intensity based, multivariate point process models. J. Econ. **141**, 876–912 (2007)
6. Carr, P., Wu, L.: Leverage effect, volatility feedback, and self-exciting market disruptions. J. Financ. Quant. Anal. **52**(5), 2119–2156 (2017)
7. Chavez-Demoulin, V., McGill, J.: High-frequency financial data modeling using Hawkes processes. J. Bank. Finance **36**, 3415–3426 (2012)
8. Da Fonseca, J., Zaatour, R.: Hawkes process: fast calibration, application to trade clustering, and diffusive limit. J. Futur. Mark. **34**(6), 548–579 (2014)
9. Dassios, A., Zhao, H.: A dynamic contagion process. Adv. Appl. Probab. **43**, 193–198 (2011)
10. Dassios, A., Zhao, H.: Exact simulation of Hawkes process with exponentially decaying intensity. Electron. Commun. Probab. **18**, 1–13 (2013)
11. Embrechts, P., Liniger, T., Lu, L.: Multivariate Hawkes processes: an application to financial data. J. Appl. Probab. **48**(A), 367–378 (2011)
12. Engle, R.F., Russell, J.R.: Autoregressive conditional duration: a new model for irregularly spaced transaction data. Econometrica **66**, 1127–1162 (1998)
13. Fulop, A., Li, J., Yu, J.: Self-exciting jumps, learning, and asset pricing implications. Rev. Financ. Stud. **28**, 876–912 (2015)
14. Hainaut, D.: A bivariate Hawkes process for interest rate modeling. Econ. Model. **57**, 180–196 (2016)
15. Hainaut, D.: A model for interest rates with clustering effects. Quant. Finance **16**, 1203–1218 (2016)
16. Hainaut, D.: Hedging of options in the presence of jump clustering. J. Comput. Finance **22**(3), 1–35 (2018)
17. Hainaut, D., Deelstra, G.: A self-exciting switching jump diffusion: properties, calibration and hitting time. Quant. Finance **19**, 407–426 (2019)
18. Hainaut, D., Moraux, F.: A switching self-exciting jump diffusion process for stock prices. Ann. Finance **15**, 267–306 (2019)
19. Hawkes, A.: Point spectra of some mutually exciting point processes. J. R. Stat. Soc. B **33**, 438–443 (1971)
20. Hawkes, A.: Spectra of some self-exciting and mutually exciting point processes. Biometrika **58**, 83–90 (1971)
21. Hawkes, A.G.: Hawkes processes and their applications to finance. Quant. Finance **18**(2), 193–198 (2018)
22. Hawkes, A., Oakes, D.: A cluster representation of a self-exciting process. J. Appl. Probab. **11**, 493–503 (1974)
23. Ogata, Y.: Space-time point-process models for earthquake occurrences. Ann. Inst. Stat. Math. **50**, 379–402 (1998)

Chapter 5
Non-Markov Models for Contagion and Spillover

We have seen in Chap. 4 that self-excited processes offer a natural way to introduce contagion between shocks in financial markets. In this approach, the occurrence of a shock depends on previous ones. In the most common specification, the intensity of jumps, that is akin to the instantaneous probability of a shock, increases as soon as a jump is observed. The influence of this jump on the intensity next decays with time according to a memory function, also called a kernel. When this memory kernel is exponential as in the SEJD, the pair jump intensity is a bivariate Markov process. In this case, the moment generating function (mgf) admits an analytical solution, found by Itô calculus. If the memory kernel is not exponential, we lose most of the analytical tractability offered by stochastic calculus. This chapter fills a gap in the literature by providing a closed form expression of the moment generating function (mgf) of non-Markov self-exciting jump processes.

   We show that when the memory kernel is the Fourier transform of a probability measure, the self-excited jump process can be reformulated as an infinite-dimensional Markov process in the complex plane. We focus on four types of memory kernels associated to Laplace, Gaussian, logistic, and Cauchy measures. Next, we approach the infinite-dimensional Markov model by a finite-dimensional process and obtain its mgf using standard stochastic calculus. The mgf of the infinite-dimensional process is obtained by considering the limit of the finite-dimensional approximation. As an illustration, we exploit these results to compute the probability density function of large positive and negative shocks in S&P 500 daily returns. Note that a similar procedure is applied in Chap. 8 to study properties of a non-Markov interest rate model. Furthermore, the Laplace, Gaussian, logistic, and Cauchy measures also serve in Chap. 7 to define autocovariances in a Gaussian field model.

D. Hainaut, *Continuous Time Processes for Finance*,
Bocconi & Springer Series 12, https://doi.org/10.1007/978-3-031-06361-9_5

5.1 The Multivariate Processes

In this chapter, we focus exclusively on jump processes. We consider m jump processes that are mutual and self-excited. In the numerical illustration concluding this chapter, we consider for instance the joint dynamic of large negative and positive jumps of a stock index. In the same manner, results in this chapter may be applied to model joint large negative shocks for multiple stocks. The framework proposed in this chapter may also be combined with the SEJD for option pricing.

The kth aggregate jump process (in absolute value) is denoted by $\left(L_t^k\right)_{t\geq 0}$ for $k = 1, \ldots, m$. This is the sum of N_t^k shocks, denoted $J_j^k \in \mathbb{R}^+$:

$$\left(L_t^k\right)_{t\geq 0} = \sum_{j=1}^{N_t^k} J_j^k \, , k = 1, \ldots, m. \tag{5.1}$$

The counting process $\left(N_t^k\right)_{t\geq 0}$ is a point process with an intensity denoted by $\left(\lambda_t^k\right)_{t\geq 0}$. All processes are defined on a probability space Ω endowed with a measure $\mathbb{P}$ and a filtration $(\mathcal{F}_t)_{t\geq 0}$. The negative shocks in absolute value $\left(J_j^k\right)$ are independent copies of a random variable J^k. The probability density function (pdf) of J^k is $\zeta_k(z)$ for $z \in \mathbb{R}^+$. The intensities of counting processes $\left(N_t^k\right)_{k=1,\ldots,m}$ are akin to the instantaneous probabilities of observing a new shock in the kth process. To introduce memory and contagion into the dynamic of jumps, the arrival intensities are driven by the following equations:

$$\lambda_t^k = \theta_k + \left(\lambda_0^k - \theta_k\right) g_{kk}(t) + \sum_{j=1}^{m} \eta_{kj} \int_0^t g_{kj}(t-s)\mathrm{d}L_s^j \quad , k = 1, \ldots, m \, , \tag{5.2}$$

where $\left(g_{kj}(t)\right)_{k,j=1,\ldots,m}$ are positive definite functions such that $g_{kj}(0) = 1$ and $\lim_{t\to\infty} g_{kj}(t) = 0$. The parameters θ_k and $\eta_{k,j}$ are positive. In this framework, the occurrence of a shock J^j of type j immediately raises the probability of observing a new shock across all processes, proportionally to $\left(\eta_{k,j} J^j\right)_{k=1,\ldots,m}$. Between two successive jumps, the intensities revert to $(\theta_k)_{k=1,\ldots,m}$. The speed of reversion depends upon the decay rate of $g_{kk}(\cdot)$. The functions $\left(g_{kj}(t)\right)_{k,j=1,\ldots,m}$ are kernel functions defining the "memory" of jump processes. If they are fast decreasing, the contagion effect between jumps is limited in time. In the opposite case, the processes remember the history of shocks for a long period.

Let us adopt the following vector and matrix notations: $\boldsymbol{\lambda}_t = \left(\lambda_t^1, \ldots, \lambda_t^m\right)^\top$, $\boldsymbol{\theta} = \left(\theta^1, \ldots, \theta^m\right)^\top$, $\boldsymbol{L}_t = \left(L_t^1, \ldots, L_t^m\right)^\top$, $\boldsymbol{N}_t = \left(N_t^1, \ldots, N_t^m\right)^\top$, $G(t) = \left(g_{kj}(t)\right)_{k,j=1,\ldots,m}$, and $H = (\eta_{kj})_{k,j=1,\ldots,m}$. We reformulate the dynamic of intensity in a concise way

$$\boldsymbol{\lambda}_t = \boldsymbol{\theta} + (\boldsymbol{\lambda}_0 - \boldsymbol{\theta})^\top \operatorname{diag}(G(t)) + H \int_0^t G(t - s) \, \mathrm{d}\boldsymbol{L}_s \,,$$

where $\operatorname{diag}(G(t))$ is the matrix of diagonal elements of $G(t)$. The vector of expected jump sizes is denoted by $\boldsymbol{\mu}_J = (\mathbb{E}(J^1), \ldots, \mathbb{E}(J^m))^\top$. Let

$$||\Xi|| = \left(\int_0^\infty |\eta_{kj} g_{kj}(s) \mathbb{E}(J^j)| \mathrm{d}s \right)_{k,j=1,\ldots,m}$$

be the $m \times m$ matrix of L_1-norms of the elementwise product of H, $G(t)$ and expected jump sizes. To ensure the stationarity of N_t, the spectral radius of the matrix $||\Xi||$ must be strictly lower than one. We assume that this is satisfied in the remainder of this chapter. In this setting, the long-term expectation, denoted $\bar{\boldsymbol{\lambda}} = \lim_{t \to \infty} \mathbb{E}(\boldsymbol{\lambda}_t)$, is such that $\bar{\boldsymbol{\lambda}} = \boldsymbol{\theta} + ||\Xi|| \bar{\boldsymbol{\lambda}}$ and therefore is equal to $\bar{\boldsymbol{\lambda}} = (I - ||\Xi||)^{-1} \boldsymbol{\theta}$.

We consider functions $g_{kj}(\cdot)$ that admit an integral representation. Bochner showed that positive definite functions may be defined by a spectral measure, as stated in the next proposition.

Proposition 5.1 (Bochner) *A continuous function $g(h)$ from $\mathbb{R}^+$ to the complex plane $\mathbb{C}$ is positive definite if and only if it may be represented as the Fourier transform of a measure $v(\cdot)$ on $\mathbb{R}$:*

$$g(h) = \int_{\mathbb{R}} e^{ihu} v(\mathrm{d}u) \,, \tag{5.3}$$

where $v(\cdot)$ is a bounded, real-valued function such that $\int_A v(\mathrm{d}u) \geq 0$ for all $A \subset \mathbb{R}$.

The function $v(\cdot)$ is called the *spectral distribution function* for g. In the sequel to this chapter, we assume that the following holds:

Assumption The kernel functions $g_{kj}(\cdot)$ are positive definite, and their spectral measure $v_{kj}(\cdot)$ is a probability measure, i.e., $\int_{\mathbb{R}} v_{kj}(\mathrm{d}u) = 1$.

By definition, an alternative representation of a positive definite function $g(\cdot)$ is

$$g(h) = \int_{\mathbb{R}} (\cos(hu) + i \sin(hu)) \, v(\mathrm{d}u) \,.$$

Therefore, if $\boldsymbol{\lambda}_t$ is $\mathbb{R}$-valued, $g(h)$ is also $\mathbb{R}$-valued and the imaginary part in this last equation must be null: $\int_{\mathbb{R}} \sin(hu) \, v(\mathrm{d}u) = 0$. This implies that the spectral distribution is symmetric around the origin. If $\nu(\cdot)$ is continuous, then $\nu(du) = \nu(-du)$ and

$$g(h) = \int_{\mathbb{R}} \cos{(hu)}\,\nu(du),$$

$$= 2 \int_0^\infty \cos{(|h|u)}\,\nu(du).$$

We consider four spectral distributions for $\nu(\cdot)$: the Laplace, Gaussian, logistic, and Cauchy measures. These are detailed below:

(1) The first measure is the *Laplace* or *double exponential measure*. Let $\alpha \in \mathbb{R}^+$ be constant. A symmetric exponential measure is defined by the following relation:

$$\nu_L(du) = \frac{1}{2}\left(\alpha e^{-\alpha u}\,\mathbb{1}_{\{u \geq 0\}} + \alpha e^{\alpha u}\,\mathbb{1}_{\{u \leq 0\}}\right) du. \tag{5.4}$$

The variance of $\nu_L(\cdot)$ is $\sigma_L^2(\alpha) = \frac{2}{\alpha^2}$. This spectral distribution generates the following memory kernel:

$$g_L(h) = \frac{1}{2}\left[\int_{\mathbb{R}^+} \alpha e^{ihu} e^{-\alpha u}\,du + \int_{\mathbb{R}^-} \alpha e^{ihu} e^{\alpha u}\,du\right] \tag{5.5}$$

$$= \frac{\alpha^2}{\alpha^2 + h^2},$$

which is a power decreasing function. If $m = 1$, as $\int_0^\infty \eta\,\mu_J\,g_L(s)\,ds = \frac{\pi \eta \mu_J \alpha}{2}$, we infer the condition $\alpha < \frac{2}{\pi \eta \mu_J}$, to ensure the stability of λ_t in long term.

(2) The second category of functions $g(\cdot)$ is obtained with a *Gaussian spectral measure*,

$$\nu_G(du) = \sqrt{\frac{\alpha}{\pi}} e^{-\alpha u^2}\,du. \tag{5.6}$$

The variance of $\nu_G(\cdot)$ is $\sigma_G^2(\alpha) = \frac{1}{2\alpha}$. Given that $\int_{\mathbb{R}^+} e^{-\alpha u^2} \cos{(hu)}\,du = \frac{1}{2}\sqrt{\frac{\pi}{\alpha}} e^{-\frac{h^2}{4\alpha}}$, the kernel is in this case equal to

$$g_G(h) = 2\sqrt{\frac{\alpha}{\pi}} \int_{\mathbb{R}^+} e^{-\alpha u^2} \cos{(|h|u)}\,du \tag{5.7}$$

$$= e^{-\frac{h^2}{4\alpha}}.$$

In the one-dimensional case ($m = 1$), $\int_0^\infty \eta \mu_J e^{-\frac{h^2}{4\alpha}}\,ds = \eta \mu_J \sqrt{\pi}\sqrt{\alpha}$, and we infer that $\sqrt{\alpha} < \frac{1}{\eta \mu_J \sqrt{\pi}}$ ensures the stability of λ_t.

(3) The third distribution is the *logistic* one. For $\alpha > 0$, the spectral measure is in this case

$$\nu_{\log}(\mathrm{d}u) = \frac{e^{-u/\alpha}}{\alpha \left(1 + e^{-u/\alpha}\right)^2} \mathrm{d}u \,. \tag{5.8}$$

The variance of $\nu_{\log}(\cdot)$ is $\sigma^2_{\log}(\alpha) = \frac{\alpha^2 \pi^2}{3}$, and the memory kernel is equal to

$$g_{\log}(h) = \begin{cases} \frac{\pi \alpha h}{\sinh(\pi \alpha h)} & h > 0 \\ 1 & h = 0 \end{cases} \,.$$

If $m = 1$, as $\int_0^\infty \eta \mu_J g(s) \, \mathrm{d}s = \frac{\eta \mu_J \pi}{4\alpha}$, we infer that $\frac{\eta \mu_J \pi}{4} < \alpha$ ensures the stability of λ_t.

(4) The last type of spectral measure that we consider is the *Cauchy distribution*:

$$\nu(\mathrm{d}u) = \frac{1}{\pi} \frac{\alpha}{\alpha^2 + u^2} \mathrm{d}u \,, \tag{5.9}$$

where $\alpha \in \mathbb{R}^+$. By construction, this measure is symmetric, and the function $g(h)$ is an exponential decreasing function:

$$g(h) = \frac{2\alpha}{\pi} \int_{\mathbb{R}+} \frac{\cos(|h|u)}{\alpha^2 + u^2} \mathrm{d}u$$
$$= e^{-\alpha |h|} \,. \tag{5.10}$$

If $m = 1$, as $\int_0^\infty \eta \mu_J e^{-\alpha s} \, \mathrm{d}s = \frac{\eta \mu_J}{\alpha}$, we infer that $\eta \mu_J < \alpha$ to ensure stability of λ_t.

This fourth spectral measure encompasses the classical Hawkes process. This kind of process has already been widely studied in the literature and will serve as a benchmark in this work. If the functions $(g_{kj})_{k,j=1,\ldots,m}$ are defined by Eq. (5.10) with parameters $\alpha_{kj} = \alpha_k$ for $k = 1, \ldots, m$, the vector of densities is a solution of the stochastic differential equations (SDEs):

$$\mathrm{d}\boldsymbol{\lambda}_t = \mathrm{diag}\left((\alpha_k)_{k=1,\ldots,m}\right)(\boldsymbol{\lambda}_0 - \boldsymbol{\theta}) \, \mathrm{d}t + H \, \mathrm{d}\boldsymbol{L}_t \,,$$

and the multivariate process $(\boldsymbol{\lambda}_t, \boldsymbol{L}_t)_{t \geq 0}$ is Markov. This means that for a well-behaved function $f(\boldsymbol{\lambda}_t, \boldsymbol{L}_t)$, the conditional expectation $\mathbb{E}\left(f(\boldsymbol{\lambda}_s, \boldsymbol{L}_s)|\mathcal{F}_t\right)$, for $s \geq t$, depends exclusively on $\boldsymbol{\lambda}_t$ and $\boldsymbol{L}_t$. Furthermore, we can rely on Itô calculus to determine, for example, the moment generating functions (mgf's) of jump processes. The mgf is of particular importance since it may be numerically inverted by a discrete Fourier transform to retrieve the pdf of $\boldsymbol{L}_t$. Notice that when the condition $\alpha_{kj} = \alpha_k$ is not fulfilled, the process $(\boldsymbol{\lambda}_t, \boldsymbol{L}_t)_{t \geq 0}$ is no longer Markov.

In order to compare memory kernels, we adjust the parameters α in order to match the variances of spectral densities. For a given $\sigma \in \mathbb{R}^+$, we set the parameters of Laplace, Gaussian, and logistic kernels to

$$\alpha_L = \sqrt{\frac{2}{\sigma^2}}, \ \alpha_G = \frac{1}{2\sigma^2}, \ \alpha_{\log} = \sqrt{\frac{3\sigma^2}{\pi^2}}. \tag{5.11}$$

For the Cauchy distribution, the variance is not defined. Therefore, we simply find an α by minimizing the sum of least squares between the Log and Cauchy memory kernels. The result of this procedure for $\sigma^2 = 16$ is illustrated in Fig. 5.1. This graph emphasizes that the Laplace kernel has a slower decay than the Gaussian and logistic kernels.

For most memory kernels, the process $(\lambda_s, L_s)_{s \geq 0}$ is not Markov, and we have very little information on its behavior conditionally to the filtration $\mathcal{F}_t$ for $t \leq s$. The next sections propose an alternative formulation used later to find the joint moment generating function of jump processes.

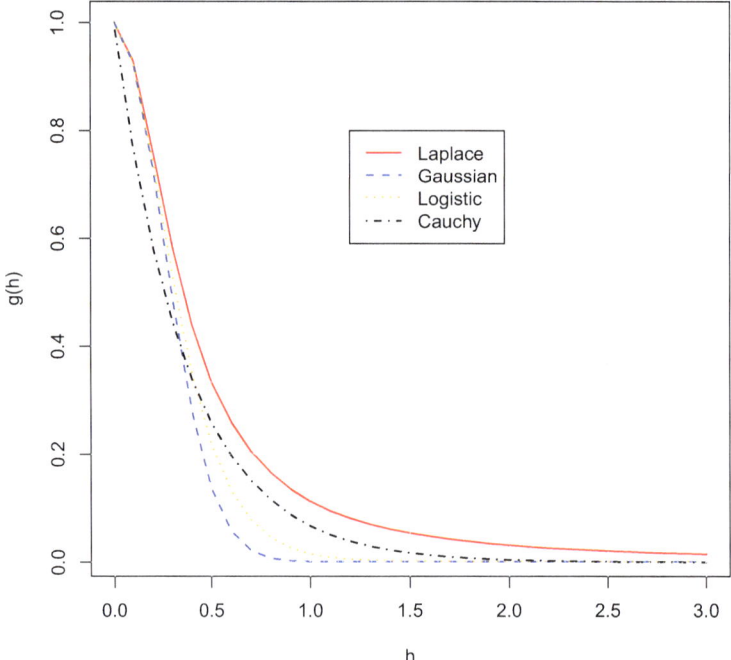

Fig. 5.1 Comparison of Laplace, Gaussian, and logistic kernels for $\sigma_L^2(\alpha_L) = \sigma_G^2(\alpha_G) = \sigma_{\log}^2(\alpha_{\log}) = 16$. The Cauchy kernel is computed with $\alpha_C = 2.71$

5.2   Infinite-Dimensional Reformulation

Except for the Cauchy kernel with $\alpha_{kj} = \alpha_k$, the process $(\boldsymbol{L}_t, \boldsymbol{\lambda}_t)_{t \geq 0}$ is not Markov. Nevertheless, if the functions $g_{kj}(t)$ admit a spectral representation with a probability measure $\nu_{kj}(z)$, we can reformulate the model as an infinite-dimensional Markov process in the complex plane. Indeed, from the Bochner equation (5.3) and after changing the order of integration, the intensity process may be rewritten as

$$\lambda_t^k = \theta_k + \left(\lambda_0^k - \theta_k \right) g_{kk}(t) + \sum_{j=1}^m \eta_{kj} \int_0^t \int_{\mathbb{R}} e^{i(t-s)\xi} \nu_{kj}(\mathrm{d}\xi) \mathrm{d}L_s^j$$

$$= \theta_k + \left(\lambda_0^k - \theta_k \right) g_{kk}(t) + \sum_{j=1}^m \eta_{kj} \int_{\mathbb{R}} \left(\int_0^t e^{i(t-s)\xi} \mathrm{d}L_s^j \right) \nu_{kj}(\mathrm{d}\xi),$$

for $k = 1, \ldots, m$. Let us now define a family of processes $Y_t^{(j,\xi)}$ indexed by $\xi \in \mathbb{R}$ and $j = 1, \ldots, m$,

$$Y_t^{(j,\xi)} = \int_0^t e^{i(t-s)\xi} \, \mathrm{d}L_s^j \, .$$

These processes are defined on the complex plane $\mathbb{C}$, and their differential is provided by the SDE

$$\mathrm{d}Y_t^{(j,\xi)} = i \, \xi \, Y_t^{(j,\xi)} \mathrm{d}t + \mathrm{d}L_t^j \, . \tag{5.12}$$

The intensities can then be rewritten as a sum of integrals with respect to $Y_t^{(j,\xi)}$,

$$\lambda_t^k = \theta_k + \left(\lambda_0^k - \theta_k \right) g_{kk}(t) + \sum_{j=1}^m \eta_{kj} \int_{\mathbb{R}} Y_t^{(j,\xi)} \nu_{kj}(\mathrm{d}\xi) \, .$$

As $\nu_{kj}(\cdot)$ is a probability measure, we can check that $\lim_{\xi \to \infty} Y_t^{(j,\xi)} \nu_{kj}(\mathrm{d}\xi) = 0$ and $\lim_{\xi \to -\infty} Y_t^{(j,\xi)} \nu_{kj}(\mathrm{d}\xi) = 0$. The differential of intensities is given by

$$\mathrm{d}\lambda_t^k = \left(\lambda_0^k - \theta_k \right) \frac{\mathrm{d}g_{kk}(t)}{\mathrm{d}t} \mathrm{d}t + \sum_{j=1}^m \eta_{kj} \int_{\mathbb{R}} \mathrm{d}Y_t^{(j,\xi)} \nu_{kj}(\mathrm{d}\xi) \, .$$

It is worth noting that λ_t is real, whereas $Y_t^{(j,\xi)}$ is complex. To check this, let us denote by τ_l the time of the lth jump of type j. We can then rewrite processes $Y_t^{(j,\xi)}$

and $Y_t^{(j,-\xi)}$ as the sums:

$$Y_t^{(j,\xi)} = \sum_{l=1}^{N_t^j} e^{i(t-\tau_l)\xi} J_l^j = \sum_{l=1}^{N_t^j} [\cos((t-\tau_l)\xi) + i\sin((t-\tau_l)\xi)] J_l^j,$$

$$Y_t^{(j,-\xi)} = \sum_{l=1}^{N_t^j} e^{-i(t-\tau_l)\xi} J_l^j = \sum_{l=1}^{N_t^j} [\cos((t-\tau_l)\xi) - i\sin((t-\tau_l)\xi)] J_l^j.$$

Given that the measures $\nu_{kj}(\cdot)$ are symmetric, the sum of $Y_t^{(j,\xi)}$ and $Y_t^{(j,-\xi)}$ weighted by $\nu_{kj}(d\xi)$ and $\nu_{kj}(-d\xi)$,

$$Y_t^{(j,\xi)}\nu_{kj}(d\xi) + Y_t^{(j,-\xi)}\nu_{kj}(-d\xi) = 2\sum_{l=1}^{N_t^j} \cos((t-\tau_l)\xi) J_l^j \nu_{kj}(d\xi),$$

and so $\int_{\mathbb{R}} dY_t^{(j,\xi)} \nu_{kj}(d\xi)$ are real. From the rewriting of $Y_t^{(j,\xi)}$ as $\sum_{l=1}^{N_t^j} e^{i(t-\tau_l)\xi} J_l^j$, we immediately infer that the quadratic variation of $Y_t^{(j,\xi)}$ is

$$[Y^{(j,\xi)}, Y^{(j,\xi)}]_t = \sum_{l=1}^{N_t^j} e^{2i(t-\tau_l)\xi} \left(J_l^j\right)^2,$$

whereas the quadratic covariation of $Y^{(j_1,\xi_1)}$ and $Y^{(j_2,\xi_2)}$ is equal to

$$[Y^{(j_1,\xi_1)}, Y^{(j_2,\xi_2)}]_t = \begin{cases} \sum_{l=1}^{N_t^{j_1}} e^{i(t-\tau_l)(\xi_1+\xi_2)} \left(J_l^{j_1}\right)^2 & j_1 = j_2, \\ 0 & j_1 \neq j_2. \end{cases}$$

We have reformulated the process (λ_t, L_t) as an infinite-dimensional Markov process $(\lambda_t, L_t, \left(Y_t^{(j,\xi)}\right)_{\xi\in\mathbb{R}, j=1,\ldots,m})$. We propose a discrete approximation in the next section and infer the moment generating function of jump processes in this setting. By considering the limit, in Sect. 5.4, we will retrieve the mgf of the original process.

5.3   Finite-Dimensional Approximation

Instead of considering an infinity of processes $\left(Y_t^{(j,\xi)}\right)_{\xi\in\mathbb{R}}$, we approach them with a finite number of equivalent processes. This method presents several advantages. First, it makes it possible to implement our model. Second, we can use the Itô calculus to find the mgf's of $\left(\lambda_t^k\right)_{k=1,\ldots,m}$. The key step consists in approaching $v_{kj}(\cdot)$ by a symmetric discrete measure with a finite number of atoms. For this purpose, we consider a symmetric partition $\mathcal{E}^{(n)} := \{-\infty < \xi_{-n}^{(n)} < \ldots < \xi_0^{(n)} = 0 < \ldots < \xi_n^{(n)} < \infty\}$, with $\xi_{-l}^{(n)} = -\xi_l^{(n)}$. The midpoint of each interval $(\xi_l^{(n)}, \xi_{l+1}^{(n)})$ is denoted by

$$b_{l+1} = \frac{\xi_l^{(n)} + \xi_{l+1}^{(n)}}{2}, l \in \{0, \ldots, n-1\}, \tag{5.13}$$

$$b_l = \frac{\xi_l^{(n)} + \xi_{l+1}^{(n)}}{2}, l \in \{-n, \ldots, -1\}.$$

The mass of corresponding atoms for the measure $v_{kj}(\cdot)$ is defined as the measure of intervals:

$$w_{l+1}^{(k,j)} = \int_{\xi_l^{(n)}}^{\xi_{l+1}^{(n)}} v_{kj}(\mathrm{d}z), l \in \{0, \ldots, n-1\}, \tag{5.14}$$

$$w_l^{(k,j)} = \int_{\xi_l^{(n)}}^{\xi_{l+1}^{(n)}} v_{kj}(\mathrm{d}z), l \in \{-n, \ldots, -1\}$$

for $l = \in \{-n, \ldots, n\}\backslash\{0\}$ and $k, j = 1, \ldots, m$. Due to the symmetry of $v_{kj}(\cdot)$, $w_{-l}^{(k,j)} = w_l^{(k,j)}$ and $b_{-l} = -b_l$. The discrete measure for a partition of size n is defined as follows:

$$\tilde{v}_{kj}(z) = \sum_{l=-n}^{n} w_l^{(k,j)} \delta_{b_l}(z), \tag{5.15}$$

where $\delta_{b_l^{(k,j)}}(z)$ is the Dirac measure located at point b_l. Note that we have used $\sum_{l=-n}^{n}$ to denote $\sum_{l \in \{-n,\ldots,n\}\backslash\{0\}}$, to lighten the notation. We assume that the following holds for the partition $\mathcal{E}^{(n)}$. First, $\xi_{-n}^{(n)} \to -\infty$ and $\xi_n^{(n)} \to \infty$ when $n \to \infty$. Second, $\max |\xi_{i+1}^{(n)} - \xi_i^{(n)}| \to 0$ when $n \to \infty$. Third, $\mathcal{E}^{(n)} \subset \mathcal{E}^{(n+1)}$. In this case, for any function $f(\cdot)$, integrable with respect to $v_{kj}(\cdot)$, we have that $\lim_{n\to\infty} \int_{-\infty}^{\infty} f(z) \tilde{v}_{kj}(\mathrm{d}z) = \int_{-\infty}^{\infty} f(z) v_{kj}(\mathrm{d}z)$. To further lighten the notation, we write $\tilde{Y}_t^{(j,l)} := Y_t^{(j,b_l)}$ for $j = 1, \ldots m$ and $l = -n, \ldots, n$. Each of the $m \times (2n+1)$ processes $\tilde{Y}_t^{(k,l)}$ is ruled by the SDE

$$\mathrm{d}\tilde{Y}_t^{(j,l)} = i\, b_l \tilde{Y}_t^{(j,l)} \mathrm{d}t + \mathrm{d}\tilde{L}_t^j,$$

where $\tilde{L}_t^j = \sum_{l=1}^{\tilde{N}_t^j} J_l^j$ is the jth jump process in the discretized model. Its intensity is denoted by $\tilde{\lambda}_t^k$ and is such that

$$\tilde{\lambda}_t^k = \theta_k + \left(\tilde{\lambda}_0^k - \theta_k\right) g_{kk}(t) + \sum_{j=1}^{m} \eta_{kj} \sum_{l=-n}^{n} w_l^{(k,j)} \tilde{Y}_t^{(j,l)} .$$

The symmetry of $w_l^{(k,j)}$ (which follows from the symmetry of the partition $\mathcal{E}^{(n)}$ and of v_{kj}) ensures that λ_t^k is real. Its differential is given by

$$d\tilde{\lambda}_t^k = \left(\tilde{\lambda}_0^k - \theta_k\right) \frac{dg_{kk}(t)}{dt} dt + \sum_{j=1}^{m} \eta_{kj} \sum_{l=-n}^{n} w_l^{(k,j)} d\tilde{Y}_t^{(j,l)} \qquad (5.16)$$

$$= \left(\tilde{\lambda}_0^k - \theta_k\right) \frac{dg_{kk}(t)}{dt} dt + \sum_{j=1}^{m} \sum_{l=-n}^{n} \eta_{kj} w_l^{(k,j)} i \, b_l \tilde{Y}_t^{(j,l)} dt$$

$$+ \sum_{j=1}^{m} \sum_{l=-n}^{n} \eta_{kj} w_l^{(k,j)} d\tilde{L}_t^j .$$

We reformulate the dynamic of intensities in matrix form. For this purpose, we adopt the following notations: $\tilde{\boldsymbol{\lambda}}_t = \left(\tilde{\lambda}_t^1, \dots, \tilde{\lambda}_t^m\right)^\top$, $\tilde{\boldsymbol{L}}_t = \left(\tilde{L}_t^1, \dots, \tilde{L}_t^m\right)^\top$, $\tilde{\boldsymbol{Y}}_t^l = \left(\tilde{Y}_t^{(1,l)}, \dots, \tilde{Y}_t^{(m,l)}\right)^\top$, and $W_l = (w_l^{(k,j)})_{k,j=1,\dots,m}$ for $l = -n, \dots, n$. Notice that by construction, we have the two limits,

$$\lim_{n \to \infty} \sum_{l=-n}^{n} w_l^{(k,j)} \tilde{Y}_t^{(j,l)} = \int_{\mathbb{R}} Y_t^{(j,\xi)} v_{kj}(d\xi) ,$$

$$\lim_{n \to \infty} \sum_{l=-n}^{n} W_l \tilde{\boldsymbol{Y}}_t^l = \left(\sum_{j=1}^{m} \int_{\mathbb{R}} Y_t^{(j,\xi)} v_{kj}(d\xi) \right)_{k=1,\dots,m} .$$

If $\odot$ is the Hadamard (elementwise) product, we reformulate the dynamic of intensities, Eq. (5.16), as follows:

$$d\tilde{\boldsymbol{\lambda}}_t = \left[\left(\tilde{\boldsymbol{\lambda}}_0 - \boldsymbol{\theta}\right)^\top \mathrm{diag}\left(\frac{dG(t)}{dt}\right) + i \sum_{l=-n}^{n} b_l \left(H \odot W_l\right) \tilde{\boldsymbol{Y}}_t^l \right] dt$$

$$+ \sum_{l=-n}^{n} \left(H \odot W_l\right) d\tilde{\boldsymbol{L}}_t ,$$

whereas $d\tilde{\boldsymbol{Y}}_t^l = i\, b_l \tilde{\boldsymbol{Y}}_t^l\, dt + d\tilde{\boldsymbol{L}}_t$. Therefore, the differential of any real function $f\left(t, \tilde{\boldsymbol{\lambda}}_t, \left(\tilde{\boldsymbol{Y}}_t^l\right)_{l=-n,\ldots,n}, \tilde{\boldsymbol{L}}_t\right)$ differentiable once with respect to its arguments can be obtained with Itô's formula,

$$
df(\cdot) = \partial_t f(\cdot)\, dt + i \sum_{l=-n}^{n} b_l \left(\partial_{\tilde{\boldsymbol{Y}}^l} f(\cdot)\right)^\top \tilde{\boldsymbol{Y}}_t^l\, dt + \left(\partial_{\tilde{\boldsymbol{\lambda}}} f(\cdot)\right)^\top
$$

$$
\times \left[\left(\tilde{\boldsymbol{\lambda}}_0 - \boldsymbol{\theta}\right)^\top \operatorname{diag}\left(\frac{dG(t)}{dt}\right) + i \sum_{l=-n}^{n} b_l\, (H \odot W_l)\, \tilde{\boldsymbol{Y}}_t^l \right] dt
$$

$$
+ \int_{\mathbb{R}^{m,+}} f\left(t, \tilde{\boldsymbol{\lambda}}_t + \sum_{l=-n}^{n} (H \odot W_l)\, z, \left(\tilde{\boldsymbol{Y}}_t^l + z\right)_{l=-n,\ldots,n}, \tilde{\boldsymbol{L}}_t + z\right)
$$

$$
- f(\cdot)\, \tilde{\boldsymbol{L}}(dt, dz),
$$

where $\tilde{\boldsymbol{L}}(\cdot, \cdot)$ is the random measure such that $\tilde{\boldsymbol{L}}_t = \int_0^t \int_{\mathbb{R}^{m,+}} z\, d\tilde{\boldsymbol{L}}(du, dz)$. Here $\partial_t f$, $\partial_{\tilde{\boldsymbol{Y}}^l} f$ and $\partial_{\tilde{\boldsymbol{\lambda}}} f$ are the partial derivatives of $f(\cdot)$ with respect to time, $\tilde{\boldsymbol{Y}}_t^l$ and $\tilde{\boldsymbol{\lambda}}$.

As already mentioned, the mgf of the jump process is of particular importance since it may be numerically inverted by a discrete Fourier transform to retrieve the pdf of $\boldsymbol{L}_t$. The next proposition provides an analytical expression of this function.

Proposition 5.2 *The moment generating function of jump processes for* $\boldsymbol{\omega} = (\omega_1, \ldots, \omega_m)^\top \in \mathbb{C}^m$ *is equal to*

$$
\mathbb{E}\left(e^{\boldsymbol{\omega}^\top \tilde{\boldsymbol{L}}_s} \mid \mathcal{F}_t\right) \tag{5.17}
$$

$$
= \exp\left(q_0(t,s) + \boldsymbol{q}_\lambda(t,s)^\top \tilde{\boldsymbol{\lambda}}_t + \sum_{l=-n}^{n} \boldsymbol{q}_l(t,s)^\top (H \odot W_l)\, \tilde{\boldsymbol{Y}}_t^l + \boldsymbol{\omega}^\top \tilde{\boldsymbol{L}}_t\right),
$$

where $q_0(t,s) : \left(\mathbb{R}^+\right)^2 \to \mathbb{R}$, $\boldsymbol{q}_\lambda(t,s) : \left(\mathbb{R}^+\right)^2 \to \mathbb{R}^m$ *and* $\boldsymbol{q}_l(t,s) : \left(\mathbb{R}^+\right)^2 \to \mathbb{R}^m$. *We denote by* $\boldsymbol{e}_j = (0, \ldots, 1, \ldots, 0)$ *the* jth *vector of the orthonormal basis of* $\mathbb{R}^n$. *The elements of vectors* $\boldsymbol{q}_\lambda(t,s)$ *and* $\boldsymbol{q}_l(t,s)$ *are denoted by* $q_\lambda^k(t,s)$ *and* $q_l^k(t,s)$. *These functions are solutions of the system of ordinary differential equations (ODEs)*

$$
\partial_t q_0(t,s) = -\boldsymbol{q}_\lambda(t,s)^\top \operatorname{diag}\left(\frac{dG(t)}{dt}\right)\left(\tilde{\boldsymbol{\lambda}}_0 - \boldsymbol{\theta}\right), \tag{5.18}
$$

$$
\partial_t q_\lambda^j(t,s) = -\int_0^\infty e^{\left(\sum_{l=-n}^{n}(\boldsymbol{q}_\lambda(t,s)^\top + \boldsymbol{q}_l(t,s)^\top)(H \odot W_l) + \boldsymbol{\omega}^\top\right)\boldsymbol{e}_j z} - 1\, \zeta_j(dz),
$$

$$
\partial_t \boldsymbol{q}_l(t,s) = -i\, b_l \left(\boldsymbol{q}_l(t,s) + \boldsymbol{q}_\lambda(t,s)\right)
$$

for $l = -n, \ldots, n$ *and* $j = 1, \ldots, m$. *The terminal conditions are* $q_0(s, s) = 0$, $q_\lambda^j(s, s) = 0$ *and* $q_l^j(s, s) = 0$.

Proof Let us denote by $f\left(t, \tilde{\lambda}_t, \left(\tilde{Y}_t^l\right)_{l=-n,\ldots,n}, \tilde{L}_t\right)$ the moment generating function of jump processes: $\mathbb{E}\left(e^{\omega^\top \tilde{L}_s} \mid \mathcal{F}_t\right)$. As $f(\cdot)$ is a conditional expectation, it is also a martingale, and therefore, $f(\cdot)$ is a solution of the following differential equation:

$$0 = \partial_t f(\cdot) + i \sum_{l=-n}^{n} b_l \left(\partial_{\tilde{Y}^l} f(\cdot)\right)^\top \tilde{Y}_t^l + \left(\partial_{\tilde{\lambda}} f(\cdot)\right)^\top \tag{5.19}$$

$$\times \left[\operatorname{diag}\left(\frac{dG(t)}{dt}\right)\left(\tilde{\lambda}_0 - \boldsymbol{\theta}\right) + i \sum_{l=-n}^{n} b_l \left(H \odot W_l\right) \tilde{Y}_t^l \right] + \sum_{j=1}^{m} \tilde{\lambda}_t^j$$

$$\times \int_0^\infty f\left(t, \tilde{\lambda}_t + \sum_{l=-n}^{n} \left(H \odot W_l\right) e_j z, \left(\tilde{Y}_t^l + e_j z\right)_{l=-n,\ldots,n}, \tilde{L}_t + e_j z\right)$$

$$- f(\cdot)\,\zeta_j(dz).$$

We make the ansatz that $f(\cdot)$ is an exponential affine function of the form

$$f(\cdot) = \exp\left(q_0(t, s) + \boldsymbol{q}_\lambda(t, s)^\top \tilde{\lambda}_t + \sum_{l=-n}^{n} \boldsymbol{q}_l(t, s)^\top \left(H \odot W_l\right) \tilde{Y}_t^l + \omega^\top \tilde{L}_t\right).$$

The mgf is real and therefore $\sum_{l=-n}^{n} \boldsymbol{q}_l(t, s)^\top \left(H \odot W_l\right) \tilde{Y}_t^l$ must be real. Given that the $\tilde{Y}_t^{(j,l)}$ are random and

$$\sum_{l=-n}^{n} \boldsymbol{q}_l(t, s)^\top \left(H \odot W_l\right) \tilde{Y}_t^l = \sum_{l=-n}^{n} \sum_{k=1}^{m} \sum_{j=1}^{m} q_l^k(t, s)\eta_{kj} w_l^{(k,j)} \tilde{Y}_t^{(j,l)},$$

the sum $\sum_{l=-n}^{n} q_l^k(t, s) w_l^{(k,j)} \tilde{Y}_t^{(j,l)}$ is real. This necessary implies that $\operatorname{Re}(q_l^k(\cdot)) = \operatorname{Re}(q_{-l}^k(\cdot))$ and $-\operatorname{Im}(q_l^k(\cdot)) = \operatorname{Im}(q_{-l}^k(\cdot))$. This relation can be checked once that $q_l^k(\cdot)$ is found. The partial derivatives of $f(\cdot)$ with respect to state variables and times are given by

$$\partial_{\tilde{Y}^l} f(\cdot) = f(\cdot)\,(H \odot W_l)^\top \boldsymbol{q}_l(t, s),$$

$$\partial_{\tilde{\lambda}} f(\cdot) = f(\cdot)\boldsymbol{q}_\lambda(t, s)$$

and

$$\partial_t f(\cdot) = f(\cdot) \left(\partial_t q_0(t,s) + \partial_t \boldsymbol{q}_\lambda(t,s)^\top \tilde{\boldsymbol{\lambda}}_t + \sum_{l=-n}^{n} \partial_t \boldsymbol{q}_l(t,s)^\top (H \odot W_l) \, \tilde{\boldsymbol{Y}}_t^l \right).$$

Under the assumption that $f(\cdot)$ is an exponential affine function, the jump term in Eq. (5.19) becomes

$$f \left(t, \tilde{\boldsymbol{\lambda}}_t + \sum_{l=-n}^{n} (H \odot W_l) \, \boldsymbol{e}_j \, z, \left(\tilde{\boldsymbol{Y}}_t^l + \boldsymbol{e}_j \, z \right)_{l=-n,\dots,n}, \tilde{\boldsymbol{L}}_t + \boldsymbol{e}_j \, z \right) - f(\cdot)$$

$$= f(\cdot) \left(e^{\boldsymbol{q}_\lambda(t,s)^\top \sum_{l=-n}^{n} (H \odot W_l) \, \boldsymbol{e}_j \, z + \sum_{l=-n}^{n} \boldsymbol{q}_l(t,s)^\top (H \odot W_l) \, \boldsymbol{e}_j \, z + \boldsymbol{\omega}^\top \boldsymbol{e}_j \, z} - 1 \right).$$

Combining the previous equations allows us to infer that $q_0(t,s)$, $\boldsymbol{q}_\lambda(t,s)$, and $\boldsymbol{q}_l(t,s)$ satisfy the relation

$$0 = \partial_t q_0(t,s) + \partial_t \boldsymbol{q}_\lambda(t,s)^\top \tilde{\boldsymbol{\lambda}}_t + \sum_{l=-n}^{n} \partial_t \boldsymbol{q}_l(t,s)^\top (H \odot W_l) \, \tilde{\boldsymbol{Y}}_t^l$$

$$+ i \sum_{l=-n}^{n} b_l \left((H \odot W_l)^\top \boldsymbol{q}_l(t,s) \right)^\top \tilde{\boldsymbol{Y}}_t^l + \boldsymbol{q}_\lambda(t,s)^\top$$

$$\times \left[\text{diag} \left(\frac{\mathrm{d}G(t)}{\mathrm{d}t} \right) \left(\tilde{\boldsymbol{\lambda}}_0 - \boldsymbol{\theta} \right) + i \sum_{l=-n}^{n} b_l \, (H \odot W_l) \, \tilde{\boldsymbol{Y}}_t^l \right]$$

$$+ \sum_{j=1}^{m} \tilde{\lambda}_t^j \int_0^\infty e^{\boldsymbol{q}_\lambda(t,s)^\top \sum_{l=-n}^{n} (H \odot W_l) \, \boldsymbol{e}_j \, z + \sum_{l=-n}^{n} \boldsymbol{q}_l(t,s)^\top (H \odot W_l) \, \boldsymbol{e}_j \, z + \boldsymbol{\omega}^\top \boldsymbol{e}_j \, z} - 1 \, \zeta_j(\mathrm{d}z).$$

Regrouping terms, we infer the following ODEs:

$$0 = \partial_t q_0(t,s) + \boldsymbol{q}_\lambda(t,s)^\top \text{diag} \left(\frac{\mathrm{d}G(t)}{\mathrm{d}t} \right) \left(\tilde{\boldsymbol{\lambda}}_0 - \boldsymbol{\theta} \right),$$

$$0 = \int_0^\infty \left[e^{\boldsymbol{q}_\lambda(t,s)^\top \sum_{l=-n}^{n} (H \odot W_l) \boldsymbol{e}_j z + \sum_{l=-n}^{n} \boldsymbol{q}_l(t,s)^\top (H \odot W_l) \boldsymbol{e}_j z + \boldsymbol{\omega}^\top \boldsymbol{e}_j z} - 1 \right] \zeta_j(\mathrm{d}z)$$

$$+ \partial_t q_{\lambda_j}(t,s),$$

$$0 = \sum_{l=-n}^{n} \partial_t \boldsymbol{q}_l(t,s)^\top (H \odot W_l) + i \sum_{l=-n}^{n} b_l \, \boldsymbol{q}_l(t,s)^\top (H \odot W_l)$$

$$+ i \sum_{l=-n}^{n} b_l \, \boldsymbol{q}_\lambda(t,s)^\top (H \odot W_l).$$

The last equation is fulfilled if

$$\partial_t q_j(t, s) = -i\, b_l\,\big(q_j(t, s) + q_\lambda(t, s)\big)$$

for $j = 1, \ldots, m$, from which we infer equations (5.18). Given that $b_l = -b_l$ and $q_\lambda \in \mathbb{R}$, we have that $\mathrm{Re}(q_l^k(\cdot)) = \mathrm{Re}(q_{-l}^k(\cdot))$ and $-\mathrm{Im}(q_l^k(\cdot)) = \mathrm{Im}(q_{-l}^k(\cdot))$. $\square$

5.4 Moment Generating Function

We establish the moment generating function (mgf) of jump processes by considering the limit of the mgf of the previous discrete model. First, we provide the analytical expression of $q_l(\cdot)$ in the next corollary:

Corollary 5.3 *The function* $q_l(t, s) : (\mathbb{R}^+)^2 \to \mathbb{R}^m$ *is equal to*

$$q_l(t, s) = \int_t^s i\, b_l e^{i\, b_l\,(u-t)} q_\lambda(u, s)\, du\,. \tag{5.20}$$

This result is proved by checking that the proposed expression of $q_l(\cdot, \cdot)$ is the solution of the ODEs (5.18),

$$\partial_t q_l(t, s) = -i\, b_l \underbrace{\left(\int_t^s i\, b_l e^{i\, b_l\,(u-t)} q_\lambda(u, s)\, du\right)}_{q_l(t,s)} - i\, b_l q_\lambda(t, s)\,.$$

As the process L_t is the limit of $\tilde{L}_t$ when $n \to \infty$, we find its moment generating function by considering the limit of Eq. (5.17). The next proposition presents the main result of this chapter.

Proposition 5.4 *The moment generating function of the jump process for* $\boldsymbol{\omega} \in \mathbb{C}^{m-}$ *is*

$$\mathbb{E}\left(e^{\boldsymbol{\omega}^\top \boldsymbol{L}_s} \mid \mathcal{F}_t\right) = \exp\left(q_0(t, s) + q_\lambda(t, s)^\top \boldsymbol{\lambda}_t + \boldsymbol{\omega}^\top \boldsymbol{L}_t\right) \tag{5.21}$$

$$\times \exp\left(\int_0^t \int_t^s q_\lambda(u, s)^\top \left(H \odot \frac{dG(u - v)}{du}\right) du\, d\boldsymbol{L}_v\right),$$

where $q_0(t, s) : \left(\mathbb{R}^+\right)^2 \to \mathbb{R}$ and $\boldsymbol{q}_\lambda(t, s) : \left(\mathbb{R}^+\right)^2 \to \mathbb{R}^m$ are solutions of the system of ordinary differential equations (ODEs)

$$\partial_t q_0(t, s) = -\boldsymbol{q}_\lambda(t, s)^\top \mathrm{diag}\left(\frac{\mathrm{d}G(t)}{\mathrm{d}t}\right)(\boldsymbol{\lambda}_0 - \boldsymbol{\theta}) \,, \tag{5.22}$$

$$\partial_t q_\lambda^j(t, s) = -\int_0^\infty \left(\mathrm{e}^{\left(\boldsymbol{q}_\lambda(t,s)^\top H + \int_t^s \boldsymbol{q}_\lambda(u,s)^\top \left(H \odot \frac{\mathrm{d}G(u-t)}{\mathrm{d}u}\right)\mathrm{d}u + \boldsymbol{\omega}^\top\right)\boldsymbol{e}_j z} - 1 \right) \zeta_j(\mathrm{d}z)$$

for $j = 1, \ldots, m$. The terminal conditions are $q_0(s, s) = 0$ and $q_\lambda^j(s, s) = 0$.

Proof By construction, $\mathbb{E}\left(\mathrm{e}^{\boldsymbol{\omega}^\top L_s} \mid \mathcal{F}_t\right) = \lim_{n\to\infty} \mathbb{E}\left(\mathrm{e}^{\boldsymbol{\omega}^\top \tilde{L}_s} \mid \mathcal{F}_t\right)$. We expand the sum in Eq. (5.17) as

$$\lim_{n\to\infty} \sum_{l=-n}^n \boldsymbol{q}_l(t, s)^\top (H \odot W_l)\, \tilde{\boldsymbol{Y}}_t^l = \sum_{k=1}^m \sum_{j=1}^m \lim_{n\to\infty} \sum_{l=-n}^n q_l^k(t, s)\eta_{kj} w_l^{(k,j)} \tilde{Y}_t^{(j,l)} \,.$$

If we remember the definition of $w_l^{(k,j)}$ and $Y_t^{(j,\xi)}$, we obtain after a change of order of integration that

$$\lim_{n\to\infty} \sum_{l=-n}^n q_l^k(t, s)\eta_{kj} w_l^{(k,j)} \tilde{Y}_t^{(j,l)} = \int_{\mathbb{R}} q_\xi^k(t, s) Y_t^{(j,\xi)} \eta_{kj} \nu_{kj}(\mathrm{d}\xi) \tag{5.23}$$

$$= \int_0^t \int_{\mathbb{R}} q_\xi^k(t, s)\mathrm{e}^{i(t-v)\xi} \eta_{kj} \nu_{kj}(\mathrm{d}\xi)\, \mathrm{d}L_v^j \,.$$

From the previous corollary, the function $q_l^k(t, s)$ is given by

$$q_l^k(t, s) = \int_t^s i\, b_l \mathrm{e}^{i\, b_l\,(u-t)} q_\lambda^k(u, s)\, \mathrm{d}u \,.$$

Therefore, the limit of this function when the size of the partition tends to infinity is

$$\lim_{n\to\infty} q_l^k(t, s) = q_\xi^k(t, s) = \int_t^s i\, \xi \mathrm{e}^{i\, \xi\,(u-t)} q_\lambda^k(u, s)\, \mathrm{d}u \,,$$

for some $\xi \in \mathbb{R}$. Injecting this expression into the integrand of Eq. (5.23) leads to

$$\int_{\mathbb{R}} q_{\xi}^{k}(t,s)e^{i(t-v)\xi}\eta_{kj}v_{kj}(\mathrm{d}\xi) \tag{5.24}$$

$$= \int_{\mathbb{R}} \int_{t}^{s} i\,\xi e^{i\,\xi\,(u-t)}q_{\lambda}^{k}(u,s)\,\mathrm{d}u\,e^{i(t-v)\xi}\eta_{kj}v_{kj}(\mathrm{d}\xi)$$

$$= \int_{t}^{s} q_{\lambda}^{k}(u,s)\eta_{kj} \int_{\mathbb{R}} i\,\xi e^{i\,\xi\,(u-v)}v_{kj}(\mathrm{d}\xi)\,\mathrm{d}u \ .$$

As $g_{kj}(\cdot)$ admits a spectral representation, its derivative is also equal to

$$\frac{\mathrm{d}g_{kj}(u-v)}{\mathrm{d}u} = \int_{\mathbb{R}} i\,\xi\,e^{i\,\xi\,(u-v)}v_{kj}(\mathrm{d}\xi) \ . \tag{5.25}$$

Combining Eqs. (5.24) and (5.25) gives us

$$\int_{\mathbb{R}} q_{\xi}^{k}(t,s)e^{i(t-v)\xi}\eta_{kj}v_{kj}(\mathrm{d}\xi) = \int_{t}^{s} q_{\lambda}^{k}(u,s)\eta_{kj}\frac{\mathrm{d}g_{kj}(u-v)}{\mathrm{d}u}\,\mathrm{d}u \ .$$

It is then possible to write Eq. (5.23) as follows:

$$\lim_{n\to\infty}\sum_{l=-n}^{n} q_{l}^{k}(t,s)\eta_{kj}w_{l}^{(k,j)}\tilde{Y}_{t}^{(j,l)} = \int_{0}^{t}\int_{t}^{s} q_{\lambda}^{k}(u,s)\eta_{kj}\frac{\mathrm{d}g_{kj}(u-v)}{\mathrm{d}u}\,\mathrm{d}u\,\mathrm{d}L_{v}^{j} \ ,$$

and the limit of $\sum_{l=-n}^{n} q_{l}(t,s)^{\top}(H \odot W_{l})\,\tilde{Y}_{t}^{l}$ when $n \to \infty$ can then be rewritten in matrix form:

$$\sum_{k=1}^{m}\sum_{j=1}^{m}\lim_{n\to\infty}\sum_{l=-n}^{n} q_{l}^{k}(t,s)\eta_{kj}w_{l}^{(k,j)}\tilde{Y}_{t}^{(j,l)}$$

$$= \int_{0}^{t}\int_{t}^{s}\sum_{k=1}^{m} q_{\lambda}^{k}(u,s)\sum_{j=1}^{m}\eta_{kj}\frac{\mathrm{d}g_{kj}(u-v)}{\mathrm{d}u}\,\mathrm{d}u\mathrm{d}L_{v}^{j}$$

$$= \int_{0}^{t}\int_{t}^{s} q_{\lambda}(u,s)^{\top}\left(H \odot \frac{\mathrm{d}G(u-v)}{\mathrm{d}u}\right)\,\mathrm{d}u\,\mathrm{d}L_{v} \ .$$

The expression (5.21) immediately follows from this last result. The derivative of q_{λ}^{j} with respect to time is

$$\partial_{t}q_{\lambda}^{j}(t,s) =$$

$$- \lim_{n\to\infty}\int_{0}^{\infty} e^{q_{\lambda}(t,s)^{\top}\sum_{l=-n}^{n} H\odot W_{l}\,e_{j}\,z + \sum_{l=-n}^{n} q_{l}(t,s)^{\top} W_{l}\,e_{j}\,z + \omega^{\top} e_{j}\,z} - 1\,\zeta_{j}(\mathrm{d}z) \ .$$

Using a similar reasoning as previously, we infer that

$$
\lim_{n \to \infty} \sum_{l=-n}^{n} \boldsymbol{q}_l(t,s)^\top \left(H \odot W_l\right) \boldsymbol{e}_j = \lim_{n \to \infty} \sum_{k=1}^{m} \sum_{l=-n}^{n} q_l^k(t,s) \eta_{kj} w_l^{(k,j)}
$$

$$
= \sum_{k=1}^{m} \int_{\mathbb{R}} \int_t^s i\,\xi e^{i\,\xi\,(u-t)} q_\lambda^k(u,s)\,\mathrm{d}u\;\eta_{kj}\,v_{kj}(\mathrm{d}\xi)
$$

$$
= \sum_{k=1}^{m} \int_t^s q_\lambda^k(u,s)\eta_{kj} \int_{\mathbb{R}} i\,\xi e^{i\,\xi\,(u-t)}\,v_{kj}(\mathrm{d}\xi)\,\mathrm{d}u
$$

$$
= \sum_{k=1}^{m} \int_t^s q_\lambda^k(u,s)\eta_{kj} \frac{\mathrm{d}g_{kj}(u-t)}{\mathrm{d}u}\,\mathrm{d}u
$$

$$
= \int_t^s \boldsymbol{q}_\lambda(u,s)^\top \left(H \odot \frac{\mathrm{d}G(u-t)}{\mathrm{d}u}\right)\mathrm{d}u\,\boldsymbol{e}_j\,.
$$

As $v_{kj}(\cdot)$ is a probability measure, we have that $\lim_{n \to \infty} \sum_{l=-n}^{n} w_l^{(k,j)} = 1$ and

$$
\lim_{n \to \infty} \sum_{l=-n}^{n} \boldsymbol{q}_\lambda(t,s)^\top \left(H \odot W_l\right) \boldsymbol{e}_j = \lim_{n \to \infty} \sum_{l=-n}^{n} \sum_{k=1}^{m} q_\lambda^k(t,s) \eta_{kj}\, w_l^{(k,j)}
$$

$$
= \sum_{k=1}^{m} q_\lambda^k(t,s)\eta_{kj}
$$

$$
= \boldsymbol{q}_\lambda(t,s)^\top H \boldsymbol{e}_j\,.
$$

These last two expressions lead to Eq. (5.22). $\qquad\qquad\qquad\qquad\qquad\square$

By numerically differentiating the mgf (5.21), we can estimate the expectation and variance of the jump process. Figure 5.2 illustrates this for a one-dimensional process with $\theta = 10$, $\eta = 180$, $\lambda_0 = 5$, and exponential jumps of average equal to 1%. The α parameters tuning the Laplace, Gaussian, and logistic kernels are set in order to match variances of spectral densities (see Eq. (5.11) with $\sigma^2 = 16$) as done in the illustration concluding Sect. 5.1. The densities are approached by a discrete distribution with 100 atoms covering a 95% confidence interval. The Laplace process has a longer term memory, and therefore, the intensity of shock arrivals is on average higher than those of Gauss and logistic processes. This explains why the expectation and volatility of the jump process with this memory kernel dominate in long term those obtained with Gauss and logistic kernels. In contrast, the Gauss jump process has a shorter memory and thus lower expectation–variance.

Another interesting use of the mgf is the computation of joint and marginal probability density functions of aggregated negative jumps. This point is developed

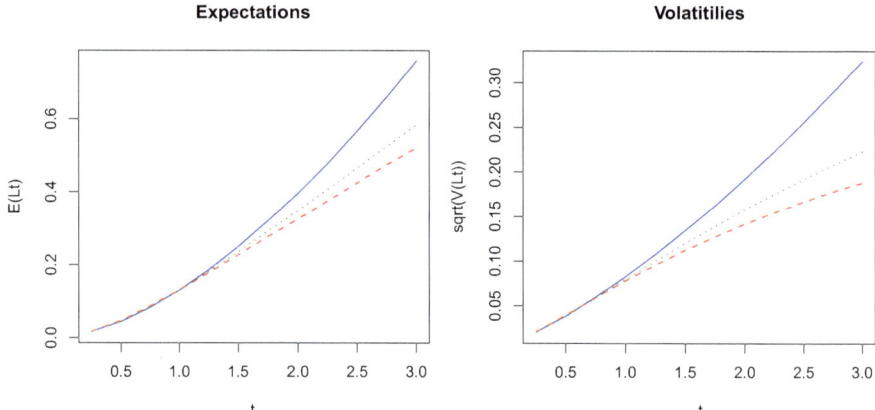

Fig. 5.2 Comparison of $\mathbb{E}\left(\tilde{L}_t \mid \mathcal{F}_0\right)$ and $\sqrt{\mathbb{V}\left(\tilde{L}_t \mid \mathcal{F}_0\right)}$ of a one-dimensional jump process with Laplace, Gauss, and logistic kernels

in the case study concluding the chapter. First, we focus on a sub-category of Cauchy memory kernels for which $(\boldsymbol{\lambda}_t, \boldsymbol{L}_t)_{t\geq 0}$ is Markov.

5.5 Cauchy Memory Kernels

As already mentioned in Sect. 5.1, the process $(\boldsymbol{\lambda}_t, \boldsymbol{L}_t)_{t\geq 0}$ is Markov with Cauchy memory kernels and $\alpha_{kj} = \alpha_k$ for $k = 1, \dots, m$. This model has been extensively studied in the literature. For such kernels, the moment generating function is obtained by taking advantage of the Markov properties of $(\boldsymbol{\lambda}_t, \boldsymbol{L}_t)_{t\geq 0}$ and Itô calculus. However, the expression of the mgf obtained in this way differs from the one presented in Proposition 5.4. The aim of this section is to reconcile these results. We start by briefly reviewing the expression of the mgf inferred by Itô calculus. If $g_{kj}(t) = \mathrm{e}^{-\alpha_k t}$ with $\alpha_k \in \mathbb{R}^+$ for $k, j \in \{1, \dots, m\}$, the multivariate process $(\boldsymbol{\lambda}_t, \boldsymbol{L}_t)_{t\geq 0}$ is Markov and

$$d\boldsymbol{\lambda}_t = \mathrm{diag}(\boldsymbol{\alpha}) \left(\boldsymbol{\theta} - \boldsymbol{\lambda}_t\right) dt + H\, d\boldsymbol{L}_t\,,$$

where $\boldsymbol{\alpha} = (\alpha_1, \dots, \alpha_m)^\top$. The matrix $G(t) = \left(\mathrm{e}^{-\alpha_k t}\right)_{k,j=1,\dots,m}$ has identical columns and

$$\frac{dG(t)}{dt} = \left(-\alpha_k \mathrm{e}^{-\alpha_k t}\right)_{k,j=1,\dots,m} = -\mathrm{diag}(\boldsymbol{\alpha})\, G(t)\,.$$

We write $\mathrm{diag}(\boldsymbol{\alpha} e^{-\boldsymbol{\alpha} t}) = \mathrm{diag}\left((\alpha_k e^{-\alpha_k t})_{k=1...m}\right)$ to lighten the notation. According to Proposition 5.4, the mgf of the cumulated jumps process is given by

$$\mathbb{E}\left(e^{\boldsymbol{\omega}^\top L_s} \mid \mathcal{F}_t\right) = \exp\left(q_0(t, s) + \boldsymbol{q}_\lambda(t, s)^\top \boldsymbol{\lambda}_t + \boldsymbol{\omega}^\top \boldsymbol{L}_t\right) \tag{5.26}$$

$$\times \exp\left(-\int_0^t \int_t^s \boldsymbol{q}_\lambda(u, s)^\top \left(H \odot (\mathrm{diag}(\boldsymbol{\alpha})\, G(u - v))\right) \mathrm{d}u \, \mathrm{d}\boldsymbol{L}_v\right),$$

where $q_0(t, s) : (\mathbb{R}^+)^2 \to \mathbb{R}$, $\boldsymbol{q}_\lambda(t, s) : (\mathbb{R}^+)^2 \to \mathbb{R}^m$ are solutions of the system of ODEs

$$\partial_t q_0(t, s) = \boldsymbol{q}_\lambda(t, s)^\top \mathrm{diag}(\boldsymbol{\alpha} e^{-\boldsymbol{\alpha} t})\,(\boldsymbol{\lambda}_0 - \boldsymbol{\theta}) \tag{5.27}$$

$$\partial_t q_\lambda^j(t, s) = -\int_0^\infty e^{\boldsymbol{q}_\lambda(t,s)^\top H e_j z - \int_t^s \boldsymbol{q}_\lambda(u,s)^\top (H \odot (\mathrm{diag}(\boldsymbol{\alpha})\, G(u-t)))\mathrm{d}u\, e_j z + \boldsymbol{\omega}^\top e_j z}$$

$$-1\,\zeta_j(\mathrm{d}z)$$

for $j = 1, \ldots, m$. The terminal conditions are $q_0(s, s) = 0$ and $q_\lambda^j(s, s) = 0$. This formulation of the mgf is not standard in the literature. The most common expression for the mgf is obtained with Itô's lemma and recalled in the next proposition.

Proposition 5.5 *When the memory kernels are exponential decreasing with $\alpha_{kj} = \alpha_k$ for $k = 1, \ldots, m$, the moment generating function of cumulated jump processes for $\boldsymbol{\omega} \in \mathbb{C}^{m-}$ is*

$$\mathbb{E}\left(e^{\boldsymbol{\omega}^\top L_s} \mid \mathcal{F}_t\right) = \exp\left(p_0(t, s) + \boldsymbol{p}_\lambda(t, s)^\top \boldsymbol{\lambda}_t + \boldsymbol{\omega}^\top \boldsymbol{L}_t\right), \tag{5.28}$$

where $p_0(t, s) : (\mathbb{R}^+)^2 \to \mathbb{R}$, $\boldsymbol{p}_\lambda(t, s) : (\mathbb{R}^+)^2 \to \mathbb{R}^m$ are solutions of the system of ordinary differential equations (ODEs)

$$\begin{cases} \partial_t p_0(t, s) = -\boldsymbol{p}_\lambda(t, s)^\top \mathrm{diag}\,(\boldsymbol{\alpha})\, \boldsymbol{\theta} \\ \partial_t p_\lambda^j(t, s) = \alpha_j p_\lambda^j(t, s) - \int_0^\infty e^{\boldsymbol{p}_\lambda(t,s)^\top H e_j z + \boldsymbol{\omega}^\top e_j z} - 1\, \zeta_j(\mathrm{d}z) \end{cases} \tag{5.29}$$

for $j = 1, \ldots, m$. The terminal conditions are $p_0(s, s) = 0$ and $p_\lambda^j(s, s) = 0$.

Proof Let us temporarily denote by $f(t, \boldsymbol{\lambda}_t, \boldsymbol{L}_t)$ the mgf $\mathbb{E}\left(e^{\boldsymbol{\omega}^\top L_s} \mid \mathcal{F}_t\right)$. Since $f(\cdot)$ is a conditional expectation, it is also a martingale. This is then a solution of the following Itô equation:

$$0 = \partial_t f(\cdot) + \partial f_\lambda(\cdot)^\top \mathrm{diag}(\boldsymbol{\alpha})(\boldsymbol{\theta} - \boldsymbol{\lambda}_t)$$

$$+ \sum_{j=1}^m \lambda_t^j \int_0^\infty f(t, \boldsymbol{\lambda}_t + H\, e_j z, \boldsymbol{L}_t + e_j z) - f(\cdot)\, \zeta_j(\mathrm{d}z).$$

If we consider the ansatz $f(t, \boldsymbol{\lambda}_t, \boldsymbol{L}_t) = \exp(p_0(t, s) + \boldsymbol{p}_\lambda(t, s)^\top \boldsymbol{\lambda}_t + \boldsymbol{\omega}^\top \boldsymbol{L}_t)$, then

$$0 = \partial_t p_0(t, s) + \partial_t \boldsymbol{p}_\lambda(t, s)^\top \boldsymbol{\lambda}_t + \boldsymbol{p}_\lambda(t, s)^\top \mathrm{diag}(\boldsymbol{\alpha})(\boldsymbol{\theta} - \boldsymbol{\lambda}_t)$$

$$+ \sum_{j=1}^m \lambda_t^j \int_0^\infty e^{\boldsymbol{p}_\lambda(t,s)^\top H \boldsymbol{e}_j\, z + \boldsymbol{\omega}^\top \boldsymbol{e}_j\, z} - 1\, \zeta_j(\mathrm{d}z)\, .$$

Regrouping terms leads to the result. □

The next proposition reconciliates Propositions 5.5 and 5.4.

Proposition 5.6 *When memory kernels are exponential decreasing with $\alpha_{kj} = \alpha_k$ for $k = 1, \ldots, m$, the functions $p_0(t, s)$, $\boldsymbol{p}_\lambda(t, s)$ defined by Eqs. (5.29) are related to the functions $q_0(t, s)$, $\boldsymbol{q}_\lambda(t, s)$ of Eqs. (5.27) by the relations:*

$$\begin{cases} p_0(t, s) & = q_0(t, s) + \int_t^s \boldsymbol{q}_\lambda(u, s)^\top \mathrm{diag}(e^{-\alpha(u-t)})\, \mathrm{d}u \\ & \quad \times \left(\mathrm{diag}(\boldsymbol{\alpha})\, \boldsymbol{\theta} + \mathrm{diag}(\boldsymbol{\alpha} e^{-\alpha t})\, (\boldsymbol{\lambda}_0 - \boldsymbol{\theta})\right) \\ \boldsymbol{p}_\lambda(t, s)^\top & = \boldsymbol{q}_\lambda(t, s)^\top - \int_t^s \boldsymbol{q}_\lambda(u, s)^\top \left(H \odot (\mathrm{diag}(\boldsymbol{\alpha})\, G(u - t))\right) H^{-1}\, \mathrm{d}u \\ & = \boldsymbol{q}_\lambda(t, s)^\top - \int_t^s \boldsymbol{q}_\lambda(u, s)^\top \mathrm{diag}(\boldsymbol{\alpha} e^{-\alpha(u-t)})\, \mathrm{d}u\, . \end{cases}$$

$$(5.30)$$

Proof Let us denote by $E = (1)_{k,j=1\ldots m}$ the $m \times m$ matrix of ones. We start the proof by differentiating the second equation of (5.30) with respect to time,

$$\partial_t \boldsymbol{p}_\lambda(t, s)^\top = \partial_t \boldsymbol{q}_\lambda(t, s)^\top + \boldsymbol{q}_\lambda(t, s)^\top (H \odot (\mathrm{diag}(\boldsymbol{\alpha}) E)) H^{-1} \qquad (5.31)$$

$$- \int_t^s \boldsymbol{q}_\lambda(u, s)^\top \left(H \odot \left(\mathrm{diag}(\boldsymbol{\alpha})^2\, G(u - t)\right)\right) H^{-1}\, \mathrm{d}u\, .$$

Basic matrix algebra allows us to infer two interesting relations:

$$(H \odot (\mathrm{diag}(\boldsymbol{\alpha}) E)) H^{-1} = \mathrm{diag}(\boldsymbol{\alpha})$$

and

$$\left(H \odot \left(\mathrm{diag}(\boldsymbol{\alpha})^2\, G(u - t)\right)\right) H^{-1}$$

$$= \mathrm{diag}(\boldsymbol{\alpha})\, (H \odot (\mathrm{diag}(\boldsymbol{\alpha})\, G(u - t))) H^{-1}\, .$$

Equation (5.19) then becomes

$$\partial_t \boldsymbol{p}_\lambda(t, s)^\top = \partial_t \boldsymbol{q}_\lambda(t, s)^\top + \boldsymbol{q}_\lambda(t, s)^\top \mathrm{diag}(\boldsymbol{\alpha})$$

$$- \int_t^s \boldsymbol{q}_\lambda(u, s)^\top \mathrm{diag}(\boldsymbol{\alpha})\, (H \odot (\mathrm{diag}(\boldsymbol{\alpha})\, G(u - t))) H^{-1}\, \mathrm{d}u$$

from which we infer that

$$\partial_t p_\lambda^j(t,s) = \partial_t q_\lambda^j(t,s) + \alpha_j \, \boldsymbol{p}_\lambda(t,s)^\top \boldsymbol{e}_j . \tag{5.32}$$

By definition of $q_\lambda^j(t,s)$, we have

$$\partial_t q_\lambda^j(t,s) = -\int_0^\infty e^{\boldsymbol{p}_\lambda(t,s)^\top H \boldsymbol{e}_j z + \boldsymbol{\omega}^\top \boldsymbol{e}_j z} - 1 \, \zeta_j(dz),$$

and we check that $\partial_t p_\lambda^j(t,s)$ satisfies the ODEs in Eq. (5.29). Let us recall that $\boldsymbol{q}_\lambda(s,s) = 0$. If we denote by $\mathrm{diag}\left(\boldsymbol{\alpha}^2 e^{-\boldsymbol{\alpha}t}\right)$ the $m \times m$ diagonal matrix $\mathrm{diag}\left((\alpha_k^2 e^{-\alpha_k t})_{k=1\ldots m}\right)$, the derivative of $p_0(t,s)$ with respect to t is given by

$$\partial_t p_0(t,s) = \partial_t q_0(t,s) - \int_t^s \boldsymbol{q}_\lambda(u,s)^\top \mathrm{diag}(e^{-\boldsymbol{\alpha}(u-t)}) \, du$$

$$\times \left(\mathrm{diag}\left(\boldsymbol{\alpha}^2 e^{-\boldsymbol{\alpha}t}\right) (\boldsymbol{\lambda}_0 - \boldsymbol{\theta}) \right)$$

$$- \boldsymbol{q}_\lambda(t,s)^\top \left(\mathrm{diag}(\boldsymbol{\alpha}) \, \boldsymbol{\theta} + \mathrm{diag}(\boldsymbol{\alpha} e^{-\boldsymbol{\alpha}t}) (\boldsymbol{\lambda}_0 - \boldsymbol{\theta}) \right)$$

$$+ \int_t^s \boldsymbol{q}_\lambda(u,s)^\top \mathrm{diag}(\boldsymbol{\alpha} e^{-\boldsymbol{\alpha}(u-t)}) \, du$$

$$\times \left(\mathrm{diag}(\boldsymbol{\alpha}) \, \boldsymbol{\theta} + \mathrm{diag}(\boldsymbol{\alpha} e^{-\boldsymbol{\alpha}t}) (\boldsymbol{\lambda}_0 - \boldsymbol{\theta}) \right) .$$

From the definition of $q_0(t,s)$, its derivative is equal to

$$\partial_t q_0(t,s) = \boldsymbol{q}_\lambda(t,s)^\top \mathrm{diag}\left(\boldsymbol{\alpha} e^{-\boldsymbol{\alpha}t}\right) (\boldsymbol{\lambda}_0 - \boldsymbol{\theta}) ,$$

and therefore, $\partial_t p_0(t,s)$ satisfies the ODEs in Eq. (5.29):

$$\partial_t p_0(t,s) = - \underbrace{\left[\boldsymbol{q}_\lambda(t,s)^\top - \int_t^s \boldsymbol{q}_\lambda(u,s)^\top \mathrm{diag}(\boldsymbol{\alpha} e^{-\boldsymbol{\alpha}(u-t)}) \, du \right]}_{\boldsymbol{p}_\lambda(t,s)^\top} \mathrm{diag}(\boldsymbol{\alpha}) \, \boldsymbol{\theta} .$$

$\square$

To conclude this section, we draw the attention of the reader to the fact that Proposition 5.5 holds only if $\alpha_{kj} = \alpha_k$ for $k = 1, \ldots, m$. If this condition is not fulfilled, the process is no longer Markov, and the mgf must be evaluated with Propositions 5.2 or 5.4. Before explaining how we can retrieve the joint density function from the mgf, we briefly review the method for fitting parameters of a multivariate Hawkes process.

5.6 Estimation

Multivariate self-excited processes may be estimated by log-likelihood maximization. Before presenting an illustration, we briefly recall the expression of this
log-likelihood adapted to our framework. The time interval of observations is
denoted by $[0, \mathcal{T}]$, whereas the lth jump time of L_t^k is denoted by τ_l^k for $k = 1, \ldots, m$. If the number of jumps of $L_{\mathcal{T}}^k$ is n_k and under the assumption that $\lambda_0^k = \theta_k$,
the sample path of intensities is calculated as follows:

$$\lambda_t^k = \theta_k + \sum_{j=1}^{m} \sum_{l=1}^{N_t^k} \eta_{kj} g_{kj}(t - \tau_l^j) J_l^j, \quad k = 1, \ldots, m.$$

From Embrechts et al. [4], we know that the log-likelihood is equal to

$$\ln \mathcal{L} = -\sum_{k=1}^{m} \int_0^{\mathcal{T}} \lambda_s^k ds + \sum_{k=1}^{m} \sum_{j=1}^{n_k} \log\left(\lambda_{\tau_j^k}^k\right). \tag{5.33}$$

We illustrate this section by fitting a bivariate process to large jumps in daily
S&P 500 log-returns from 31/1/2001 to 31/1/2020 (4496 joint observations). We
consider that a jump occurs when the daily log-return in absolute value is above
the threshold of 1.5%. Figure 5.3 shows the samples of detected jumps. Figure 5.4
displays the empirical distributions of recorded shocks (in absolute value) to which
we fit exponential distributions. The parameters are provided in Table 5.1.

The model is next fitted by maximizing the log-likelihood with a standard
"R" optimization function ("optim"). The integrals of intensities in Eq. (5.33) are
approached with daily steps. As all parameters are strictly positive, we optimize

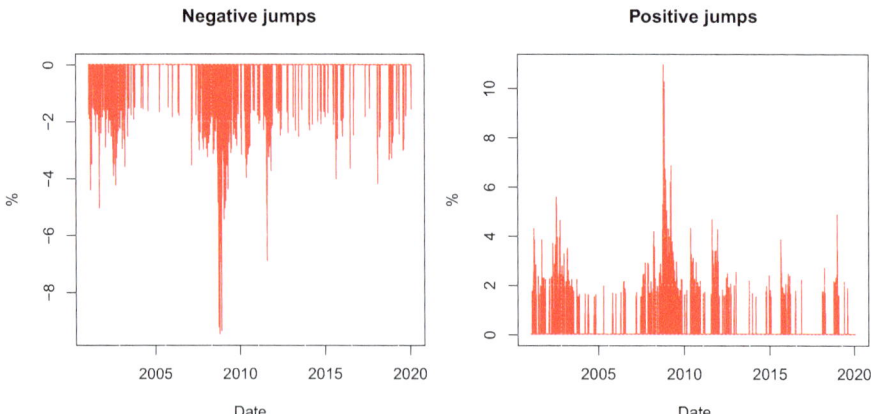

Fig. 5.3 S&P 500 negative and positive jumps above 1.5% in absolute value

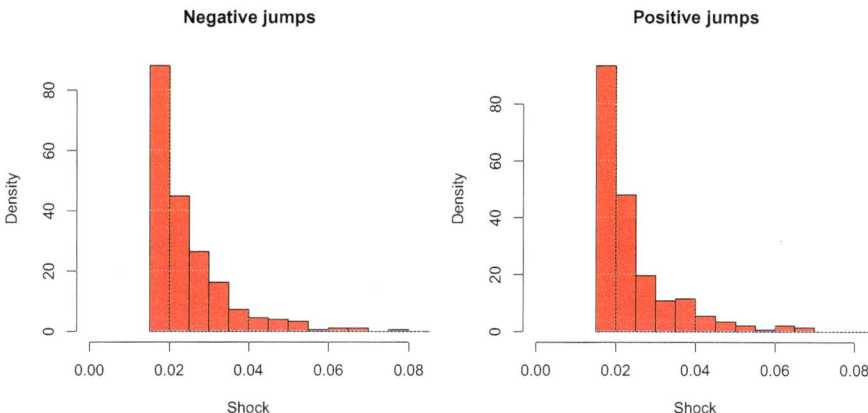

Fig. 5.4 Empirical distributions of negative and positive jumps (in absolute value)

Table 5.1 Parameter estimates of jumps distribution

	Negative jumps, $k = 1$	Positive jumps, $k = 2$
Number of jumps	356	295
Average size	2.48%	2.45%
ρ_k	40.25	40.82

Table 5.2 Parameter estimates and log-likelihoods ($\ln \mathcal{L}$)

	Kernel			
	Laplace	Gauss	Logistic	Cauchy
θ_1	7.21	9.61	8.55	8.16
θ_2	0.62	3.97	2.41	2.08
α_{11}	0.0314	$0.5942e{-}3$	14.74	22.30
α_{12}	0.0227	$0.6067e{-}3$	14.91	29.96
α_{21}	0.0399	$1.0696e{-}3$	12.60	20.99
α_{22}	0.0082	$0.1273e{-}3$	0.94	26.96
η_{11}	631.99	646.62	530.02	653.43
η_{12}	3.16	5.34	0.1194	0.1792
η_{21}	559.29	561.52	483.92	655.34
η_{22}	0.7882	0.8526	0.7163	0.4845
$\ln \mathcal{L}$	1742.80	1726.98	1735.32	1738.96

the log-likelihood with respect to their log-values to avoid introducing positivity constraints. Table 5.2 presents the results of the estimation procedure, while Fig. 5.5 compares the logistic, Laplace, and Cauchy memory functions $g_{kj}(\cdot)$ computed with the estimated parameters. The best fit is obtained with a Laplace kernel. The self-exciting memory with this kernel is slightly longer at medium term than with others (except for g_{22}). The mutual and self-excitation parameters reveal that negative jumps trigger positive shocks with a high probability ($\eta_{12}\!\gg\!0$). This phenomenon is

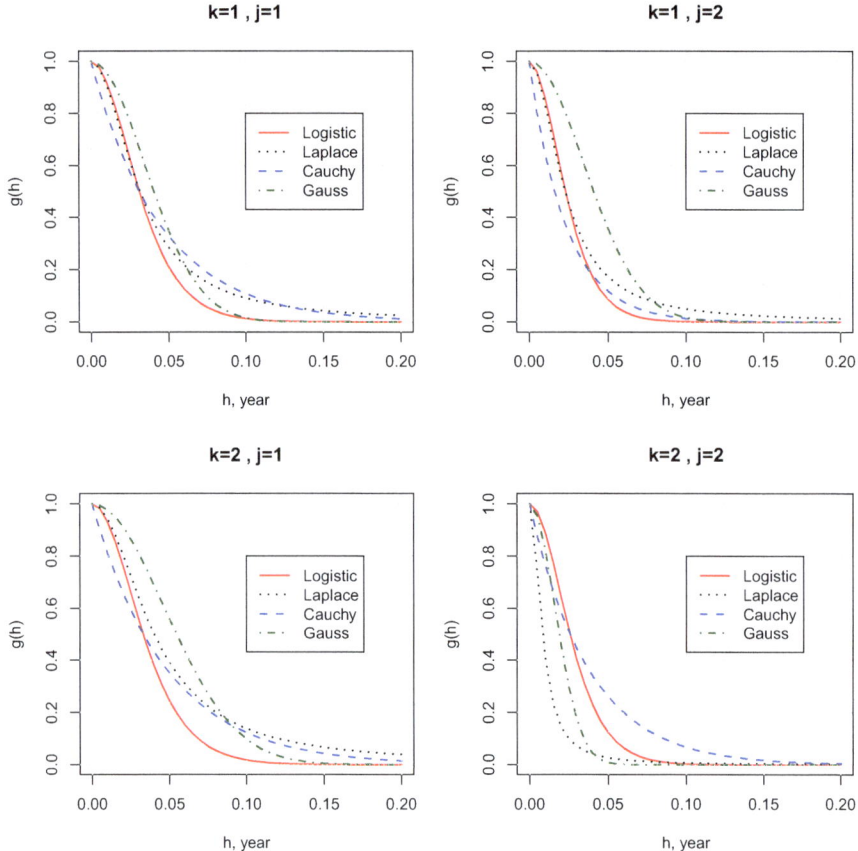

Fig. 5.5 Comparison of estimated Laplace, logistic, Cauchy, and Gauss memory kernels

well known by traders, who call it "technical bounce back." In contrast to negative jumps, positive shocks are not self-exciting (η_{22} is nearly null, whereas $\eta_{11} \gg 0$).

5.7   Probability Density Functions by Fast Fourier Transform

In Sect. 5.4, we found the moment generating function of a multivariate jump process with mutual excitation. We now exploit this result to compute the joint or the marginal probability density functions (pdf's) of this process. Our approach is based on a Discrete Fast Fourier Transform (DFFT). It roughly consists in inverting the characteristic function of the jump process, that is, the mgf valued on the imaginary axis.

For numerical reasons, we focus on the discrete version of the process, introduced in Sect. 5.3. Its characteristic function, denoted by $\Upsilon_{t,s}(i\boldsymbol{\omega}) = \mathbb{E}\left(e^{i\boldsymbol{\omega}^\top \tilde{L}_s} \mid \mathcal{F}_t\right)$, is the inverse Fourier transform of the multivariate probability density function (pdf) $f_{t,s}(\boldsymbol{x})$ of $\tilde{L}_s|\mathcal{F}_t$. Therefore, this density can be retrieved by computing the following multiple integrals (the Fourier transform of $\Upsilon_{t,s}(\cdot)$):

$$f_{t,s}(\boldsymbol{x}) = \frac{1}{(2\pi)^m} \mathcal{F}[\Upsilon_{t,s}(i\boldsymbol{\omega})](\boldsymbol{x})$$

$$= \frac{1}{(2\pi)^m} \int_{-\infty}^{+\infty} \cdots \int_{-\infty}^{+\infty} \Upsilon_{t,s}(i\boldsymbol{\omega}) e^{-i\boldsymbol{\omega}^\top \boldsymbol{x}} d\omega_1 \ldots d\omega_m . \tag{5.34}$$

The calculation of this multiple integral is computationally intensive, but it may speed up if we discretize using an expression that is compatible with the DFFT, as presented in the next proposition.

Proposition 5.7 *Let M be the number of steps used in the Discrete Fast Fourier Transform (DFFT) and $\Delta_x = \frac{x_{\max}}{M}$ be the discretization step. Let $\Delta_\omega = \frac{2\pi}{M\Delta_x}$ and*

$$\boldsymbol{\omega}_{j_1, j_2, \ldots, j_m} = \left(-\frac{M}{2}\Delta_\omega + (j_1 - 1)\Delta_\omega, \ldots, -\frac{M}{2}\Delta_\omega + (j_m - 1)\Delta_\omega \right)$$

for $j_1, \ldots, j_m \in \{1, \ldots, M+1\}$. The values of $f_{t,s}(\cdot)$ at points

$$\boldsymbol{x}_{k_1, k_2, \ldots, k_m} = ((k_1 - 1)\Delta_x, (k_2 - 1)\Delta_x, \ldots, (k_m - 1)\Delta_x)$$

for $k_1, \ldots, k_m \in \{1, \ldots, M\}$ are approached by

$$f_{t,s}(\boldsymbol{x}_{k_1, \ldots, k_m}) = (\Delta_\omega)^m \sum_{j_1=1}^{M+1} \sum_{j_2=1}^{M+1} \cdots \sum_{j_m=1}^{M+1} \delta_{j_1, j_2, \ldots, j_m} \Upsilon_{t,s}\left(i(\boldsymbol{\omega}_{j_1, j_2, \ldots, j_m})\right) \tag{5.35}$$

$$\times \exp\left(i \sum_{l=1}^{m} ((k_l - 1)\pi)\right) \times \exp\left(-i \sum_{l=1}^{m} (k_l - 1)(j_l - 1)\frac{2\pi}{M}\right),$$

where $\Upsilon_{t,s}(i\omega) = \mathbb{E}\left(e^{i\omega \tilde{L}_s} \mid \mathcal{F}_t\right)$ is defined by Eq. (5.17) and

$$\delta_{j_1, j_2, \ldots, j_m} = \left(\frac{1}{2}\right)^{(1_{\{j_1=1\}}+\ldots+1_{\{j_m=1\}})} (1 - 1_{\{j_1 \neq 1, \ldots, j_m \neq 1\}}) + 1_{\{j_1 \neq 1, \ldots, j_m \neq 1\}} .$$

Proof Using a trapezoidal approximation of integral (5.34) leads to the following discrete approximation:

$$\int_{-\infty}^{+\infty} \cdots \int_{-\infty}^{+\infty} \Upsilon_{t,s}(i\boldsymbol{\omega}) e^{-i\boldsymbol{\omega}^{\top}\boldsymbol{x}} d\omega_1 \ldots d\omega_N = \sum_{j_1=1}^{M+1} \sum_{j_2=1}^{M+1} \cdots \sum_{j_m=1}^{M+1} \delta_{j_1,j_2,\ldots,j_m}$$

$$\Upsilon_{t,s}\left(i(\boldsymbol{\omega}_{j_1,j_2,\ldots,j_m})\right) e^{-i\,\boldsymbol{\omega}_{j_1,j_2,\ldots,j_m}^{\top}\,\boldsymbol{x}_{k_1,k_2,\ldots,k_m}} (\Delta_{\omega})^m,$$

and given that $\Delta_x \Delta_{\omega} = \frac{2\pi}{M}$, the scalar product $\boldsymbol{\omega}_{j_1,j_2,\ldots,j_m}^{\top} \boldsymbol{x}_{k_1,k_2,\ldots,k_m}$ is rewritten as follows:

$$\boldsymbol{\omega}_{j_1,j_2,\ldots,j_m}^{\top} \boldsymbol{x}_{k_1,k_2,\ldots,k_m}$$

$$= \sum_{l=1}^{m} \left(-\frac{M}{2}(k_l - 1)\Delta_x\Delta_{\omega} + (k_l - 1)(j_l - 1)\Delta_x\Delta_{\omega} \right)$$

$$= \sum_{l=1}^{m} \left(-(k_l - 1)\pi + (k_l - 1)(j_l - 1)\frac{2\pi}{M} \right).$$

We immediately retrieve Eq. (5.35).   □

The expression (5.35) emphasizes that $f_{t,s}(\boldsymbol{x}_{k_1,\ldots,k_m})$ is computable with a standard DFFT algorithm applied to

$$(\Delta_{\omega})^m \, \delta_{j_1,j_2,\ldots,j_m} \, \Upsilon_{t,s}\left(i(\boldsymbol{\omega}_{j_1,j_2,\ldots,j_m})\right) \exp\left(i \sum_{l=1}^{m} ((k_l - 1)\pi) \right)$$

valued on the mesh grid $j_1, \ldots, j_m \in \{1, \ldots, M+1\}$ (eventually adjusted by $\frac{1}{(2\pi)^m}$, depending on the DFFT implementation).

In practice, computing the joint pdf of more than two processes is time consuming. Nevertheless, in high dimension, it is still possible to calculate the marginal pdf's or the pdf of aggregated jump processes with a one-dimensional Fourier transform. Let $\boldsymbol{\chi}$ be an m-vector taking its value in the set of orthonormal basis vectors $\{\boldsymbol{e}_1, \ldots \boldsymbol{e}_m\}$ of $\mathbb{R}^m$ or being equal to the unit vector $\boldsymbol{e} = (1, \ldots, 1)^{\top}$. We denote by $f_{t,s}^{\chi}(\cdot)$ the univariate pdf of the linear combination $\boldsymbol{\chi}^{\top}\tilde{\boldsymbol{L}}_s | \mathcal{F}_t$. This density is equal to the following univariate Fourier transform:

$$f_{t,s}^{\chi}(x) = \frac{1}{2\pi} \int_{-\infty}^{+\infty} \Upsilon_{t,s}(i\,\omega\,\boldsymbol{\chi}) e^{-i\omega x} d\omega.$$

From the previous proposition, this integral can be discretized in an expression that is compatible with the one-dimensional DFFT. Let us insist on the fact that the domain of $\chi^\top \tilde{L}_s | \mathcal{F}_t$ is $\mathbb{R}^+$. The next corollary is an adapted version of Proposition 1.8 of Chap. 1 taking into account this specificity.

Corollary 5.8 *Let M be the number of steps used in the Discrete Fast Fourier Transform (DFFT) and* $\Delta_x = \frac{x_{\max}}{M}$ *be the discretization step. Let* $\Delta_\omega = \frac{2\pi}{M \Delta_x}$ *and* $\omega_j = -\frac{M}{2}\Delta_\omega + (j-1)\Delta_\omega$ *for* $j = 1, \dots, M+1$. *The values of* $f_{t,s}^\chi(\cdot)$ *at points* $x_k = (k-1)\Delta_x$ *for* $k = 1, \dots, M$ *are approached by*

$$f_{t,s}^\chi(x_k) = \Delta_\omega \sum_{j=1}^{M+1} \delta_j \Upsilon_{t,s}\left(i(\omega_j\,\chi)\right) \exp\left(i\left((k-1)\pi\right)\right) \tag{5.36}$$

$$\times \exp\left(-i(k-1)(j-1)\frac{2\pi}{M}\right),$$

where $\delta_j = \left(\frac{1}{2}\right)^{1_{\{j_1=1\}}} + 1_{\{j\neq 1\}}$.

Marginal and aggregated pdf's are then computable in a reasonable time (less than 1 min for $M = 2^8$ on a standard laptop) with a 1D DFFT applied to

$$\Delta_\omega \delta_j \Upsilon_{t,s}\left(i(\omega_j\,\chi)\right) \exp\left(i\left((k-1)\pi\right)\right).$$

For any other linear combination $\chi^\top \tilde{L}_s | \mathcal{F}_t$ defined on the whole real axis, we can use Proposition 1.8 of Chap. 1.

Figure 5.7 presents the pdf of negative, positive and sum of cumulated jumps over a period of 1 year for the model with Laplace kernels and parameter estimates of Table 5.2. These pdf's are obtained with a 1D DFFT and 2^8 discretization steps. The corresponding probability measures are discretized on a grid of 100 atoms ($n = 50$) covering the largest 95% confidence interval of the four measures. Figure 5.6 presents these atomic measures. The system of ODEs (5.18) is numerically solved by backward iterations[1] with a time step of $1/200$. We also set $\lambda_0 = \theta$. The right plot of Fig. 5.7 reveals that the distribution of the sum of negative and positive jumps is left tailed with a negative mean.

5.8 Further Reading

In its standard version, the jump-intensity process is Markov. In this case, the moment generating function (mgf) admits an analytical solution, found by Itô's calculus. This approach is applied in Dassios and Jang [3] to price catastrophe

---

[1] The $q_l(\cdot)$ are computed with the relation $q_l(t - \Delta, s) = i\, b_l e^{i\, b_l\, \Delta} q_\lambda(t, s)\, \Delta + e^{i\, b_l\, \Delta} q_l(t, s)$ coming from Eq. (5.20) instead of using the related ODEs.

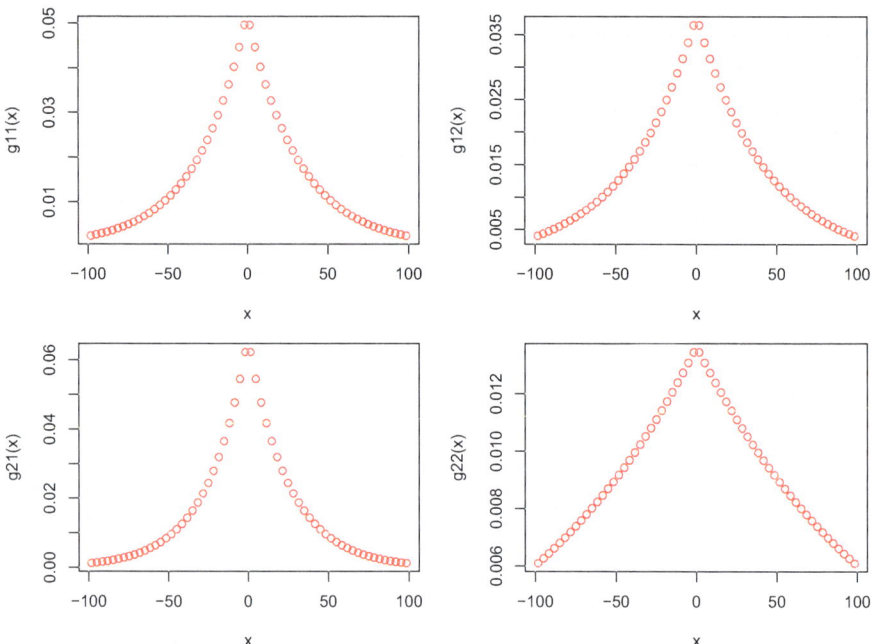

Fig. 5.6 Plots of atomic measures approaching the Laplace distribution defining memory kernels g_{11}, g_{12}, g_{21}, and g_{22}

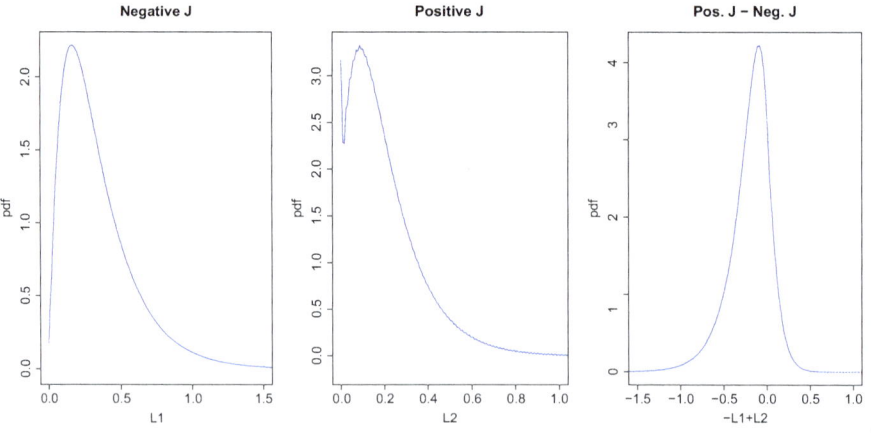

Fig. 5.7 From left to right: 1 year pdf of negative, positive, and sum of cumulated jumps

reinsurance, in Jang and Dassios [6] to a bivariate model, and in Jang et al. [7] to the calculation of moments of shot noise processes. In a general setting, we lose most of the analytical tractability offered by stochastic calculus mainly because the jump-intensity process is not Markov anymore. Apart from moments or asymptotic properties, very few results are available in the literature. For instance, Muzy et al. [8] find the first moments of stationary processes and their limit behavior. Stabile and Torrisi [9] study the asymptotic behavior of infinite and finite horizon ruin probabilities of non-stationary Hawkes jump arrivals. The material in this chapter is inspired by the article of Hainaut [5] in which non-Markov self-exciting processes are used for modeling contagion between insurance claims. The application to cyber-criminality in the USA confirms the results of Bessy-Roland et al. [2], who detect self-excitation and contagion of cyber-attacks. Another interesting application is provided by Barsotti et al. [1], who use a shot noise process for modeling the lapse risk in life insurance.

References

1. Barsotti, F., Milhaud, X., Salhi, Y.: Lapse risk in life insurance: correlation and contagion effects among policyholders' behaviors. Insurance Math. Econ. **71**, 317–331 (2016)
2. Bessy-Roland, Y., Boumezoued, A., Hillairet, C.: Multivariate Hawkes process for cyber insurance. Ann. Actuarial Sci. **15**(1), 14–39 (2020)
3. Dassios, A., Jang, J.: Pricing of catastrophe reinsurance and derivatives using the Cox process with shot noise intensity. Finance Stochast. **7**(1), 73–95 (2003)
4. Embrechts, P., Liniger, T., Lu, L.: Multivariate Hawkes processes: an application to financial data. J. Appl. Probab. **48**(A), 367–378 (2011)
5. Hainaut, D.: Moment generating function of non-Markov self-excited claims processes. Insurance Math. Econ. **101**, 406–424 (2021)
6. Jang, J., Dassios, A.: A bivariate shot noise self-exciting process for insurance. Insurance Math. Econ. **53**(3), 524–532 (2013)
7. Jang, J., Dassios, A., Hongbiao, Z.: Moments of renewal shot-noise processes and their applications. Scand. Actuarial J. **8**, 727–752 (2018)
8. Muzy, J.-F., Delattre, S., Hoffmann, M., Bacry, E.: Some limit theorems for Hawkes processes and application to financial statistics. Stoch. Process. Appl. **123**(7), 2475–2499 (2013)
9. Stabile, G., Torrisi, G.L.: Risk processes with non-stationary Hawkes claims arrivals. Methodol. Comput. Appl. Probab. **12**(3), 415–429 (2010)

Chapter 6
Fractional Brownian Motion

The previous chapters provide empirical evidence that models with stochastic volatility outperform their deterministic counterpart. In Chap. 1, the multifractal process competes with GARCH models, whereas the Heston model of Chap. 3 achieves a better likelihood than the Black and Scholes model. On the other hand, models based on fractional Brownian motion (fBm) have emerged in recent years. For instance, in the rough Heston model, the volatility is mean reverting and ruled by an fBm. This model belongs to the class of Volterra processes that are introduced in Chap. 9. Fractional Brownian motion is also a particular type of Gaussian field. These Gaussian fields are developed in Chap. 7. This motivates us to review the main properties of fBm.

Fractional Brownian motion is comparable to a continuous fractal random walk. However, unlike regular Brownian motion, fBm has dependent increments, which means that the current "step" of an fBm is dependent on previous "steps." This dependence is measured on a scale from zero to one by the Hurst index, $H \in (0, 1)$, named after hydrologist Harold Edwin Hurst for his work in the field of hydrology. The Hurst index describes the raggedness of the path of the fBm. A value of 1/2 corresponds to the Brownian motion with independent increments. A value of H greater (resp., lower) than 1/2 corresponds to positive (resp., negative) correlation between increments. Processes with $H > 1/2$ have a long-term memory that is observed in time series of stock returns or interest rates, while fBm's with $H < 1/2$ explain well the volatility.

© The Author(s), under exclusive license to Springer Nature Switzerland AG 2022   143
D. Hainaut, *Continuous Time Processes for Finance*,
Bocconi & Springer Series 12, https://doi.org/10.1007/978-3-031-06361-9_6

6.1 Definition and Properties

Fractional Brownian motion is one of the simplest stochastic processes that exhibits long-range dependence. It was introduced by Kolmogorov [10] and studied by Mandelbrot and van Ness [11]. In this section, we review its definition and its main properties.

Definition 6.1 Let $(\Omega, \mathcal{F}, \mathbb{P})$ denote a probability space and $H, 0 \leq H \leq 1$, be referred to as the Hurst parameter. The stochastic process $\left(B_t^H\right)_{t \geq 0}$ defined on this probability space is a *fractional Brownian motion (fBm)* of order H if:

1. $\mathbb{P}(B_0^0 = 0) = 1$.
2. For each $t \in \mathbb{R}^+$, B_t^H is an $\mathcal{F}_t$-measurable Gaussian process such that $\mathbb{E}\left(B_t^H\right) = 0$.
3. For $t, s \in \mathbb{R}^+$, the autocovariance is

$$\mathbb{E}\left(B_t^H B_s^H\right) = R_H(t, s) = \frac{1}{2}\left(t^{2H} + s^{2H} - |t - s|^{2H}\right). \qquad (6.1)$$

It follows from (6.1) and from Kolmogorov's continuity criterion that, for $H \geq 1/2$, the sample paths of B_t^H are continuous with probability one, but nowhere differentiable. From the definition, we infer that the variance of the fBm is equal to

$$\mathbb{V}\left(B_t^H\right) = t^{2H}.$$

We also observe that increments are stationary:

$$B_{t+s}^H - B_s^H \sim B_t^H \quad , s, t \geq 0.$$

The fractional process is also self-similar, i.e., the process $a^{-H} B_{at}^H$ has the same law as B_t^H. The increments of the fBm are defined by $X_n = B_{n+1}^H - B_n^H$ for $n \in \mathbb{N}$. $(X_n)_{n \in \mathbb{N}}$ are normal random variables with unit variance. The covariance between X_k and X_{k+n}, denoted $\rho(n)$, is

$$\rho_H(n) = \frac{1}{2}\left((n + 1)^{2H} + (n - 1)^{2H} - 2n^{2H}\right). \qquad (6.2)$$

Using a Taylor series of second order for x^{2H} allows us to infer that

$$\rho_H(n) \approx H(2H - 1)n^{2H-2},$$

as n tends to infinity. In particular, we have the following limit:

$$\lim_{n \to \infty} \frac{\rho_H(n)}{H(2H - 1)n^{2H-2}} = 1.$$

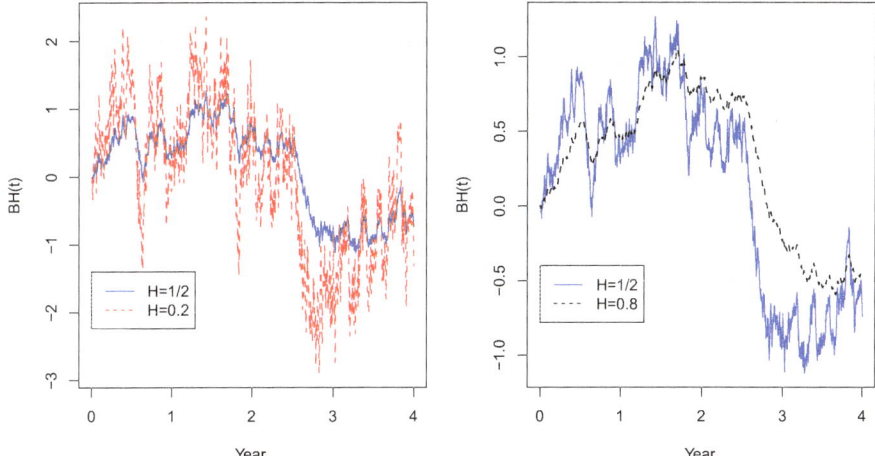

Fig. 6.1 Left and right plots: comparison of fBm sample paths for various Hurst indexes

From these last equations, we infer that for $H > \frac{1}{2}$, the sum $\sum_{n=1}^{\infty} \rho_H(n) = \infty$. The process presents in this case an aggregation behavior and may describe systems with memory and persistence of past events. This point is emphasized by the right plot of Fig. 6.1, which shows fBm sample paths computed with $H = \frac{1}{2}$ and $H = 0.8$. The curve with the highest Hurst index is clearly smoother due to the aggregation of previous variations. In contrast, the right plot of the same figure reveals that the fBm sample path is more chaotic than that of a Brownian motion when $H < \frac{1}{2}$.

Fractional Brownian motion can be rewritten as an integral with respect to a standard Brownian motion, denoted $(W_t)_{t \in \mathbb{R}^+}$, which highlights the memory of the process.

Proposition 6.2 *Fractional Brownian motion admits a Brownian representation:*

$$B_t^H = \frac{1}{C_H} \left[\int_{-\infty}^0 \left((t-u)^{H-\frac{1}{2}} - (-u)^{H-\frac{1}{2}} \right) dW_u + \int_0^t (t-u)^{H-\frac{1}{2}} dW_u \right],$$
(6.3)

where C_H is a function of the Hurst index:

$$C_H = \left(\int_{-\infty}^0 \left((1-u)^{H-\frac{1}{2}} - (-u)^{H-\frac{1}{2}} \right)^2 du + \frac{1}{2H} \right)^{\frac{1}{2}}.$$

Proof Let us denote by Z_t the right-hand side of Eq. (6.3). First, notice that Z_t is a stochastic integral with respect to a standard Brownian motion. Hence, it is Gaussian with a null expectation. We must check that the variance and autocovariance are equal to those of an fBm. The variance of Z_t is equal to

$$
\mathbb{E}\left(Z_t^2\right) = \frac{1}{C_H^2}\mathbb{E}\left(\int_{-\infty}^t (t-u)^{H-\frac{1}{2}}\,\mathrm{d}W_u - \int_{-\infty}^0 (-u)^{H-\frac{1}{2}}\,\mathrm{d}W_u\right)^2
$$

$$
= \frac{1}{C_H^2}\int_{-\infty}^{+\infty}\left((t-u)_+^{H-\frac{1}{2}} - (-u)_+^{H-\frac{1}{2}}\right)^2\mathrm{d}u
$$

$$
= \frac{t^{2H}}{C_H^2}\int_{-\infty}^{+\infty}\left((1-v)_+^{H-\frac{1}{2}} - (-v)_+^{H-\frac{1}{2}}\right)^2\mathrm{d}v = t^{2H},
$$

where $u = tv$. This is also the variance of an fBm. The autocovariance of Z_t is equal to $\mathbb{E}(Z_t Z_s)$. To prove that it is equal to Eq. (6.1), we start from

$$
\mathbb{E}\left((Z_t - Z_s)^2\right) = \mathbb{E}\left((Z_t)^2\right) - 2\mathbb{E}(Z_t Z_s) + \mathbb{E}\left((Z_s)^2\right).
$$

The random variable in the left-hand term is developed as follows:

$$
(Z_t - Z_s)^2 = \frac{1}{C_H^2}\left(\int_{-\infty}^0\left((t-u)^{H-\frac{1}{2}} - (s-u)^{H-\frac{1}{2}}\right)\mathrm{d}W_u\right.
$$

$$
\left. + \int_0^t (t-u)^{H-\frac{1}{2}}\,\mathrm{d}W_u - \int_0^s (s-u)^{H-\frac{1}{2}}\,\mathrm{d}W_u\right)^2
$$

$$
= \frac{1}{C_H^2}\left(\int_{-\infty}^t (t-u)^{H-\frac{1}{2}}\,\mathrm{d}W_u - \int_{-\infty}^s (s-u)^{H-\frac{1}{2}}\,\mathrm{d}W_u\right)^2.
$$

The expectation of the integral in this last equation becomes

$$
\mathbb{E}\left(\left(\int_{-\infty}^{+\infty} (t-u)_+^{H-\frac{1}{2}} - (s-u)_+^{H-\frac{1}{2}}\,\mathrm{d}W_u\right)^2\right)
$$

$$
= \int_{-\infty}^{+\infty}\left((t-u)_+^{H-\frac{1}{2}} - (s-u)_+^{H-\frac{1}{2}}\right)^2\mathrm{d}u
$$

$$
= \int_{-\infty}^{+\infty}\left((t-s-u)_+^{H-\frac{1}{2}} - (-u)_+^{H-\frac{1}{2}}\right)^2\mathrm{d}u.
$$

If we change of variable, $u = |t - s|v$, we get that

$$\int_{-\infty}^{+\infty} \left((t - s - u)_+^{H-\frac{1}{2}} - (-u)_+^{H-\frac{1}{2}} \right)^2 du$$

$$= |t - s|^{2H} \int_{-\infty}^{+\infty} \left((1 - v)_+^{H-\frac{1}{2}} - (-v)_+^{H-\frac{1}{2}} \right)^2 dv$$

$$= |t - s|^{2H} \int_{-\infty}^{0} \left((1 - v)_+^{H-\frac{1}{2}} - (-v)_+^{H-\frac{1}{2}} \right)^2 dv + \int_0^1 (1 - v)^{2H-1} dv$$

$$= |t - s|^{2H} C_H^2 .$$

We retrieve the covariance of the fBm. □

Several other stochastic integral representations are proposed in the literature. We just mention that the covariance function may be rewritten as a double integral for $H > \frac{1}{2}$:

$$R_H(t, s) = H(H - 2) \int_0^t \int_0^s |r - u|^{2H-2} du \, dr .$$

For $H < \frac{1}{2}$, a similar decomposition also exists. The definition of the Itô integral is a direct consequence of the martingale property of Brownian motion. But fBm does not exhibit this property. In fact, fBm is not even a semi-martingale. In order to show this, we first define the p-variation of B_t^H.

Definition 6.3 The *p-variation* of a random process X_t over the interval $[0, T]$ is defined as

$$V_p(X, [0, T]) = \sup_\Pi \sum_{i=1}^n \left| X_{t_i} - X_{t_i+1} \right|^p ,$$

where Π is a finite partition over $[0, T]$. The *index* of the p-variation is defined to be

$$I(X, [0, T]) = \inf \left\{ p > 0 ; \ V_p(X, [0, T]) < \infty \right\} .$$

We admit without proof the following result (we refer the interested reader to Biagini et al. [1] for details):

Proposition 6.4 *The index of p-variation of fractional Brownian motion is*

$$I\left(B^H, [0, T] \right) = \frac{1}{H} .$$

Moreover, $V_p\left(B^H, [0, T]\right) = 0$ *when* $pH > 1$ *and* $V_p\left(B^H, [0, T]\right) = \infty$ *when* $pH < 1$.

When $H = \frac{1}{2}$, the index of p-variation is 2 and the quadratic variation is $V_2\left(B^{1/2}, [0, T]\right) = T$. This result allows us to show that B_t^H is not a semi-martingale when $H \neq \frac{1}{2}$.

Proposition 6.5 *Fractional Brownian motion is not a semi-martingale if $H \neq \frac{1}{2}$.*

Proof A process $(X_t)_{t \geq 0}$ is called a semi-martingale if it admits the Doob–Meyer decomposition

$$X_t = X_0 + M_t + A_t,$$

where M_t is an $\mathcal{F}_t$ local martingale with $M_0 = 0$, A_t is a càdlàg adapted process of locally bounded variation, and X_0 is $\mathcal{F}_0$-measurable. Moreover, any semi-martingale has locally bounded quadratic variation. Now, if $H \in (0, 1/2)$, then B_t^H cannot even be a martingale since it has infinite quadratic variation; hence, it is not a semi-martingale.

If $H \in (1/2, 1)$, then the quadratic variation of B_t^H is zero. So, let us suppose that it is a semi-martingale. Then, $M_t = B_t^H - A_t$ has quadratic variation equal to zero. So, $M_t = 0$ for all t a.s. In this case, $B_t^H = A_t$, but this is not possible since B_t^H has unbounded variation. Hence, B_t^H is not a semi-martingale for any $H \neq \frac{1}{2}$. $\qquad\square$

As mentioned earlier, a consequence of Propositions 6.4 and 6.5 is the difficulty of defining the integral of a function or an adapted process with respect to the fBm. We may think to define it as the limit of a Riemann sum:

$$\int_0^t f(s)\mathrm{d}B_s^H = \lim_{\|\Pi\| \to 0} \sum_i f(t_{i-1})(B_{t_i}^H - B_{t_{i-1}}^H),$$

where Π is a partition of $[0, t]$ and $\|\Pi\|$ is the length of its largest sub-interval. However, if $H < \frac{1}{2}$, the convergence of the Riemann sum is warranted only if $f(\cdot)$ satisfies restrictive conditions on its variations. We will come back to this point in Sect. 6.3, but first we explain how to estimate the Hurst index of a time series.

6.2 Estimation of H by Rescaled Range Analysis

There exist several estimation methods of the Hurst index of a time series. We present in this section the rescaled range analysis (R/S) popularized by Mandelbrot and Wallis [12]. We first estimate the dependence of the rescaled range on the time span n among N observations. A time series, $(x_k)_{k=1,\ldots N}$ of full length N, is divided into shorter time series of length $n = N, N/2, \ldots, N/2^j, \ldots$. We calculate the average rescaled range according to the following steps:

1. Calculate the means and standard deviations of each time series:

$$\bar{x}(n) = \frac{1}{n}\sum_{i=1}^{n} x_i \quad , \quad s(n) = \sqrt{\frac{1}{n}\sum_{i=1}^{n}(x_i - \bar{x})^2}.$$

2. Create a mean-adjusted series:

$$y_i = x_i - \bar{x} \quad \text{for } i = 1, \ldots, n.$$

3. Calculate the cumulative deviate series of this adjusted series:

$$z_i = \sum_{k=1}^{i} y_k \quad \text{for } i = 1, \ldots, n.$$

4. Compute the range, denoted $r(n)$:

$$r(n) = \max(z_1, \ldots, z_n) - \min(z_1, \ldots, z_n).$$

5. Calculate the rescaled range $r(n)/s(n)$ and average over all the partial time series of length n.

The Hurst exponent is estimated by fitting the power law $\mathbb{E}[r(n)/s(n)] = Cn^H$ to the data. Since

$$\ln(\mathbb{E}[r(n)/s(n)]) = H \ln(n) + \ln(C) ,$$

this can be done by linearly regressing the $\ln(\mathbb{E}[r(n)/s(n)])$ on $\ln(n)$. For small n, there is a significant deviation from the 0.5 slope. Peters [14] estimated the theoretical (i.e., for white noise) values of the $r(n)/s(n)$ statistic to be

$$\mathbb{E}(r(n)/s(n)) = \begin{cases} \frac{n-\frac{1}{2}}{n}\frac{\Gamma(\frac{n-1}{2})}{\sqrt{\pi}\Gamma(\frac{n}{2})} \sum_{i=1}^{n-1}\sqrt{\frac{n-i}{i}}, & \text{for } n \le 340 \\[2ex] \frac{n-\frac{1}{2}}{n}\frac{1}{\sqrt{n\frac{\pi}{2}}} \sum_{i=1}^{n-1}\sqrt{\frac{n-i}{i}}, & \text{for } n > 340, \end{cases}$$

where Γ is the gamma function. The corrected R/S Hurst exponent is calculated as 0.5 plus the slope of $r(n)/s(n) - \mathbb{E}(r(n)/s(n))$. We have applied this procedure to daily log-return of the S&P 500 from the 31/1/01 to the 31/1/20. The left plot of Fig. 6.2 presents the ratio $r(n)/s(n)$ for $n = 4$ to 4778. We clearly observe that this curve is concave. The right plot compares the trend obtained by linear regression to $\ln(r(n)/s(n))$. The slope of this trend is $H = 0.5849$, whereas the Hurst index

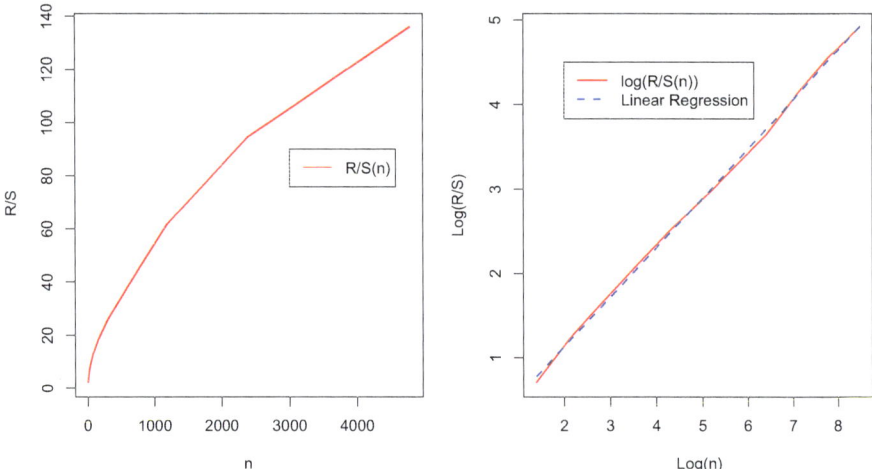

Fig. 6.2 S&P 500: The left plot shows the evolution of the ratio $r(n)/s(n)$ as a function of n. The right plot shows the same information on a logarithmic scale and compares with the linear trend line

estimated with corrected R/S is equal to $H = 0.5532$. Since $H > 0.5$, this confirms that the stock market has a longer memory than the Brownian motion. This also motivates us to focus on fBm with a Hurst index in $(1/2, 1)$. Notice that the fBm with Hurst index lower than $1/2$ explains well the stock volatility dynamics. Models based on such fBm are called "rough" and are presented in Chap. 9 as a particular case of affine Volterra processes.

6.3   Integrals of Deterministic Functions and the Wick Product

There exist multiple alternative definitions of the integral with respect to an fBm. Symmetric, backward, and forward integrals are defined pathwise in a similar manner to a Riemann integral. The Wiener integral exploits the normality of the fBm. The divergence-type integral is introduced as a dual operator of the stochastic derivative. Each of them has interesting features and limitations. In this section, we focus on a definition based on the Bochner–Minlos theorem.[1] This approach offers an easy way to define the integral of deterministic functions with respect to the fBm and admits an extension to adapted processes based on the Wick product.

---

[1] Notice that Bochner's theorem is also used in Chap. 5 to define memory kernels of self-excited processes and in Chap. 7 to construct covariance functions of Gaussian fields.

In the two next sections of this chapter, we focus on fBm with a Hurst index $H \in (1/2, 1)$, following Duncan et al. [6] and Biagini et al. [1]. This corresponds to processes with a long memory for which the correlation between increments is positive. Such memory effects are often observed in financial markets. An extension to fBm with $H \in (0, 1/2)$ exists but is outside the scope of this book.

We denote by $\Omega = C_0(\mathbb{R}_+, \mathbb{R})$ the space of real-valued continuous functions on $\mathbb{R}_+$ with initial value zero and the topology of local uniform convergence. We first define the weight function:

$$\phi(s, t) = H(2H - 1)|s - t|^{2H - 2} \quad s, t \in \mathbb{R}. \tag{6.4}$$

The integral of this function is equal to the autocovariance of an fBm:

$$\int_0^t \int_0^s \phi(u, v) \mathrm{d}u \mathrm{d}v = \frac{1}{2}\left(s^{2H} + t^{2H} - |t - s|^{2H}\right)$$

$$= R_H(t, s).$$

We denote by L_ϕ^2 the Hilbert space of real functions (deterministic and random processes) endowed with the scalar product

$$\langle f, g \rangle_H = \int_{\mathbb{R}} \int_{\mathbb{R}} f(s)g(t)\phi(s, t) \, \mathrm{d}s \, \mathrm{d}t \qquad f, g \in L_\phi^2,$$

and the associated norm

$$\|f\|_H^2 = \langle f, f \rangle_H .$$

The map $f \to \exp\left(-1/2 \|f\|_H^2\right)$ is positive definite. By the Bochner–Minlos theorem, there exist a probability measure $\mathbb{P}^H$ on Ω such that

$$\int_\Omega e^{i\langle \omega, f \rangle} \mathrm{d}\mathbb{P}^H(\omega) = e^{-\frac{1}{2}\|f\|_H^2}, \tag{6.5}$$

where $\langle \omega, f \rangle$ is the impact of ω on f (i.e., $\omega(f)$). We denote by $\mathcal{F}$ the Borel σ-algebra such that on the probability space $(\Omega, \mathcal{F}, \mathbb{P}^H)$ the coordinate process $B^H : \Omega \to \mathbb{R}$ defined as $B_t^H(\omega) = \omega(t)$ for $\omega \in \Omega$ is an fBm.

The set of deterministic functions of L_ϕ^2 is denoted by $L_H^2(\mathbb{R})$. If f is deterministic, we define the stochastic integral as

$$\int_{\mathbb{R}} f(s) \mathrm{d}B_s^H := \langle \omega, f \rangle .$$

For these functions, their integral over $\mathbb{R}$ with respect to B_t^H is defined as a Gaussian random variable with zero mean and variance $\|f\|_H^2$. By extension, we can prove that:

Proposition 6.6 *If f and g belong to $L^2_H(\mathbb{R})$, then $\int_{\mathbb{R}} f(s) \mathrm{d}B^H_s$ and $\int_{\mathbb{R}} g(s) \mathrm{d}B^H_s$ ($H > \frac{1}{2}$) are well defined zero mean Gaussian random variables with variances $\|f\|^2_H$ and $\|g\|^2_H$ and*

$$\mathbb{E}\left[\int_{\mathbb{R}} f(s) \mathrm{d}B^H_s \int_{\mathbb{R}} g(s) \mathrm{d}B^H_s\right] = \int_{\mathbb{R}} \int_{\mathbb{R}} f(s) g(t) \phi(s, t) \mathrm{d}s \, \mathrm{d}t$$

$$= \langle f, g \rangle_H .$$

This proposition can be proved by approximating f and g by piecewise constant functions and then proceeding by passage to the limit.

When the integrand f is a random process, we need to specify $\langle \omega, f \rangle = \int_{\mathbb{R}} f(s) \mathrm{d}B^H_s$ with care. This requires several intermediate steps. First, we show that every random variable $F \in L^2\left(\mathbb{P}^H\right)$ is expandable as a series (by the fractional Wiener–Itô chaos theorem). Applying this result to fBm allows us to define the *fractional white noise*, W^H_t, which is interpreted as the derivative of the fBm: $\mathrm{d}B^H_t = W^H_t \mathrm{d}t$. We next introduce the *Wick product*, denoted $\diamond$, which is a convolution of Wiener–Itô chaos expansions. Finally, we define in the next section the integral $\int Y_t \mathrm{d}B^H_t$ as $\int Y_t \diamond W^H_t \mathrm{d}t$.

Let $L^p(\mathbb{P}^H) = L^p$ be the space of all random variables $F : \Omega \to \mathbb{R}$ with a bounded L^p norm:

$$\|F\|_{L^p(\mathbb{P}^H)} = \mathbb{E}_{\mathbb{P}^H}\left(|F|^p\right)^{1/p} < \infty .$$

For any deterministic function $f \in L^2_H(\mathbb{R})$, we can define a random variable as follows:

$$\mathcal{E}(f) = \exp\left(\int_{\mathbb{R}} f(t) \mathrm{d}B^H_t - \frac{1}{2}\|f\|^2_H\right), \qquad (6.6)$$

called the *exponential function* (when $H = 1/2$, $\mathcal{E}(f)$ corresponds to the Doléans-Dade exponential). We will see later that this exponential is also the Wick exponential, denoted by $\exp^{\diamond}(f) = \mathcal{E}(f)$. It may be proved that the linear span of $\mathcal{E}(f)$ is dense in $L^p(\mathbb{P}^H)$. This Wick exponential is related to the Wick product that we introduce in the following paragraphs. This product is based on a series development of random variables in L^2_ϕ. In Biagini et al. [1], it is shown that there exists a smooth orthonormal basis of $L^2_\phi(\mathbb{R})$, denoted $\{e_i\}^\infty_{i=1}$, such that for any $f \in L^2_\phi(\mathbb{R})$

$$f(x) = \sum_{i=1}^{\infty} \langle f, e_i \rangle_H e_i(x) \quad .$$

We do not need the explicit forms of $\{e_i\}_{i=1}^{\infty}$. Let us now introduce the *Hermite polynomials*, $h_n(x)$, defined by

$$h_n(x) = (-1)^n \, e^{x^2/2} \frac{d^n}{dx^n} \left(e^{-x^2/2} \right) , \quad n = 0, 1, 2, \ldots$$

Let $\mathcal{J} = \left(\mathbb{N}_0^{\mathbb{N}} \right)_c$ denote the set of all finite multi-indexes $\alpha = (\alpha_1, \alpha_2, \ldots, \alpha_m)$ of non-negative integers; then we put

$$\tilde{\mathcal{H}}_\alpha(\omega) = h_{\alpha_1}\left(\langle \omega, e_1 \rangle \right) \times \ldots \times h_{\alpha_m}\left(\langle \omega, e_m \rangle \right) . \tag{6.7}$$

In particular, if the ith unit vector is $\epsilon^{(i)} = (0, \ldots, 0, 1, 0, \ldots, 0)$, then we have that

$$\tilde{\mathcal{H}}_{\epsilon^{(i)}}(\omega) = h_1\left(\langle \omega, e_i \rangle \right) = \langle \omega, e_i \rangle . \tag{6.8}$$

The next result presents the fractional Wiener–Itô chaos expansion theorem:

Theorem 6.7 *Let $F \in L^2 \left(\mathbb{P}^H \right)$. Then there exist constants $c_\alpha \in \mathbb{R}$ and $\alpha \in \mathcal{J}$ such that*

$$F(\omega) = \sum_{\alpha \in \mathcal{J}} c_\alpha \tilde{\mathcal{H}}_\alpha(\omega) ,$$

where the convergence holds in $L^2 \left(\mathbb{P}^H \right)$. Moreover, $\left(\tilde{\mathcal{H}}_\alpha \right)_{\alpha \in \mathcal{J}}$ are orthogonal in $L^2(\mathbb{P}^H)$ and

$$||F||_{L^2(\mathbb{P}^H)}^2 = \sum_{\alpha \in \mathcal{J}} \alpha! \, c_\alpha^2 ,$$

where $\alpha! = \alpha_1! \times \cdots \times \alpha_m!$.

The proof of this expansion is based on the fact that the linear span of functions (6.6) is dense in $L^2(\mathbb{P}^H)$ and is outside the scope of this introduction. By the orthogonality of $\left(\tilde{\mathcal{H}}_\alpha \right)_{\alpha \in \mathcal{J}}$ in $L^2(\mathbb{P}^H)$, we infer that the coefficients of the Wiener–Itô chaos expansion are

$$c_\alpha = \frac{1}{\alpha!} \mathbb{E}_{\mathbb{P}^H} \left(F \tilde{\mathcal{H}}_\alpha \right) .$$

Therefore, if we choose $f \in L_H^2(\mathbb{R})$ and set $F(\omega) = \langle \omega, f \rangle = \int_{\mathbb{R}} f(s) \mathrm{d}B_s^H$, then F and $\langle \omega, e_i \rangle$ are Gaussian, and from Proposition 6.6, we infer that

$$
\begin{aligned}
\mathbb{E}_{\mathbb{P}^H} \left(F \tilde{\mathcal{H}}_{\epsilon^{(i)}} \right) &= \mathbb{E}_{\mathbb{P}^H} \left(\langle \omega, f \rangle \langle \omega, e_i \rangle \right) \\
&= \langle f, e_i \rangle_H \\
&= \int_{-\infty}^{\infty} \int_{-\infty}^{\infty} f(u) e_i(v) \phi(u, v) \mathrm{d}u \, \mathrm{d}v .
\end{aligned}
$$

Moreover, $\mathbb{E}_{\mathbb{P}^H} \left(F \tilde{\mathcal{H}}_{\alpha} \right) = 0$ if $|\alpha| > 1$. The fractional integral may then be expanded as

$$
\int_{\mathbb{R}} f(s) \mathrm{d}B_s^H = \sum_{i=1}^{\infty} \langle f, e_i \rangle_H \, \tilde{\mathcal{H}}_{\epsilon^{(i)}}(\omega) . \tag{6.9}
$$

If we consider $f(s) = \mathbf{1}_{[0,t]}$, the indicator function equal to 1 on $[0, t]$, we can rewrite the fBm as a sum of integrals:

$$
B_t^H = \sum_{i=1}^{\infty} \left[\int_0^t \left(\int_{-\infty}^{+\infty} e_i(v) \phi(u, v) \mathrm{d}v \right) \mathrm{d}u \right] \tilde{\mathcal{H}}_{\epsilon^{(i)}}(\omega) . \tag{6.10}
$$

If we define the fractional white noise W_t^H as follows:

$$
W_t^H = \sum_{i=1}^{\infty} \left(\int_{-\infty}^{+\infty} e_i(v) \phi(t, v) \mathrm{d}v \right) \tilde{\mathcal{H}}_{\epsilon^{(i)}}(\omega) , \tag{6.11}
$$

we infer that the fBm may be rewritten as an integral of the white noise process:

$$
B_t^H = \int_0^t W_s^H \mathrm{d}s \quad , \quad \mathrm{d}B_t^H = W_t^H \mathrm{d}t . \tag{6.12}
$$

Definition 6.8 Let $F(\omega)$ and $G(\omega)$ be two random functions of $L^2 \left(\mathbb{P}^H \right)$ that admit expansions:

$$
F(\omega) = \sum_{\alpha \in \mathcal{J}} a_\alpha \tilde{\mathcal{H}}_\alpha(\omega) \quad G(\omega) = \sum_{\alpha \in \mathcal{J}} b_\alpha \tilde{\mathcal{H}}_\alpha(\omega)
$$

with bounded $L^2\left(\mathbb{P}^H\right)$ norm. The *Wick product* $F \diamond G$ is defined by

$$(F \diamond G)(\omega) = \sum_{\alpha,\beta \in \mathcal{J}} a_\alpha b_\beta \tilde{\mathcal{H}}_{\alpha+\beta}(\omega)$$

$$= \sum_{\gamma \in \mathcal{J}} \left(\sum_{\alpha+\beta=\gamma} a_\alpha b_\beta \right) \tilde{\mathcal{H}}_\gamma(\omega) .$$

The Wick product of random variables arises through the orthogonal decomposition of Theorem 6.7. This is a particular way of defining an adjusted product of a set of random variables. In the lowest order product, the adjustment corresponds to subtracting off the mean value, to leave a result whose expectation is zero. We will show this for the product of two fBm integrals, but first we respectively define the *Wick power* $X^{\diamond n}$ and *Wick exponential* by

$$X^{\diamond n} = X \diamond X \diamond \cdots \diamond X \quad \text{(n factors)},$$

$$\exp^\diamond(X) = \sum_{n=0}^{\infty} \frac{1}{n!} X^{\diamond n} .$$

By construction, the Wick power and exponential are solutions of the following differential equations:

$$\frac{\mathrm{d} X^{\diamond n}}{\mathrm{d} X} = n\, X^{\diamond n-1} ,$$

$$\frac{\mathrm{d} \exp^\diamond(X)}{\mathrm{d} X} = \exp^\diamond(X) .$$

Observe that by definition of the Wick power, we have

$$(\langle \omega, e_k \rangle)^{\diamond n} = \left(\tilde{\mathcal{H}}_{\epsilon(k)}(\omega) \right)^{\diamond n} = \tilde{\mathcal{H}}_{n\epsilon(k)}(\omega) = h_n(\langle \omega, e_k \rangle) .$$

If $c_k \in \mathbb{R}$, then the Wick exponential becomes

$$\exp^\diamond(c_k \langle \omega, e_k \rangle) = \sum_{n=0}^{\infty} \frac{c_k^n}{n!} \langle \omega, e_k \rangle^{\diamond n}$$

$$= \sum_{n=0}^{\infty} \frac{c_k^n}{n!} h_n(\langle \omega, e_k \rangle)$$

$$= \exp\left(c_k \langle \omega, e_k \rangle - \frac{1}{2} c_k^2 \right) ,$$

where the last line is a direct consequence of the definition of Hermite polynomials. Let us now consider a function $f \in L^2_H(\mathbb{R})$. The Wick exponential of $\langle \omega, f \rangle$ is

$$
\exp^\diamond \left(\langle \omega, f \rangle \right) = \exp^\diamond \left(\sum_{k=1}^{\infty} \langle f, e_k \rangle_H \langle \omega, e_k \rangle \right)
$$

$$
= \prod_{k=1}^{\infty} \diamond \exp^\diamond \left(\langle f, e_k \rangle_H \langle \omega, e_k \rangle \right)
$$

$$
= \prod_{k=1}^{\infty} \exp \left(\langle f, e_k \rangle_H \langle \omega, e_k \rangle - \frac{1}{2} \langle f, e_k \rangle_H^2 \right)
$$

$$
= \exp \left(\sum_{k=1}^{\infty} \langle f, e_k \rangle_H \langle \omega, e_k \rangle - \frac{1}{2} \sum_{k=1}^{\infty} \langle f, e_k \rangle_H^2 \right)
$$

$$
= \exp \left(\langle \omega, f \rangle - \frac{1}{2} \|f\|_H^2 \right). \tag{6.13}
$$

We have thus proven that the exponential function $\mathcal{E}(f)$ as defined in Eq. (6.6) is the Wick exponential of the function f. Furthermore, we have that

$$
\exp^\diamond \left(\langle \omega, f \rangle \right) \exp^\diamond \left(\langle \omega, g \rangle \right) = \exp^\diamond \left(\langle \omega, f + g \rangle \right)
$$

for all functions $f, g \in L^2_H(\mathbb{R})$.

As a last example, we consider the Wick product of two integrals of deterministic functions $f, g \in L^2_H(\mathbb{R})$. From Eq. (6.9), we have that

$$
\int_{\mathbb{R}} f(s) \mathrm{d}B_s^H \diamond \int_{\mathbb{R}} g(s) \mathrm{d}B_s^H
$$

$$
= \sum_{i=1}^{\infty} \langle f, e_i \rangle_H \tilde{\mathcal{H}}_{\epsilon^{(i)}}(\omega) \diamond \sum_{i=1}^{\infty} \langle g, e_i \rangle_H \tilde{\mathcal{H}}_{\epsilon^{(i)}}(\omega)
$$

$$
= \sum_{i,j=1}^{\infty} \langle f, e_i \rangle_H \langle g, e_j \rangle_H \tilde{\mathcal{H}}_{\epsilon^{(i)} + \epsilon^{(j)}}(\omega).
$$

Using Eq. (6.8), we have that $\tilde{\mathcal{H}}_{\epsilon^{(i)}+\epsilon^{(j)}}(\omega) = \langle \omega, e_i \rangle \langle \omega, e_j \rangle$ for $i \neq j$, whereas $\tilde{\mathcal{H}}_{2\epsilon^{(i)}}(\omega) = \langle \omega, e_i \rangle^2 - 1$ since $h_2(x) = x^2 - 1$. Then, we can rewrite the Wick product as

$$\int_{\mathbb{R}} f(s)\mathrm{d}B_s^H \diamond \int_{\mathbb{R}} g(s)\mathrm{d}B_s^H$$

$$= \sum_{\substack{i\neq j \\ i,j=1}}^{\infty} \langle f, e_i \rangle_H \langle g, e_j \rangle_H \langle \omega, e_i \rangle \langle \omega, e_j \rangle$$

$$+ \sum_{i=1}^{\infty} \langle f, e_i \rangle_H \langle g, e_i \rangle_H \left(\langle \omega, e_i \rangle^2 - 1 \right)$$

$$= \left(\sum_{i=1}^{\infty} \langle f, e_i \rangle_H \langle \omega, e_i \rangle \right) \left(\sum_{i=1}^{\infty} \langle g, e_j \rangle_H \langle \omega, e_j \rangle \right)$$

$$- \sum_{i=1}^{\infty} \langle f, e_i \rangle_H \langle g, e_i \rangle_H .$$

Given that $\sum_{i=1}^{\infty} \langle f, e_i \rangle_H \langle g, e_i \rangle_H = \langle f, g \rangle_H$, we infer the interesting relation between the product and the Wick product of integrals with respect to an fBm:

$$\int_{\mathbb{R}} f(s)\mathrm{d}B_s^H \diamond \int_{\mathbb{R}} g(s)\mathrm{d}B_s^H = \left(\int_{\mathbb{R}} f(s)\mathrm{d}B_s^H \right) \int_{\mathbb{R}} g(s)\mathrm{d}B_s^H \qquad (6.14)$$
$$- \langle f, g \rangle_H .$$

This last equation emphasizes that the Wick product of two fBm integrals is equal to their product adjusted by their expected value ($\langle f, g \rangle_H$ from Proposition 6.6), to leave a result whose expectation is zero. The last result of this section introduces a fractional version of the Girsanov formula.

Theorem 6.9 *Let $T > 0$ and γ be a continuous function with support on $[0, T]$. Let K be a function such that*

$$\langle K, f \rangle_H = \langle \gamma, f \rangle_{L^2(\mathbb{R})}$$

for all functions $f \in L_\phi^2$ with support in $[0, T]$, i.e.,

$$\int_{\mathbb{R}} K(s)\phi(s, t)ds = \gamma(t) \quad , \quad 0 \leq t \leq T .$$

On the natural σ-algebra $\mathcal{F}_T^H$ of $\left(B_t^H\right)_{t\in[0,T]}$, we define a probability measure:

$$\frac{d\mathbb{P}^{H,\gamma}}{d\mathbb{P}^H} = \exp^\diamond\left(-\langle\omega, K\rangle\right). \tag{6.15}$$

Then $\widehat{B}_t^H = B_t^H + \int_0^t \gamma(s)ds$ for $t \in [0, T]$ is an fBm under $\mathbb{P}^{H,\gamma}$.

Proof According to Eq. (6.5), we know that $B_t^H(\omega) = \langle\omega, \mathbf{1}_{[0,t]}\rangle$, and we must show that

$$\int_\Omega e^{\langle\omega+\gamma,f\rangle}d\mathbb{P}^{H,\gamma}(\omega) = \int_\Omega e^{\langle\omega,f\rangle}d\mathbb{P}^H(\omega) = e^{\frac{1}{2}\|f\|_H},$$

for all $f \in L_H^2(\mathbb{R})$. By definition (6.15) of the change of measure, we have that

$$\int_\Omega e^{\langle\omega+\gamma,f\rangle}d\mathbb{P}^{H,\gamma}(\omega) = \int_\Omega e^{\langle\omega+\gamma,f\rangle}\exp^\diamond\left(-\langle\omega, K\rangle\right)d\mathbb{P}^H(\omega).$$

According to Eq. (6.13), we then have that

$$\int_\Omega e^{\langle\omega+\gamma,f\rangle}d\mathbb{P}^{H,\gamma}(\omega) = \int_\Omega \exp\left(\langle\omega+\gamma, f\rangle - \langle\omega, K\rangle - \frac{1}{2}\|K\|_H^2\right)d\mathbb{P}^H(\omega)$$

$$= \int_\Omega \exp\left(\langle\omega, f - K\rangle + \langle\gamma, f\rangle_{L^2(\mathbb{R})} - \frac{1}{2}\|K\|_H^2\right)d\mathbb{P}^H(\omega)$$

$$= \exp\left(\frac{1}{2}\|f - K\|_H^2 + \langle\gamma, f\rangle_{L^2(\mathbb{R})} - \frac{1}{2}\|K\|_H^2\right)$$

$$= \exp\left(\frac{1}{2}\|f\|_H^2 - \langle K, f\rangle_H + \langle\gamma, f\rangle_{L^2(\mathbb{R})}\right)$$

$$= \exp\left(\frac{1}{2}\|f\|_H^2 - \langle K, f\rangle_H + \langle\gamma, f\rangle_{L^2(\mathbb{R})}\right)$$

$$= \exp\left(\frac{1}{2}\|f\|_H^2\right),$$

which ends the proof.   □

6.4 Fractional Integrals and Itô's Lemma

When $f \in L_H^2(\mathbb{R})$ is a deterministic function, we have seen that the integral $\langle w, f\rangle = \int_\mathbb{R} f(t)dB_t^H$ is well defined using the Bochner–Minlos theorem. It is in this case a Gaussian random variable with a null mean and a variance equal to $\|f\|_H^2$. If f is random, this definition is extended using the Wick product.

Definition 6.10 The *fractional integral* of a process $(Y_t)_{t \in \mathbb{R}}$, such that $Y_t \diamond W_t^H$ is a well-posed random variable, is given by

$$\int_{\mathbb{R}} Y_t \mathrm{d} B_t^H := \int_{\mathbb{R}} Y_t \diamond W_t^H \mathrm{d} t . \tag{6.16}$$

Why do we use the Wick product? As mentioned in the previous section, the Wick product is a product from which the mean value is subtracted. The expectation of the fBm integral (6.16) is therefore null. As an example, we consider a piecewise random process:

$$Y(t) = \sum_{i=1}^{n} F_i(\omega) 1_{[t_i, t_{i+1})}(t) ,$$

where $F_i(\omega)$ admits a fractional Wiener–Itô chaos expansion with bounded norm. According to the definition of the fractional integral, we have that

$$\int_{\mathbb{R}} Y_t \mathrm{d} B_t^H := \sum_{i=1}^{n} F_i(\omega) \diamond \left(B_{t_{i+1}}^H - B_{t_i}^H \right) ,$$

and it may be proved that

$$\mathbb{E} \left(\int_{\mathbb{R}} Y_t \mathrm{d} B_t^H \right) = \sum_{i=1}^{n} \mathbb{E} \left(F_i(\omega) \right) \diamond \mathbb{E} \left(B_{t_{i+1}}^H - B_{t_i}^H \right)$$

$$= 0 .$$

Taking the limit, this relation holds for all integral processes. As another illustration, we calculate the fractional integral of B_t^H:

$$\int_0^t B_s^H \mathrm{d} B_s^H = \int_0^t B_s^H \diamond W_s^H \mathrm{d} s$$

$$= \int_0^t B_s^H \diamond \frac{B_s^H}{\mathrm{d} s} \mathrm{d} s$$

$$= \frac{1}{2} \left(B_t^H \right)^{\diamond 2} .$$

The passage from the second to the third lines is a consequence of standard Wick calculus. Using the relation (6.14) and $\int_0^t \int_0^t \phi(u, v) \mathrm{d} u \mathrm{d} v = t^{2H}$ leads to

$$\int_0^t B_s^H \mathrm{d} B_s^H = \frac{1}{2} \left(B_t^H \right)^2 - \frac{1}{2} t^{2H} . \tag{6.17}$$

There exists an analogue of Itô's formula for fractional integrals. We refer the reader to Biagini et al. [1] or Duncan et al. [6] for a general version of this formula that involves a fractional version of Malliavin's derivative. We present a more restrictive version that is however sufficient for later developments.

Proposition 6.11 *Let* $\eta_t = \int_0^t a_s \mathrm{d}B_s^H$, *where* $a \in L_H^2(\mathbb{R})$ *and* $f(s, x) : \mathbb{R}_+ \times \mathbb{R} \to \mathbb{R}$, *a* $C^{1,2}$ *function. Then*

$$f(t, \eta_t) = f(0, 0) + \int_0^t \frac{\partial f(s, \eta_s)}{\partial s} \mathrm{d}s + \int_0^t \frac{\partial f(s, \eta_s)}{\partial x} a_s \mathrm{d}B_s^H \qquad (6.18)$$

$$+ \int_0^t \frac{\partial^2 f(s, \eta_s)}{\partial x^2} a_s \int_0^s \phi(s, v) a_v \mathrm{d}v \, \mathrm{d}s \, .$$

An alternative formal formulation is

$$\mathrm{d}f(t, \eta_t) = \frac{\partial f(t, \eta_t)}{\partial t} \mathrm{d}t + \frac{\partial f(t, \eta_t)}{\partial x} a_t \mathrm{d}B_t^H \qquad (6.19)$$

$$+ \frac{\partial^2 f(t, \eta_t)}{\partial x^2} a_t \int_0^t \phi(t, v) a_v \mathrm{d}v \, \mathrm{d}t \, .$$

As an illustration, let us consider the function

$$S_t := f\left(t, \int_0^t \sigma \mathrm{d}B_s^H\right) = S_0 \exp\left(\mu t - \frac{1}{2}\sigma^2 t^{2H} + \int_0^t \sigma \mathrm{d}B_s^H\right), \qquad (6.20)$$

where S_0, μ, and σ are in $\mathbb{R}^+$. According to Eqs. (6.18) and (6.19), the differential of this function is

$$\mathrm{d}S_t = S_t \left(\mu - \sigma^2 H t^{2H-1}\right) \mathrm{d}t + S_t \sigma \mathrm{d}B_t^H$$

$$+ S_t \sigma^2 \int_0^t \phi(t, v) \mathrm{d}v \, \mathrm{d}t \, .$$

But the integral of the function $\phi(\cdot)$ is given by

$$\int_0^t \phi(t, v) \mathrm{d}v = H(2H - 1) \int_0^t (t - v)^{2H-2} \mathrm{d}v$$

$$= H t^{2H-1} \, .$$

We have thus established that (6.20) is a solution of the following geometric fractional diffusion equation:

$$\frac{\mathrm{d}S_t}{S_t} = \mu \mathrm{d}t + \sigma \mathrm{d}B_t^H \, . \qquad (6.21)$$

Notice that $\mathbb{E}(S_t) = S_0 e^{\mu t}$. By extension, any $C^{1,2}$ function, f, of time and S_t defined by Eqs. (6.20) and (6.21) is a solution of

$$df(t, S_t) = \frac{\partial f(t, S_t)}{\partial t}dt + \frac{\partial f(t, S_t)}{\partial x}\mu S_t dt + \frac{\partial f(t, S_t)}{\partial x}\sigma S_t dB_t^H \quad (6.22)$$

$$+ \frac{\partial^2 f(t, S_t)}{\partial x^2}\sigma^2 S_t^2 H t^{2H-1} dt.$$

In view of Eq. (6.21), it is tempting to consider it for modeling the stock price in a Black and Scholes market. This approach is explored by Hu and Oksendal [9]. Despite the apparent simplicity of this model, we warn the reader that the presence of the Wick product implies a modification of the definition of a self-financing portfolio. This has important consequences on the economic interpretability of the model. This point is detailed in the next section.

The fractional Itô lemma admits a multivariate extension. Let us denote by $\mathbf{B}_t^H = \left(B_t^{H_1}, \ldots, B_t^{H_m}\right)^\top$ a vector of m fBm's with Hurst index $H_1, \ldots H_m \in (1/2, 1)$. The probability measure $\mathbb{P}^H$ is defined on $\left(L_\phi^2\right)^d$ by $\mathbb{P}^H = \mathbb{P}^{H_1} \otimes \ldots \otimes \mathbb{P}^{H_m}$ on $\Omega = C_0(\mathbb{R}_+, \mathbb{R}^m)$. Ω is the space of real-valued continuous functions on $\mathbb{R}_+$ with the initial value zero and the topology of local uniform convergence. We also define

$$\phi_{H_i}(s, t) = H_i(2H_i - 1)|s - t|^{2H_i - 2} \quad s, t \in \mathbb{R}. \quad (6.23)$$

Proposition 6.12 *Let* $X_t = \left(X_t^1, \ldots, X_t^m\right)^\top$ *with*

$$dX_t^i = \sum_{j=1}^m \sigma_{i,j}(t)dB_t^{H_j}, \quad i \in \{1, \ldots, m\},$$

where $\sigma_{i,j} \in L_H^2(\mathbb{R})$. *Let* $f : \mathbb{R}_+ \times \mathbb{R}^m \to \mathbb{R}$ *be a* $C^{1,2^m}$ *function. Then*

$$f(t, X_t) = f(0, 0) + \int_0^t \frac{\partial f(s, X_s)}{\partial s}ds + \sum_{j=1}^m \int_0^t \frac{\partial f(s, X_s)}{\partial x_j}dX_s^j \quad (6.24)$$

$$+ \int_0^t \sum_{i,j=1}^m \frac{\partial^2 f(s, X_s)}{\partial x_i \partial x_j}\left(\sum_{k=1}^m \sigma_{i,k}(s)\int_0^s \phi_{H_k}(s, v)\sigma_{j,k}(v)dv\right)ds.$$

An alternative formal formulation is

$$\mathrm{d}f(t, X_t) = \frac{\partial f(t, X_t)}{\partial t}\mathrm{d}t + \sum_{j=1}^{m} \frac{\partial f(t, X_t)}{\partial x_j}\mathrm{d}X_t^j \tag{6.25}$$

$$+ \sum_{i,j=1}^{m} \frac{\partial^2 f(s, X_s)}{\partial x_i \partial x_j}\left(\sum_{k=1}^{m} \sigma_{i,k}(t)\int_0^t \phi_{H_k}(t, v)\sigma_{j,k}(v)\mathrm{d}v\right)\mathrm{d}t .$$

As an illustration, let us consider a bivariate fractional motion, defined by

$$\begin{pmatrix} \mathrm{d}X_t^1 \\ \mathrm{d}X_t^2 \end{pmatrix} = \begin{pmatrix} \sigma_{1,1} & 0 \\ 0 & \sigma_{2,2} \end{pmatrix}\begin{pmatrix} \mathrm{d}B_t^{H_1} \\ \mathrm{d}B_t^{H_2} \end{pmatrix},$$

where $\sigma_{1,1}$ and $\sigma_{2,2} \in \mathbb{R}^+$. We calculate the differential of the following function:

$$S_t := f(t, X_t) = S_0 \exp\left(\mu t - \frac{1}{2}\sigma_{1,1}^2 t^{2H_1} - \frac{1}{2}\sigma_{2,2}^2 t^{2H_2} + X_t^1 + X_t^2\right), \tag{6.26}$$

which according to Eq. (6.25) is given by

$$\mathrm{d}S_t = S_t\left(\mu t - \sigma_{1,1}^2 H_1 t^{2H_1-1} - \sigma_{2,2}^2 H_2 t^{2H_2-1}\right)\mathrm{d}t$$

$$+ S_t\left(\sigma_{1,1}\mathrm{d}B_t^{H_1} + \sigma_{2,2}\mathrm{d}B_t^{H_2}\right)$$

$$+ S_t\sigma_{1,1}^2\int_0^t \phi_{H_1}(t, v)\mathrm{d}v + S_t\sigma_{2,2}^2\int_0^t \phi_{H_2}(t, v)\mathrm{d}v .$$

We conclude that S_t is driven by a bivariate geometric fractional diffusion:

$$\frac{\mathrm{d}S_t}{S_t} = \mu\mathrm{d}t + \sigma_{1,1}\mathrm{d}B_t^{H_1} + \sigma_{2,2}\mathrm{d}B_t^{H_2} . \tag{6.27}$$

There exists an extension of the fractional integral for fBm with a Hurst index in $(0, 1/2)$, called the Wick–Itô–Skorohod integral. However in this framework, we only have an Itô formula for functions of the form $f(t, B_t^H)$ (and not for a function of fractional integrals).

6.5   Options Pricing in a Fractional Setting

In this section, we consider a Black and Scholes market in which the stock price is ruled by Eqs. (6.20) and (6.21). This approach is explored by Hu and Oksendal [9] and in Biagini et al. [1, Chapter 7] and leads to a closed form formula for European options. However, the presence of the Wick product in the definition of the integral raises serious questions about the economic interpretability of the model. Let us consider a risky asset whose price, S_t, ruled by a geometric fractional Brownian motion:

$$S_t = S_0 \exp\left(\mu t - \frac{1}{2}\sigma^2 t^{2H} + \sigma B_t^H\right),$$

where $\mu, \sigma \in \mathbb{R}^+$, and $H \in (1/2, 1)$. We denote by $(\Omega, \mathcal{F}, \mathbb{P}^H)$ the probability space on which $(S_t)_{t \geq 0}$ is defined. The market also proposes a risk-free rate asset, S_t^0, earning a constant rate, $r \in \mathbb{R}$. The stock price is the solution of a stochastic differential equation

$$\frac{\mathrm{d}S_t}{S_t} = \mu\,\mathrm{d}t + S_t\mathrm{d}B_t^H.$$

In the standard way, the value of a portfolio, V_t, made up of h_t^0 risk-free assets and of h_t^1 stocks is defined by

$$V_t = h_t^0 S_t^0 + h_t^1 S_t. \tag{6.28}$$

The portfolio is self-financing in the sense that

$$V_t = V_0 + \int_0^t h_s^0 \mathrm{d}S_s^0 + \int_0^t h_s^1 \mathrm{d}S_s.$$

Hu and Oksendal [9] define the Wick portfolio as an extension of (6.28):

$$V_t = h_t^0 S_t^0 + h_t^1 \diamond S_t \tag{6.29}$$

and consider the Wick self-financing conditions,

$$V_t = V_0 + \int_0^t h_s^0 \mathrm{d}S_s^0 + \int_0^t h_s^1 \diamond \mathrm{d}S_s. \tag{6.30}$$

From Eq. (6.29), we infer that $h_t^0 = \frac{V_t - h_t^1 \diamond S_t}{S_t^0}$, which substituted into the differential of Eq. (6.30) gives

$$\mathrm{d}V_t = r V_t \mathrm{d}t + h_t^1 \diamond S_t (\mu - r)\,\mathrm{d}t + h_t^1 \diamond S_t \mathrm{d}B_t^H.$$

The pricing of options is done under a risk neutral measure $\mathbb{Q}$. Under this measure, the portfolio earns on average the risk-free rate, $r \in \mathbb{R}$ and $\mathbb{E}^{\mathbb{Q}}(dV_t) = rV_t dt$. From Proposition 6.6, we know that an equivalent measure is defined by the change of measure:

$$\frac{d\mathbb{Q}}{d\mathbb{P}^H} = \exp^{\diamond}\left(-\int_0^T K(s)dB_s^H\right)$$

$$= \exp\left(-\int_0^T K(s)dB_s^H - \frac{1}{2}||K\,\mathbf{1}_{[0,T]}||_H^2\right),$$

where here $K(s) \in L_\phi^2(\mathbb{R})$. If $\gamma(t) \in L^2(\mathbb{R})$ is the function defined by the following integral:

$$\int_{\mathbb{R}} K(s)\phi(s,t)ds = \gamma(t),$$

then $\widehat{B}_t^H = B_t^H + \int_0^t \gamma(s)ds$ for $t \in [0,T]$ is an fBm under $\mathbb{Q}$. From Itô's lemma for fBm, Eq. (6.19), the dynamics of $(V_t)_{t\geq0}$ under the risk neutral measure is

$$dV_t = rV_t dt + h_t^1 \diamond S_t\,(\mu - r - \sigma\gamma(t))\,dt + h_t^1 \diamond S_t d\widehat{B}_t^H.$$

The portfolio earns on average the risk-free rate under $\mathbb{Q}$ if

$$\gamma(t) = \frac{\mu - r}{\sigma}.$$

The function $K(s) = K(s,T)$ therefore satisfies the equality:

$$H\,(2H-1)\int_0^T K(s,T)|s-t|^{2H-2}ds = \frac{\mu - r}{\sigma} \quad \forall t \in [0,T].$$

Hu and Oksendal [9, Appendix] proved that

$$K(s,T) = a\left(Ts - s^2\right)^{1/2-H},$$

where a is constant and equal to

$$a = \frac{\mu - r}{2\sigma H\,(2H-1)\,\Gamma\,(2-2H)\cos\,(\pi\,(H-1/2))}.$$

Important warning: The Black and Scholes market is arbitrage-free at the condition to replace the standard portfolio definition (6.28) by the Wick portfolio (6.29). Bjork and Hult [3] have shown in their seminal paper that this assumption is hard to motivate from an economic point of view. Indeed, the Wick portfolio, $V_t(\omega)$, is defined for $\mathbb{P}^H$-almost all $\omega \in \Omega$. This means that it is not sufficient to know the observed values of h_t^0, h_t^1 and S_t to compute the Wick value. We have to plug in the entire random variables $h_t^1 : \Omega \to \mathbb{R}$ and $S_t : \Omega \to \mathbb{R}^+$. This means that if we want to buy, say, $h_t^1 = 10$ shares at time t, it is not sufficient to instruct our broker to buy ten shares of the risky asset. We must also specify the entire random variable $h_t^1 : \Omega \to \mathbb{R}$, i.e., we must let him know how many shares we would buy in $\mathbb{P}^H$-almost all possible states of the world. Otherwise he cannot compute the Wick value of the portfolio. Bjork and Hult [3] also provide an explicit example of a portfolio strategy that is obviously self-financing in the standard sense, but which is not Wick self-financing.

Nevertheless, it is possible to infer closed form solutions for European options. The price of a call option of strike c is denoted by C_0 and computed as the expected discounted payoff under the risk neutral measure.

Proposition 6.13 *The price of a call price in a fractional setting for $H \in (1/2, 1)$ is given by*

$$C_0 = e^{-rT} \mathbb{E}^{\mathbb{Q}} \left((S_T - c)_+ \right) \tag{6.31}$$
$$= S_0 \Phi \left(-d_1(T) \right) - c e^{-rT} \Phi(-d_2(T)),$$

where $\Phi(\cdot)$ is the cumulative distribution function and

$$d_2(T) = \frac{\ln \left(\frac{c}{S_0} \right) - \left(rT - \frac{1}{2}\sigma^2 T^{2H} \right)}{\sigma T^H}, \tag{6.32}$$

$$d_1(T) = d_2(T) - \sigma T^H.$$

Proof The stock price at time T is log-normal under the risk measure: $\ln \left(\frac{S_T}{S_0} \right) \sim N \left(\mu_S(T), \sigma_S^2(T) \right)$, where

$$\mu_S(T) = rT - \frac{1}{2}\sigma^2 T^{2H},$$

$$\sigma_S^2(T) = \sigma^2 T^{2H}.$$

The option price is therefore rewritten as the difference between two integrals:

$$C_0 = \frac{S_0 e^{-rT}}{\sqrt{2\pi}} \int_{S_0 e^{\mu_S(T) + \sigma_S(T)u} \geq c}^{\infty} e^{\mu_S(T) + \sigma_S(T)u - \frac{u^2}{2}} \, du$$

$$- \frac{c e^{-rT}}{\sqrt{2\pi}} \int_{S_0 e^{\mu_S(T) + \sigma_S(T)u} \geq c}^{\infty} e^{-\frac{u^2}{2}} \, du.$$

The second term of this equation is equal to

$$ce^{-rT}\int_{u\geq\frac{\ln\left(\frac{c}{S_0}\right)-\mu_S(T)}{\sigma_S(T)}}^{\infty}\frac{e^{-\frac{u^2}{2}}}{\sqrt{2\pi}}du = ce^{-rT}\Phi(-d_2(T)),$$

where $d_2(T)$ is given by Eq. (6.32), whereas the first term is

$$\frac{S_0}{\sqrt{2\pi}}\int_{u\geq\frac{\ln\left(\frac{c}{S_0}\right)-\mu_S(T)}{\sigma_S(T)}}^{\infty}e^{-\frac{1}{2}\sigma_S^2(T)+\sigma_S(T)u-\frac{u^2}{2}}du$$

$$= S_0\int_{s\geq\frac{\ln\left(\frac{c}{S_0}\right)-\mu_S(T)}{\sigma_S(T)}-\sigma_S(T)}^{\infty}\frac{e^{-\frac{1}{2}s^2}}{\sqrt{2\pi}}ds$$

$$= S_0\Phi(-d_1(T)).\qquad\qquad\square$$

In order to illustrate this section, we compare call prices in a Black and Scholes (B&S) framework (which correspond to $H = 1/2$) to their counterparts computed with a fractional model. The interest rate is set to $r = 2\%$, the asset value is $S_0 = 1$, and $\sigma = 20\%$. The left plot of Fig. 6.3 compares 1 year call prices for strikes ranging from $K = 0.90$ to $K = 1.10$. For a Hurst index $H = 0.6$, we observe that options are less expensive than those computed with the B&S model. The right plot shows the surface of implied volatilities. For a given maturity, the implied volatility is

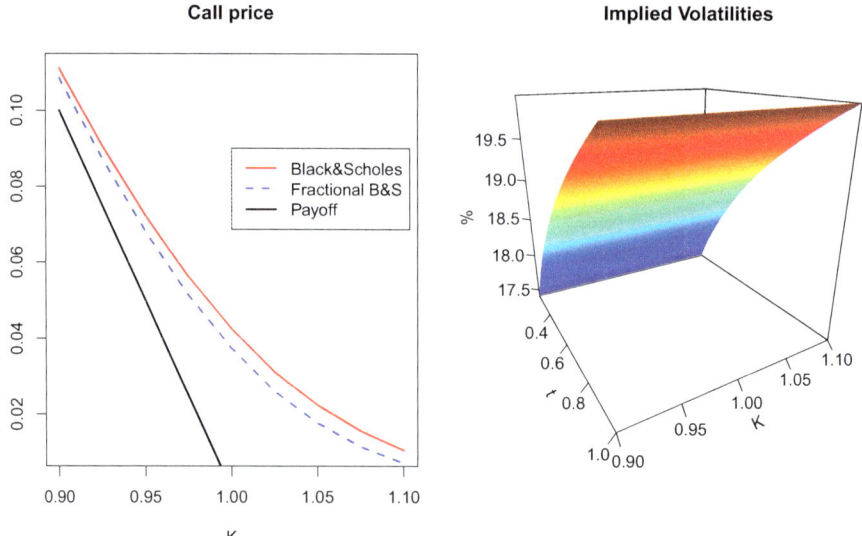

Fig. 6.3 Left plot comparison of Black and Scholes call prices versus fractional prices. Right plot: surface of implied volatilities, fractional model

nearly constant and is not able to reproduce the volatility smile observed in financial markets, but we see that implied volatilities rise with the time horizon.

However, we recall that those arbitrage-free prices are obtained under the assumption that market operators appraise a portfolio with the Wick product, as stated in Eq. (6.29). This is debatable from an economic point of view. There exist several alternative approaches to option pricing in an fBm market. For instance, we can consider the Stratonovich pathwise integral, defined in a similar manner to the Riemann integral:

$$\int_0^t f(s) \mathrm{d} B_s^H := \lim_{||\Pi|| \to 0} \sum_{i=1} \frac{f(t_{i-1}) + f(t_i)}{2} \left(B_{t_i}^H - B_{t_{i-1}}^H \right) .$$

We refer to Rostek and Schöbel [18] for a comparison of option pricing in Wick and Stratonovich frameworks.

6.6   A Fractional Interest Rate Model

A common paradigm for modeling interest rates consists in assuming that short-term rates are mean reverting processes ruled by a Brownian motion. When the time series of interest rates displays a long-range dependence, considering an fBm with a Hurst index above 1/2 instead of a Brownian motion allows us to replicate this trend. We explore in this section the properties of such a model. Notice that an alternative to the fBm is developed in Chap. 8, which studies a non-Markov model for interest rates, with a long memory.

We consider a probability space, denoted by $(\Omega, \mathcal{F}, \mathbb{Q})$, on which is defined the instantaneous rate process, $(r_t)_{t \geq 0}$. We assume that $\mathbb{Q}$ is the pricing measure. The process $(\eta_t)_{t \geq 0}$ is the following fractional integral:

$$\eta_t = \int_0^t \mathrm{e}^{-a(t-u)} \sigma \mathrm{d} B_u^H , \tag{6.33}$$

which is involved in the dynamics of the short-term rate in the following way:

$$r_t = \mathrm{e}^{-at} r_0 + b \left(1 - \mathrm{e}^{-at} \right) + \eta_t . \tag{6.34}$$

If we apply Itô's formula (6.18) for fractional Brownian motion to this function of η_t, we infer that r_t is a mean reverting process:

$$\mathrm{d} r_t = a(b - r_t) \mathrm{d}t + \sigma \mathrm{d} B_t^H \quad , \ t \geq 0 .$$

Figure 6.4 compares three sample paths of r_t with Hurst index $H = 0.5$ (Brownian motion), $H = 0.7$, and $H = 0.9$. The fractional Brownian motion is simulated as

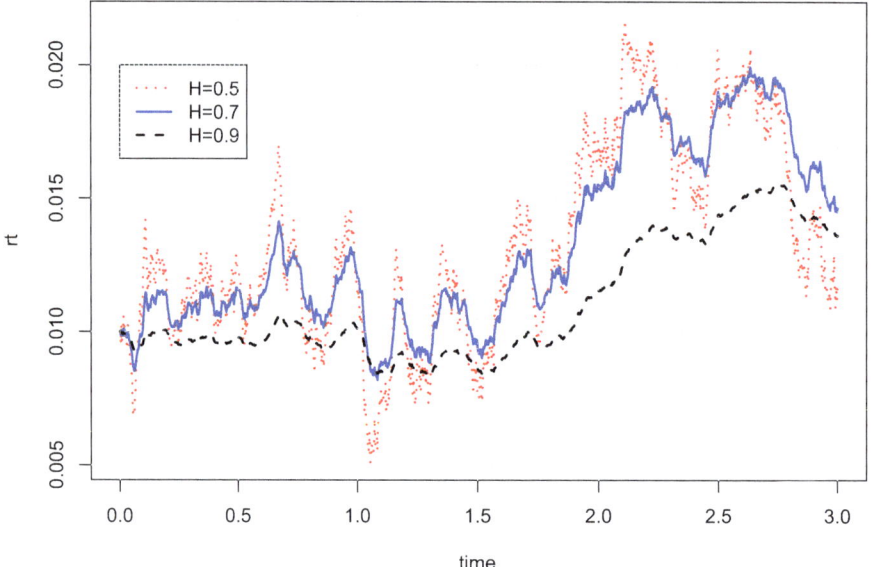

Fig. 6.4 Simulated sample paths of r_t with different Hurst indexes, $a = 0.8$, $b = 1.2\%$, $r_0 = 1.0\%$, and $\sigma = 1\%$

a multivariate normal random variable with zero mean and covariance matrix given by Eq. (6.1). We consider 1000 steps of time over a time horizon of 3 years. The seed of the random number generator is identical for each simulation. Therefore, the simulated paths are obtained from the same sample of pseudo-random numbers and comparison is possible. We clearly observe that a fractional Brownian with high Hurst index yields a smoother sample path that the one obtained with a pure Brownian motion. This smoothness is induced by the memory of the fBm.

Let us denote by

$$A(s, t) = \frac{1}{a} \left(1 - e^{-a(t-s)} \right)$$

the integral of $e^{-a(t-u)}$ over the interval $[s, t]$. For any function $\beta(u, v) : \mathbb{R} \times \mathbb{R} \to \mathbb{R}$ such that

$$\int_0^t \int_0^t \left(\int_v^t \beta(s, v) \, ds \right) \left(\int_u^t \beta(s, u) \, ds \right) \phi(u, v) \, du \, dv < \infty, \tag{6.35}$$

for all $t \geq 0$, we have a fractional version of Fubini's theorem, i.e.,

$$\int_0^t \int_0^u \beta(u, v) dB_v^H \, du = \int_0^t \int_v^t \beta(u, v) \, du \, dB_v^H .$$

This integral is a normal random variable with zero mean and variance given by Eq. (6.35). Therefore, the integral of the short-term rate over the interval $[s, t]$ is rewritten as the following sum:

$$\int_s^t r_u \, du = r_s A(s, t) + b \left((t - s) - A(s, t) \right) + \int_s^t \sigma A(v, t) \, dB_v^H . \quad (6.36)$$

By construction, this integral is a normal random variable. Let us denote by $\mathcal{F}_t^H = \{ B_v^H, \, v \in [0, t] \}$ the information about the fBm on the interval of time $[0, t]$. A zero-coupon bond of maturity t delivers one monetary unit at expiry, and its price, denoted by $P(s, t)$, is calculated as the expectation under the pricing measure $\mathbb{Q}$ of the discount factor:

$$P(s, t) = \mathbb{E} \left(e^{-\int_s^t r_u \, du} \mid \mathcal{F}_s^H \right) .$$

Since increments of the fractional Brownian motion are not independent, the conditional distribution of B_u^H for $u \geq s$ with respect to $\mathcal{F}_s^H$ is required to calculate the bond price for $s > 0$. Before developing this, the next proposition prices a bond at time 0. This is a direct consequence of the log-normality of the discount factor:

Proposition 6.14 *At time zero, a zero-coupon bond of maturity t has a price equal to*

$$P(0, t) = \exp \left(-\mu(0, t) + \frac{1}{2} \sigma(0, t)^2 \right) , \quad (6.37)$$

where $\mu(0, t)$ and $\sigma(0, t)^2$ are respectively the expectation of Eq. (6.36):

$$\mu(0, t) = r_0 A(0, t) + b (t - A(0, t)) , \quad (6.38)$$

$$\sigma(0, t)^2 = H (2H - 1) \int_0^t \int_0^t \sigma^2 A(u, t) A(v, t) |u - v|^{2H - 2} \, du \, dv \quad (6.39)$$

$$= \| \sigma A(\cdot, t) \|_H^2 .$$

The numerical calculation of the norm $\| \sigma A(\cdot, t) \|_H^2 = \sigma^2 \langle A(\cdot, t), A(\cdot, t) \rangle_H$ is challenging given the integrand is not defined for $u = v$. An alternative formulation, more stable from a numerical point of view, is provided in the next proposition.

Proposition 6.15 *The norm $\| \sigma A(\cdot, t) \|_H^2$ is equal to the following double integral:*

$$\sigma^2 \langle A(\cdot, t) \mathbf{1}_{[s,t]}, A(\cdot, t) \mathbf{1}_{[s,t]} \rangle_H \quad (6.40)$$

$$= \sigma^2 H \int_s^t A(v, t) \left(\int_v^t e^{-a(t-u)} (u - v)^{2H-1} \, du \right.$$

$$\left. + A(s, t) v^{2H-1} - \int_s^v e^{-a(t-u)} (v - u)^{2H-1} \, du \right) dv .$$

Proof The double integral in Eq. (6.39) is decomposed as follows:

$$\int_s^t \int_s^t A(u,t)A(v,t)|u-v|^{2H-2} \, du \, dv$$

$$= \int_s^t A(v,t) \left(\int_v^t A(u,t) (u-v)^{2H-2} \, du \right) dv$$

$$+ \int_s^t A(v,t) \left(\int_s^v A(u,t) (v-u)^{2H-2} \, du \right) dv \, .$$

Integrating by parts allows us to rewrite the first integrand,

$$\int_v^t A(u,t) (u-v)^{2H-2} \, du = \int_v^t \frac{e^{-a(t-u)}}{2H-1} (u-v)^{2H-1} \, du \, ,$$

and the second integrand becomes

$$\int_s^v A(u,t) (v-u)^{2H-2} \, du = \frac{A(s,t)}{2H-1} (v-s)^{2H-1}$$

$$- \int_s^v \frac{e^{-a(t-u)}}{2H-1} (v-u)^{2H-1} \, du \, . \qquad \qquad \square$$

The integrand in Eq. (6.40) does not display any irregularities, and the integral can be computed numerically by discretization. An alternative approach, based on an isometry presented in Lemma 6.21, is developed at the end of this section and used in Fink et al. [7]. Figure 6.5 shows the impact of the Hurst index on the yield curve of zero-coupon bonds. A high Hurst index tends to decrease long-term yields, compared to the Brownian motion ($H = \frac{1}{2}$).

Let us recall that we denote by $\mathcal{F}_t^H = \{B_v^H \, , \, v \in [0,t]\}$ the information about the fBm on the interval of time $[0,t]$. The conditional expectation of the fBm with respect to this filtration, which is also the best predictor based on the available information, is provided in the next proposition.

Proposition 6.16 *Let $0 \le s \le t \le T$, $H \in (1/2, 1)$ and $\kappa = H - \frac{1}{2}$; then*

$$\mathbb{E}\left(B_t^H \mid \mathcal{F}_s^H \right) = B_s^H + \int_0^s g_H(s,t,u) dB_u^H \, ,$$

where for $u \in (0,s)$,

$$g_H(s,t,u) = \frac{\sin(\pi\kappa)}{\pi} u^{-\kappa} (s-u)^{-\kappa} \int_s^t \frac{z^\kappa (z-s)^\kappa}{z-u} dz$$

and for $u \in \{0, s\}$, $g_H(s,t,u) = 0$.

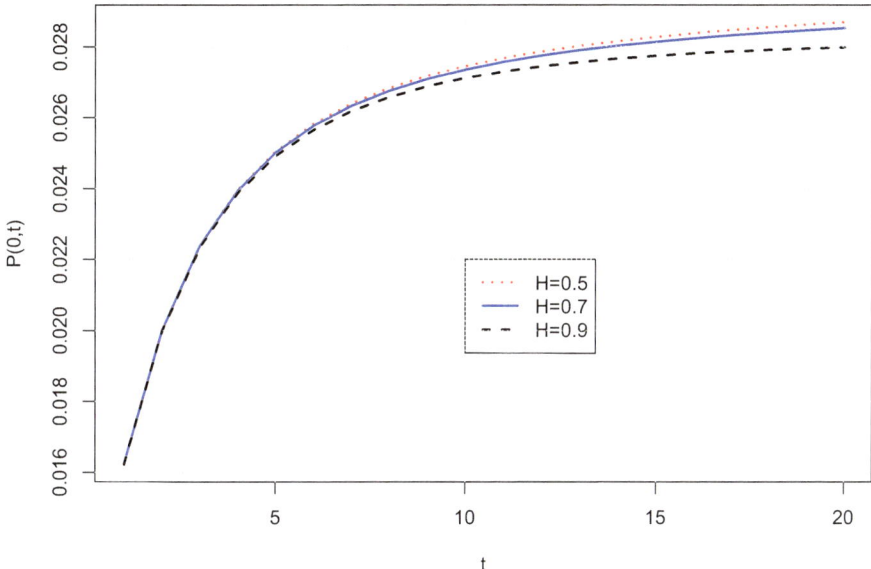

Fig. 6.5 Comparison of yield curves for different Hurst indexes: $a = 0.8$, $b = 1\%$, $\sigma = 1\%$ and $r_0 = 1\%$

The proof may be found in Gripenberg and Norros [8, Theorem 3.1] and mainly consists of finding the function g such that $\int_0^s g_H(s, t, u)\mathrm{d}B_u^H$ is orthogonal to $B_t^H - B_s^H$ with respect to the scalar product $\langle \cdot, \cdot \rangle_H$. The next corollary is a direct consequence of Proposition 6.16.

Corollary 6.17 *For $0 \leq s \leq t \leq T$, $H \in (1/2, 1)$, and a function $f \in L_\phi^2(\mathbb{R})$, we have*

$$\mathbb{E}\left(\int_0^t f(u)\mathrm{d}B_u^H \mid \mathcal{F}_s^H \right) = \int_0^s f(u)\mathrm{d}B_u^H + \int_0^s g_{f,H}(s, t, u)\mathrm{d}B_u^H , \quad (6.41)$$

where for $u \in (0, s)$ and $\kappa = H - \frac{1}{2}$,

$$g_{f,H}(s, t, u) = \frac{\sin(\pi\kappa)}{\pi}u^{-\kappa}(s - u)^{-\kappa}\int_s^t \frac{z^k(z - s)^\kappa}{z - u}f(z)\mathrm{d}z ,$$

and for $u \in \{0, s\}$, $g_{f,H}(s, t, u) = 0$.

Fink et al. [7] go a step further and prove that $\int_0^t f(u)\mathrm{d}B_u^H \mid \mathcal{F}_s^H$ is a normal random variable. To understand this, let us recall the following lemma:

Lemma 6.18 *Let Z be a multivariate normal random variable of dimension d such that $Z \sim N(\mu, \Sigma)$, where $\mu \in \mathbb{R}^d$ and $\Sigma \in \mathbb{R}^{d \times d}$. For $k \in \{1, \ldots, d - 1\}$, we* *define* $X = (Z_1, \ldots, Z_k)^\top$ *and* $Y = (Z_{k+1}, \ldots, Z_d)^\top$. *We partition the mean and covariance matrix as follows:*

$$\mu = \begin{pmatrix} \mu_1 \\ \mu_2 \end{pmatrix}, \quad \Sigma = \begin{pmatrix} \Sigma_{11} & \Sigma_{12} \\ \Sigma_{21} & \Sigma_{22} \end{pmatrix},$$

where $\mu_1 \in \mathbb{R}^k$, $\mu_2 = \mathbb{R}^{d-k}$, $\Sigma_{11} \in \mathbb{R}^k \times \mathbb{R}^k$, $\Sigma_{22} \in \mathbb{R}^{d-k} \times \mathbb{R}^{d-k}$, $\Sigma_{12}, \Sigma_{21}^\top \in \mathbb{R}^k \times \mathbb{R}^{d-k}$. *Then,*

$$X|Y = y \sim N\left(\mu_1 + \Sigma_{12}\Sigma_{22}^{-1}(y - \mu_2), \Sigma_{11} - \Sigma_{12}\Sigma_{22}^{-1}\Sigma_{21}\right).$$

The next result is Proposition 3.1 from Fink et al. [7]. Notice that Biagini et al. [2] obtain the same result using the properties of the Wick exponential.

Proposition 6.19 $\int_0^t f(u)\mathrm{d}B_u^H \mid \mathcal{F}_s^H$ *is normally distributed with an expectation given by Eq. (6.41) and a variance equal to*

$$\mathbb{V}\left(\int_0^t f(u)\mathrm{d}B_u^H \mid \mathcal{F}_s\right)$$
$$= \langle f\mathbf{1}_{[s,t]}, f\mathbf{1}_{[s,t]}\rangle_H - \langle g_{f,H}(s,t,\cdot)\mathbf{1}_{[0,s]}, g_{f,H}(s,t,\cdot)\mathbf{1}_{[0,s]}\rangle_H$$
$$= \|f\mathbf{1}_{[s,t]}\|_H^2 - \|g_{f,H}(s,t,\cdot)\mathbf{1}_{[0,s]}\|_H^2.$$

Proof Let $0 \leq s \leq t$. According to Proposition 6.6, the covariance of fBm integrals of functions $f, g \in L_\phi^2(\mathbb{R})$ is given by

$$\mathbb{E}\left(\int_0^t f(u)\mathrm{d}B_u^H \int_0^s g(u)\mathrm{d}B_u^H\right) = \langle f\mathbf{1}_{[0,t]}, g\mathbf{1}_{[0,s]}\rangle_H.$$

We choose a partition Π_n of $[0, s]$ such that $0 = s_0 \leq s_1 \leq \ldots \leq s_n = s$. According to Lemma 6.18, we have that

$$\mathbb{E}\left(\int_s^t f(u)\mathrm{d}B_u^H \mid B_{s_i}^H - B_{s_{i-1}}^H, i = 1, \ldots, n\right) = \Sigma_{12}^n \left(\Sigma_{22}^n\right)^{-1} \begin{pmatrix} \vdots \\ B_{s_i}^H - B_{s_{i-1}}^H \\ \vdots \end{pmatrix},$$

where Σ_{22}^n is the covariance matrix of fBm increments over Π_n:

$$\Sigma_{22}^n = \left(\mathbb{C}\left(B_{s_i}^H - B_{s_{i-1}}^H, B_{s_j}^H - B_{s_{j-1}}^H\right)_{i,j=1,\ldots,n}\right),$$

and Σ_{21}^n is the covariance vector between the fBm integral from s to t and fBm increments over Π_n between 0 and s:

$$\Sigma_{12}^{n\top} = \Sigma_{21}^n = \begin{pmatrix} \vdots \\ \mathbb{C}\left(\int_s^t f(u)\mathrm{d}B_u^H, \ B_{s_i}^H - B_{s_{i-1}}^H\right) \\ \vdots \end{pmatrix}.$$

According to Lemma 6.18, the variance of the conditional integral is equal to

$$\mathbb{V}\left(\int_s^t f(u)\mathrm{d}B_u^H \mid B_{s_i}^H - B_{s_{i-1}}^H, \ i = 1, \ldots, n\right)$$
$$= \Sigma_{11}^n - \Sigma_{12}^n \left(\Sigma_{22}^n\right)^{-1} \Sigma_{21}^n,$$

where Σ_{11}^n is the variance of $\int_s^t f(u)\mathrm{d}B_u^H$:

$$\Sigma_{11}^n = \mathbb{V}\left(\int_s^t f(u)\mathrm{d}B_u^H\right) = \langle f\mathbf{1}_{[s,t]}, \ f\mathbf{1}_{[s,t]}\rangle_H .$$

As $n \to \infty$, the random variable $\int_s^t f(u)\mathrm{d}B_u^H \mid B_{s_i}^H - B_{s_{i-1}}^H, \ i = 1, \ldots, n$, converges to $\int_0^t f(u)\mathrm{d}B_u^H \mid \mathcal{F}_s^H$, which is Gaussian. Therefore, from Corollary 6.17, we infer the following convergence:

$$\Sigma_{12}^n \left(\Sigma_{22}^n\right)^{-1} \begin{pmatrix} \vdots \\ B_{s_i}^H - B_{s_{i-1}}^H \\ \vdots \end{pmatrix} = \sum_{i=1}^n \left(\Sigma_{12}^n \left(\Sigma_{22}^n\right)^{-1}\right)_i \left(B_{s_i}^H - B_{s_{i-1}}^H\right)$$

$$\to \int_0^s g_{f,H}(s,t,u)\mathrm{d}B_u^H ,$$

and

$$\sum_{i=1}^n \left(\Sigma_{12}^n \left(\Sigma_{22}^n\right)^{-1}\right)_i \mathbf{1}_{[s_{i-1},s_i]}(\cdot) \to g_{f,H}(s,t,\cdot)\mathbf{1}_{[0,s]}(\cdot) ,$$

as $n \to \infty$. The next step consists in calculating the limit of the conditional variance, $\Sigma_{11}^n - \Sigma_{12}^n \left(\Sigma_{22}^n\right)^{-1} \Sigma_{21}^n$. The second term converges toward

$$\Sigma_{12}^n \left(\Sigma_{22}^n\right)^{-1} \Sigma_{21}^n = \Sigma_{12}^n \left(\Sigma_{22}^n\right)^{-1} \begin{pmatrix} \vdots \\ \mathbb{C}\left(\int_{2s}^t f(u)\mathrm{d}B_u^H, \ B_{s_i}^H - B_{s_{i-1}}^H\right) \\ \vdots \end{pmatrix}$$

$$= \sum_{i=1}^{n} \left(\Sigma_{12}^{n} \left(\Sigma_{22}^{n}\right)^{-1}\right)_{i} \left\langle f\mathbf{1}_{[s,t]}, \mathbf{1}_{[s_{i-1},s_i]}\right\rangle_{H}$$

$$= \left\langle f\mathbf{1}_{[s,t]}, \sum_{i=1}^{n} \left(\Sigma_{12}^{n} \left(\Sigma_{22}^{n}\right)^{-1}\right)_{i} \mathbf{1}_{[s_{i-1},s_i]}\right\rangle_{H}$$

$$\rightarrow \left\langle f\mathbf{1}_{[s,t]}, g_{f,H}(s,t,\cdot)\mathbf{1}_{[0,s]}\right\rangle_{H} .$$

By isometry, this scalar product may be rewritten as the following expectation:

$$\left\langle f\mathbf{1}_{[s,t]}, g_{f,H}(s,t,\cdot)\mathbf{1}_{[0,s]}\right\rangle_{H} = \mathbb{E}\left(\int_{s}^{t} f(u)\mathrm{d}B_{u}^{H} \int_{0}^{s} g_{f,H}(s,t,u)\mathrm{d}B_{u}^{H}\right)$$

$$= \mathbb{E}\left(\left(\int_{s}^{t} f(u)\mathrm{d}B_{u}^{H} - \mathbb{E}\left(\int_{s}^{t} f(u)\mathrm{d}B_{u}^{H}|\mathcal{F}_{s}^{H}\right)\right) \mathbb{E}\left(\int_{s}^{t} f(u)\mathrm{d}B_{u}^{H}|\mathcal{F}_{s}^{H}\right)\right)$$

$$+ \mathbb{E}\left(\mathbb{E}\left(\int_{s}^{t} f(u)\mathrm{d}B_{u}^{H}|\mathcal{F}_{s}^{H}\right)^{2}\right) .$$

But $\int_{0}^{s} g_{f,H}(s,t,u)\mathrm{d}B_{u}^{H}$ is orthogonal to $\int_{s}^{t} f(u)\mathrm{d}B_{u}^{H}$ with respect to the scalar product $\langle \cdot, \cdot \rangle_{H}$; therefore, the first term in this last equation is null. This implies the equality

$$\left\langle f\mathbf{1}_{[s,t]}, g_{f,H}(s,t,\cdot)\mathbf{1}_{[0,s]}\right\rangle_{H} = \mathbb{E}\left(\mathbb{E}\left(\int_{s}^{t} f(u)\mathrm{d}B_{u}^{H}|\mathcal{F}_{s}^{H}\right)^{2}\right)$$

$$= \mathbb{E}\left(\left(\int_{0}^{s} g_{f,H}(s,t,u)\mathrm{d}B_{u}^{H}\right)^{2}\right)$$

$$= \left\langle g_{f,H}(s,t,\cdot)\mathbf{1}_{[0,s]}, g_{f,H}(s,t,\cdot)\mathbf{1}_{[0,s]}\right\rangle_{H} . \qquad \square$$

Using this last proposition, we can find the price of a zero coupon at time $s > 0$.

Proposition 6.20 *At time $s > 0$, a zero-coupon bond of maturity $t > s$ has a price equal to*

$$P(s,t) = \exp\left(-\mu(s,t) + \frac{1}{2}\sigma(s,t)^{2}\right) , \qquad (6.42)$$

where $\mu(0,t)$ and $\sigma(0,t)^{2}$ are respectively the expectation of Eq. (6.36)

$$\mu(s,t) = r_{s}A(s,t) + b\left((t-s) - A(s,t)\right) + \int_{0}^{s} g_{P,H}(s,t,u)\mathrm{d}B_{u}^{H} , \qquad (6.43)$$

where, for $0 < u < s$, the function $g_{P,H}(s, t, u)$ is

$$g_{P,H}(s, t, u) = \frac{\sin(\pi\kappa)}{\pi} u^{-\kappa} (s - u)^{-\kappa} \int_s^t \frac{z^k (z - s)^k}{z - u} \sigma A(z, t) dz,$$

and

$$\sigma(s, t)^2 = \left\| \sigma A(\cdot, t) \mathbf{1}_{[s,t]} \right\|_H^2 - \left\| g_{P,H}(s, t, \cdot) \mathbf{1}_{[0,s]} \right\|_H^2. \tag{6.44}$$

Notice that the same result is obtained in Theorem 4.7 of Biagini et al. [2]. When $H = 1/2$ (Brownian motion), Eq. (6.42) still holds, but the expectation (6.43) and variance (6.44) are replaced by

$$\mu(s, t) = r_s A(s, t) + b((t - s) - A(s, t)),$$

$$\sigma(s, t)^2 = \int_s^t \sigma^2 A(u, t)^2 du.$$

The norm $\left\| \sigma A(\cdot, t) \mathbf{1}_{[s,t]} \right\|$ is computed numerically using its alternative formulation provided in Proposition 6.15. Figure 6.6 shows the function $g_{P,H}(s, t, u)$ for different values of the Hurst index. This function tends to infinity when $u \to s$ and $u \to t$ and makes unstable the numerical calculation of its norm with respect to $\phi(\cdot)$.

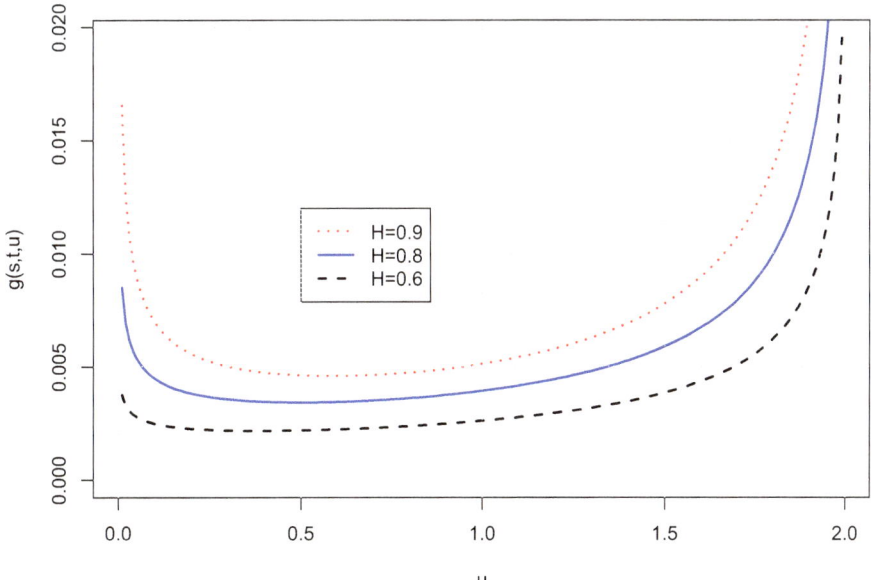

Fig. 6.6 Plot of $g_{P,H}(s, t, \cdot)$ for different Hurst indexes: $s = 2, t = 5, a = 0.8$, and $\sigma = 1\%$

For this reason, the numerical calculation of the norm $\left\|g_{P,H}(s,t,\cdot)\mathbf{1}_{[0,s]}\right\|_H^2$ is done with an isometry. We denote by $L_\phi^2([0,s])$ the space of real functions endowed with the scalar product

$$\left\langle f\mathbf{1}_{[0,s]}, g\mathbf{1}_{[0,s]}\right\rangle_H = \int_0^s \int_0^s f(u)g(v)\phi(u,v)\,du\,dv .$$

The following lemma relates the fractional Riemann–Liouville integral to the scalar product $\left\langle .\mathbf{1}_{[0,s]}, .\mathbf{1}_{[0,s]}\right\rangle_H$:

Lemma 6.21 *Let us define the fractional integral*

$$\left(I_{s-}^\alpha f\right)(v) = \frac{1}{\Gamma(\alpha)} \int_v^s (u-v)^\alpha f(u)\,du .$$

We define $\kappa = H - 1/2$ and denote by Λ_s^κ the following class of functions:

$$\Lambda_s^\kappa = \left\{ f : [0,s] \to \mathbb{R} \mid \int_0^s \left[v^{-\kappa}\left(I_{s-}^\kappa u^\kappa f(u)\right)(v)\right]^2 dv < \infty \right\} .$$

If we adopt the following notation

$$c_\kappa^2 = \frac{\pi\kappa(2\kappa+1)}{\Gamma(1-2\kappa)\sin(\pi\kappa)} ,$$

then the scalar product $(f,g)_{\Lambda_s^\kappa}$,

$$(f,g)_{\Lambda_s^\kappa} = c_\kappa^2 \int_0^s v^{-2\kappa}\left(I_{s-}^\alpha u^\kappa f(u)\right)(v)\left(I_{s-}^\alpha u^\kappa g(u)\right)(v)\,dv ,$$

is an isometry from $L_\phi^2([0,s])$ to Λ_s^κ, i.e.,

$$\left\langle f\mathbf{1}_{[0,s]}, g\mathbf{1}_{[0,s]}\right\rangle_H = (f,g)_{\Lambda_s^\kappa} .$$

We refer the reader to Pipiras and Taqqu [15] for a discussion and a proof of this isometry. We infer from this lemma that

$$\left\|g_{P,H}(s,t,\cdot)\mathbf{1}_{[0,s]}\right\|_H^2 = c_\kappa^2 \int_0^s \left[v^{-\kappa}\left(I_{s-}^\kappa u^\kappa g_{P,H}(s,t,u)\right)(v)\right]^2 dv, \quad (6.45)$$

where

$$\left(I_{s-}^\kappa u^\kappa g_{P,H}(s,t,u)\right)(v) = \frac{1}{\Gamma(\kappa)} \int_v^s (u-v)^\kappa u^\kappa g_{P,H}(s,t,u)\,du .$$

Despite this apparent complexity, the integral in this last equation can be estimated numerically by discretization and is less exposed to numerical instabilities. Details about the discretization procedure are available in Fink et al. [7].

6.7 Further Reading

For an fBm with a Hurst index $H \in (\frac{1}{2}, 1)$, there also exist a pathwise integral, as detailed in Dai and Heyde [5], and a corresponding Itô's lemma. Unlike the Wick integral, this integral does not have zero mean. Moreover, it has been shown (e.g., Cheridito [4]; Rogers [16]) that, using the pathwise integral concept, the Black–Scholes model (and related models) based on FBM is not free of arbitrage. Using the Wick integral, Hu and Oksendal [9] suggested fractional Black–Scholes models that are "free of arbitrage" in the sense induced by the use of this integral. However, the concept of self-financing portfolios is completely different from their standard counterparts and difficult to motivate from an economic point of view. In a Brownian setting, option values are solutions of a backward Itô equation, called the pricing equation. In a fractional framework, particular attention must be granted to the definition of the integral in this pricing because it may lead to serious misunderstandings. For a discussion of this point, we refer to Nualart [13] or to Rostek [17, Chapter 4].

References

1. Biagini, F., Hu, Y., Yaozhong, H., Oksendal, B., Zhang, T.: Stochastic Calculus for Fractional Brownian Motion and Applications. Springer, Berlin (2010)
2. Biagini, F., Fink, H., Klüppelberg, C.: A fractional credit model with long range dependent default rate. Stoch. Process. Appl. **123**, 1319–1347 (2013)
3. Bjork, T., Hult, H.: A note on Wick products and the fractional Black–Scholes model. Financ. Stoch. **9**(2), 197–209 (2005)
4. Cheridito, P.: Arbitrage in fractional Brownian motion models. Financ. Stoch. **7**, 533–553 (2003)
5. Dai, W., Heyde, C.C.: Itô's formula with respect to fractional Brownian motion and its application. J. Appl. Math. Stoch. Analy. **9**(4), 439–448 (1996)
6. Duncan, T.E., Hu, Y., Pasik-Duncan, B.: Stochastic calculus for fractional Brownian motion I, Theory. SIAM J. Control Optimiz. **38**, 582–612 (2000)
7. Fink, H., Klüppelberg, C., Zähle, M.: Conditional distributions of processes related to fractional Brownian motion. J. Appl. Probab. **50**, 166–183 (2013)
8. Gripenberg, G., Norros, I.: On the prediction of fractional Brownian motion. J. Appl. Probab. **33**, 400–410 (1996)
9. Hu, Y., Oksendal, B.: Fractional white noise calculus and applications to finance. Infin. Dimens. Analy. Quantum Probab. Relat. Top. **6**(1), 1–32 (2003)
10. Kolmogorov, A.: Wienersche Spiralen und einige andere interessante kurven im Hilbertschen raum. C. R. (Doklady) Acad. Sci. URSS (N.S.) **26**, 115–118 (1940)

11. Mandelbrot, B., Van Ness, J.W.: Fractional Brownian motions, fractional noises and applications. SIAM Rev. **10**(4), 422–437 (1968)
12. Mandelbrot, B.B., Wallis, J.R.: Robustness of the rescaled range R/S in the measurement of noncyclic long run statistical dependence. Water Resour. Res. **5**(5), 967–988 (1969)
13. Nualart, D.: Stochastic calculus with respect to fractional Brownian motion. Annales de la Faculté des sciences de Toulouse : Mathématiques Série **15**(1), 63–78 (2006)
14. Peters, E.E.: Fractal Market Analysis. Wiley, New York (994)
15. Pipiras, V., Taqqu, M.S.: Are classes of deterministic integrands for fractional Brownian motion on an interval complete? Bernoulli **7**(6), 873–897 (2001)
16. Rogers, L.C.G.: Arbitrage from fractional Brownian motion. Math. Financ. **7**, 95–105 (1997)
17. Rostek, S.: Option Pricing in Fractional Brownian Markets. Lecture Notes in Economics and Mathematical Systems. Springer, Berlin (2009)
18. Rostek, S., Schöbel, R.: A note on the use of fractional Brownian motion for financial modeling. Econ. Model. **30**, 30–35 (2013)

Chapter 7
Gaussian Fields for Asset Prices

The pricing of exotic options with a payoff involving asset prices at different times requires a model capable of explaining the covariance of underlying securities. Assuming that asset returns are ruled by a Brownian motion with drift is convenient for mathematical developments. However, this model does not replicate the time dependence observed for some asset classes, as underlined by Willinger et al. [22]. This point is one of the main motivations justifying the study of fractional Brownian motion (fBm) seen in Chap. 6. Gaussian fields offer a natural extension of fBm in which the marginal distribution is Gaussian with various covariance structures.

This chapter has a twofold objective. Firstly, we propose a Gaussian field model for asset prices that are analytically tractable, with three covariance structures. In a similar manner to Chap. 5, the covariance functions are characterized by their spectral distribution. Secondly, we aim to emphasize the importance of the covariance in the valuation of exotic derivatives. For this purpose, we derive closed form expressions for calendar exchange spread and Asian calendar exchange spread options. Next we numerically evaluate the bias induced by a misspecification of the covariance structure. Notice that the fractional Brownian motion introduced in Chap. 6 is nothing else than a particular type of Gaussian field.

7.1 Conditional Gaussian Fields

We consider a probability space $(\Omega, \mathcal{F}, \mathbb{Q})$ endowed with the filtration $(\mathcal{F}_t)_{t \geq 0}$ on which is defined a random field X_t indexed by time. Before detailing the features of asset prices, we recall in this section the main properties of Gaussian random fields. A *real-valued random field* is a family of random variables X_t indexed by $t \in \mathbb{R}^+$ with a collection of distribution functions of the form $F_{t_1,\dots,t_n}$,

$$F_{t_1,\dots,t_n}(b_1,\dots,b_n) = \mathbb{Q}\left(X_{t_1} \leq b_1, \dots, X_{t_n} \leq b_n\right),$$

© The Author(s), under exclusive license to Springer Nature Switzerland AG 2022
D. Hainaut, *Continuous Time Processes for Finance*,
Bocconi & Springer Series 12, https://doi.org/10.1007/978-3-031-06361-9_7

where $b_1, \ldots, b_n \in \mathbb{R}$. In this chapter, we consider Gaussian random fields for which the distribution $F_{t_1,\ldots,t_n}(\cdot)$ is a multivariate normal distribution with zero mean, $\mathbb{E}[X_t] = 0$. Note that most expectations in this chapter are calculated under the risk neutral measure $\mathbb{Q}$. For this reason, to lighten the notation, we use $\mathbb{E}(\cdot)$ in place of $\mathbb{E}^{\mathbb{Q}}(\cdot)$. This distribution is fully characterized by its covariance matrix. The covariance between X_t and X_s is denoted by $C(t, s)$. This is a non-negative definite function on $(\mathbb{R}^+ \times \mathbb{R}^+)$. Given that the mean of the Gaussian field is null, the covariance function is equal to its cross expectation:

$$C(t, s) = \mathbb{C}[X_t, X_s] = \mathbb{E}[X_t X_s] .$$

If the covariance function is equal to $C(t, s) = t \wedge s$, the Gaussian field is a pure Brownian motion with independent increments. However, in general, a Gaussian field does not have independent increments and is not a Markov process. On the other hand, a Gaussian field is not systematically a semi-martingale. If the Gaussian field is not a semi-martingale, we cannot rely on stochastic differential calculus for evaluating options. Nevertheless, working with Gaussian fields allows us to reproduce the autocovariance function of asset prices with greater accuracy than processes with independent increments. We will see later that this feature is particularly important for pricing calendar options. In this chapter, we consider continuous and differentiable covariance functions, $C(t, s) \in C^\infty$, and focus on homogeneous Gaussian random fields. A random field is *homogeneous* or *stationary* if $\mathbb{E}[X_t^2]$ is finite for all t and

1. $\mathbb{E}[X_t]$ is constant and independent of $t \in \mathbb{R}^+$.
2. $C(t, s)$ solely depends on the difference $|s - t|$.

When the Gaussian field is homogeneous, the covariance function may be rewritten as a function $g : \mathbb{R} \to \mathbb{R}^+$,

$$g(h) = C(t, t + h),$$
$$= \mathbb{C}[X_{t+h}, X_t] .$$

The construction of the function $g(\cdot)$ is detailed in the next section. For homogeneous fields, increments in the t direction are correlated. For $t, h \in \mathbb{R}^+$, the time covariance is equal to

$$\mathbb{C}[X_{t+u} - X_t, X_t] = \mathbb{C}[X_{t+u}, X_t] - \mathbb{C}[X_t, X_t] ,$$
$$= g(u) - g(0) .$$

This is a clear difference with Brownian motion, which has independent increments. Furthermore, the conditional expectation of X_s with respect to the filtration $\mathcal{F}_t$

depends upon the whole sample path of the process and not exclusively on X_t. In general, X_t is then not a Markov process.[1]

In financial applications, X_0 is known at time 0 and will determine asset prices. Therefore, we must determine the conditional distribution of X_t with respect to X_0. Given that $(X_t, X_0)^\top$ is a bivariate Gaussian distribution,

$$\begin{pmatrix} X_t \\ X_0 \end{pmatrix} \overset{d}{\sim} N\left(\begin{pmatrix} 0 \\ 0 \end{pmatrix}, \begin{pmatrix} g(0) & g(t) \\ g(t) & g(0) \end{pmatrix} \right),$$

the random variable $Z_t := X_t | X_0 = x$ is Gaussian $\overset{d}{\sim} N\left(\mu_{t|x}, \sigma_{t|x}^2 \right)$ with a mean and variance given by

$$\mu_{t|x} = 0 + \frac{g(t)}{g(0)}(x - 0),$$

$$\sigma_{t|x}^2 = g(0) - \frac{g(t)^2}{g(0)}.$$

Using the properties of the multivariate normal distribution, in the next proposition, we directly infer the covariance between Z_t and Z_s, which will be used later.

Proposition 7.1 *The joint distribution of $Z_t = X_t | X_0 = x$ and $Z_s = X_s | X_0 = x$ for $0 \le t \le s$ is a bivariate normal random variable with a mean and covariance matrix given by*

$$\begin{pmatrix} Z_s \\ Z_t \end{pmatrix} \overset{d}{\sim} N\left(\begin{pmatrix} \frac{g(s)}{g(0)}x \\ \frac{g(t)}{g(0)}x \end{pmatrix}, \begin{pmatrix} g(0) - \frac{g(s)^2}{g(0)} & g(s-t) - \frac{g(s)g(t)}{g(0)} \\ g(s-t) - \frac{g(s)g(t)}{g(0)} & g(0) - \frac{g(t)^2}{g(0)} \end{pmatrix} \right).$$

In the next section, we consider covariance functions of homogeneous fields that are decreasing functions of h such that $\lim_{h \to \infty} g(h) = 0$. From the previous proposition, we infer the asymptotic variance and covariance of Z_t:

$$\lim_{s \to \infty} \mathbb{V}[Z_s | \mathcal{F}_0] = g(0) \tag{7.1}$$

and

$$\lim_{s \to \infty} \mathbb{C}[Z_t, Z_s | \mathcal{F}_0] = 0.$$

The variance being bounded by a constant, the process Z_t reverts around its mean.

---

[1] Except for the Ornstein–Uhlenbeck process, which is a Markov process that may be reformulated as a homogeneous Gaussian field. But in general, a time Gaussian field is not a Markov process.

If X_t is a Brownian field, the covariance between X_t and X_s for $0 \leq t \leq s$ is equal to $\mathbb{C}\,[X_t, X_s] = s \wedge t = t$. Conditionally to X_0, $Z_t = X_t|X_0 = x$ and $Z_s = X_s|X_0 = x$ are distributed according to a bivariate normal distribution with mean $(x, x)^\top$ and covariance matrix $\begin{pmatrix} s & t \\ t & t \end{pmatrix}$. In this particular case, when $t \to \infty$, the variance tends to infinity and the covariance is equal to t:

$$\lim_{s \to \infty} \mathbb{V}\,[Z_s|\mathcal{F}_0] = \infty \tag{7.2}$$

$$\lim_{s \to \infty} \mathbb{C}\,[Z_t, Z_s|\mathcal{F}_0] = t\,.$$

Therefore, homogeneous fields behave very differently to Brownian motion. As their variance is asymptotically constant, homogeneous fields revert to their mean and are not adapted to model stock prices which usually have a variance increasing with the time horizon. This property makes them more suitable for the modeling of interest rates or any assets with a mean reverting yield. The proposed approach is also appropriate for commodities, mainly because their value reverts to a long run equilibrium level with a constant asymptotic variance, as illustrated in Bessembinder et al. [6]. Gibson and Schwartz [12], Cortazar and Schwartz [10], Schwartz [21], and Hainaut [14] model commodities with an Ornstein–Uhlenbeck (OU) process presenting the same features as those of Eq. (7.1).

7.2 Market Model

The filtration $(\mathcal{F}_t)_{t \geq 0}$ is generated by d independent continuous and homogeneous random fields $X_t^{(k)}$ for $k = 1$ to d. They have a null mean and the covariance function of $X_t^{(k)}$ is denoted by $g_k(h) := C_k(t, t + h)$, where $g_k : \mathbb{R} \to \mathbb{R}^+$ and $g_k(h) = g_k(-h)$. The processes $X_t^{(k)}$ are homogeneous in time, but increments of the random field in the t direction are correlated: for $t, h \in \mathbb{R}^+$, we have that $\mathbb{C}\left[X_{t+u}^{(k)} - X_t^{(k)}, X_t^{(k)}\right] = g_k(u) - g_k(0)$. Furthermore, $g_k(0) = 1$ and $X_0^{(k)} = 0$. The probability measure $\mathbb{Q}$ is here the probability measure under which the pricing of financial derivatives is done. As in the previous section, we denote by $Z_t^{(k)}$ the expectation of $X_t^{(k)}$ conditionally to $X_0^{(k)} = 0$:

$$Z_t^{(k)} := X_t^{(k)}|X_0^{(k)} = 0 \quad k = 1, \ldots, d\,.$$

The risk-free rate is assumed constant and is denoted by r. We assume that the financial market counts two securities, $S_t^{(1)}$ and $S_t^{(2)}$, with a homogeneous covariance structure. The yields[2] of these assets are denoted by $y_t^{(1)}$ and $y_t^{(2)}$. The relation between yields and prices is $S_t^{(j)} = S_0^{(j)} \exp\left(y_t^{(j)}\right)$ for $j = 1, 2$. We denote the vector of conditional Gaussian fields by $\mathbf{Z}_t = \left(Z_t^{(1)}, \ldots, Z_t^{(d)}\right)^\top$. In our model, the yields are linear functions of Gaussian fields:

$$\begin{pmatrix} y_t^{(1)} \\ y_t^{(2)} \end{pmatrix} = \begin{pmatrix} (r + c_1)t - \alpha_1(t) \\ (r + c_2)t - \alpha_2(t) \end{pmatrix} + \underbrace{\begin{pmatrix} \sigma_{11} & \sigma_{12} & \cdots & \sigma_{1d} \\ \sigma_{21} & \sigma_{22} & \cdots & \sigma_{2d} \end{pmatrix}}_{\Sigma} \mathbf{Z}_t,$$

where $\alpha_j(t)$ for $j = 1, 2$ are functions from $\mathbb{R}^+$ to $\mathbb{R}^+$ equal to

$$\alpha_j(t) := \frac{1}{2} \sum_{k=1}^d \left(\sigma_{jk}\right)^2 \left(1 - g_k(t)^2\right). \tag{7.3}$$

Σ is a $2 \times d$ real matrix.[3] c_1 and c_2 are adjustments of the drift for non-financial assets like commodities. Under the risk neutral measure, commodities earn on average the risk-free rate plus the cost of carry, less the convenience yield. If $S_t^{(1)}$ and $S_t^{(2)}$ are financial assets, these factors are null: $c_1 = c_2 = 0$. According to the usual financial theory, discounted financial security prices must be martingale processes under the pricing measure $\mathbb{Q}$. More precisely, discounted prices are such that $\mathbb{E}\left[e^{-rT} S_T^{(j)} | \mathcal{F}_t\right] = S_0^{(j)} \mathbb{E}\left[e^{-\alpha_j(T) + \Sigma_{j,:} \mathbf{Z}_T} | \mathcal{F}_t\right]$ for any times $t \leq T$. Except in the case of Brownian motion, a time Gaussian field depends in general upon its sample path in a complex manner and the process is not Markovian. Therefore, the expectation $\mathbb{E}\left[e^{\Sigma_{j,:} \mathbf{Z}_T} | \mathcal{F}_t\right]$ does not admit a closed form solution and the drift $\alpha_j(T)$ that ensures that discounted prices are always martingale is unknown.[4] However at time $t = 0$, the condition $\mathbb{E}\left[e^{-rT} S_T^{(j)} | \mathcal{F}_0\right] = S_0^{(j)}$ is fulfilled if $\alpha_j(t)$ are defined by Eq. (7.3). If $S_t^{(1)}$ and $S_t^{(2)}$ are non-financial assets, their future price is equal to $\mathbb{E}\left[S_T^{(j)} | \mathcal{F}_0\right] = S_0^{(j)} e^{(r + c_j)T}$ for $j = 1, 2$ and this relation holds only if $\alpha_j(t)$ are defined by Eq. (7.3).

---

[2] Here the *yield* $y_t^{(j)}$ is defined as the cumulative return up to time t of the jth asset.
[3] Note that Σ may eventually be replaced by a time-dependent matrix $\Sigma(t)$ in order to replicate seasonality effects in the covariance. In this framework, the pricing by simulations would still be possible.
[4] Notice that the same problem arises for models based on fractional Brownian motions which are not arbitrage-free (see, e.g., Cheridito [9] and Bender et al. [5]).

If we remember Proposition 7.1, given that $g_k(0) = 1$, the autocovariance of $y_t^{(j)}$ is

$$\mathbb{C}\left[y_t^{(j)}, y_s^{(j)}|\mathcal{F}_0\right] = \sum_{k=1}^{d} \sigma_{jk}^2 \left(g_k(s-t) - g_k(s)g_k(t)\right) \text{ for } j = 1, 2,$$

whereas the covariance of $y_t^{(1)}$ and $y_s^{(2)}$ is equal to

$$\mathbb{C}\left[y_t^{(1)}, y_s^{(2)}|\mathcal{F}_0\right] = \mathbb{E}\left[y_t^{(1)}y_s^{(2)}|\mathcal{F}_0\right] - \mathbb{E}\left[y_t^{(1)}|\mathcal{F}_0\right]\mathbb{E}\left[y_s^{(2)}|\mathcal{F}_0\right],$$

$$= \sum_{k=1}^{d} \sigma_{1k}\sigma_{2k}\left(g_k(s-t) - g_k(s)g_k(t)\right).$$

This is an important difference with existing models using Gaussian fields. For example, Goldstein [13] and Kennedy [16] develop a model for interest rates in which time increments are independent. In this particular case, the Gaussian field is a semi-martingale and we can use the Itô calculus to deduce the main properties of asset dynamics.

7.3 Choice of Autocovariance Functions

We consider three types of autocovariance functions that are built using Bochner's theorem. A function $g(h)$ is eligible as an autocovariance function if it is positive definite. We have already seen in Proposition 5.1 of Chap. 5 that a positive definite function is defined by its spectral measure. We recall this result, due to Bochner. A continuous function $g(h)$ from $\mathbb{R}^+$ to the complex plane $\mathbb{C}$ is positive definite if and only if it may be represented as the Fourier transform of a measure $\nu(\cdot)$ on $\mathbb{R}$:

$$g(h) = \int_{\mathbb{R}} e^{ihu} \nu(du),$$

where $\nu(\cdot)$ is a bounded, real-valued function such that $\int_A \nu(du) \geq 0$ for all $A \subset \mathbb{R}$. The function $\nu(\cdot)$ is called the *spectral distribution function* for g. An alternative representation is

$$g(h) = \int_{\mathbb{R}} \left(\cos(hu) + i\sin(hu)\right) \nu(du).$$

Then if Z_t is $\mathbb{R}$-valued, $g(h)$ is also $\mathbb{R}$-valued and the imaginary part in this last equation is zero: $\int_{\mathbb{R}} \sin(hu)\, v(du) = 0$. This implies that the spectral distribution is symmetric around the origin. If $v(\cdot)$ is continuous, we then have $v(du) = v(-du)$ and

$$g(h) = \int_{\mathbb{R}} \cos(hu)\, v(du),$$

$$= 2 \int_0^{\infty} \cos(|h|u)\, v(du).$$

We consider three spectral distributions $v(\cdot)$: a Laplace, a Cauchy, and a Gaussian measure. These are detailed below:

(1) The first measure considered is the *Laplace spectral measure*. Let $\lambda \in \mathbb{R}^+$ be a constant. A symmetric exponential measure is defined by the following relation:

$$v(du) = \frac{1}{2}\left(\lambda e^{-\lambda u} 1_{\{u \geq 0\}} + \lambda e^{\lambda u} 1_{\{u \leq 0\}}\right) du. \tag{7.4}$$

If we use this function as the spectral distribution of the covariance, we obtain the following function $g(h)$:

$$g(h) = \frac{1}{2}\left[\int_{\mathbb{R}^+} \lambda e^{ihu} e^{-\lambda u} du + \int_{\mathbb{R}^-} \lambda e^{ihu} e^{\lambda u} du\right]$$

$$= \frac{\lambda^2}{\lambda^2 + h^2},$$

which is a power decreasing function. In order to define autocovariance functions of Gaussian fields $Z_t^{(k)}$ involved in the dynamics of the financial market, we consider d spectral density functions (7.4) parameterized by a vector $\lambda = \{\lambda_1, \ldots, \lambda_d\} \in \mathbb{R}^{d+}$. The corresponding autocovariance functions are in this case denoted by $g_k^L(\cdot)$ and defined by

$$g_k^L(h) := \frac{(\lambda_k)^2}{(\lambda_k)^2 + h^2} \quad k = 0, 1, \ldots, d. \tag{7.5}$$

(2) The second type of spectral measure that we study is the *Cauchy spectral measure*:

$$v(du) = \frac{1}{\pi} \frac{\lambda}{\lambda^2 + u^2} du, \tag{7.6}$$

where $\lambda \in \mathbb{R}^+$. By construction this measure is symmetric and the function $g(h)$ is an exponential decreasing function:

$$g(h) = \frac{2\lambda}{\pi} \int_{\mathbb{R}^+} \frac{\cos(|h|u)}{\lambda^2 + u^2} du$$

$$= e^{-\lambda|h|}.$$

It may be shown that this autocovariance function is closely related to that of an Ornstein–Uhlenbeck process as detailed in the next proposition.

Proposition 7.2 *If the autocovariance functions of $X_t^{(k)}$ are exponential, then $\mathbb{C}\left[X_t^{(k)}, X_{t+h}^{(k)}\right] = e^{-\lambda_k|h|}$, and if $X_0^{(k)}$ are $N(0,1)$ random variables, then $X_t^{(k)}$ are Ornstein–Uhlenbeck (OU) processes that obey to the following dynamics:*

$$dX_t^{(k)} = -\lambda_k X_t^{(k)} dt + \sqrt{2\lambda_k} dW_t^{(k)} \quad k = 1, \dots, d, \tag{7.7}$$

where $W_t^{(k)}$ are d independent Brownian motions.

Proof Let us consider OU processes that are ruled by Eq. (7.7). We apply Itô's lemma to the process $U_t = -e^{\lambda_k t} X_t^{(k)}$, then

$$dU_t = -\lambda_k e^{\lambda_k t} X_t^{(k)} dt - e^{\lambda_k t} dX_t^{(k)}$$

$$= -e^{\lambda_k t} \sqrt{2\lambda_k} dW_t^{(k)},$$

and

$$U_s = U_t - \sqrt{2\lambda_k} \int_t^s e^{\lambda_k u} dW_u^{(k)}.$$

Therefore we infer that

$$X_s^{(k)} = e^{-\lambda_k(s-t)} X_t^{(k)} + \sqrt{2\lambda_k} \int_t^s e^{-\lambda_k(s-u)} dW_u^{(k)},$$

and $X_0^{(k)}$ is distributed according to a Gaussian law, with mean zero and variance equal to 1. We have that $\sqrt{2\lambda_k} \int_0^t e^{-\lambda_k u} dW_u^{(k)}$ is a normal random of null mean and variance equal to

$$2\lambda_k \int_0^t e^{-2\lambda_k u} du = 1 - e^{-2\lambda_k t},$$

and $e^{-\lambda_k t} X_0^{(k)}$ has a variance equal to $e^{-2\lambda_k t}$. The variance of $X_t^{(k)}$ is then $e^{-2\lambda_k t} + 1 - e^{-2\lambda_k t} = 1$, whereas the covariance between $X_s^{(k)}$ and $X_t^{(k)}$ is equal to

$$\mathbb{C}\left[X_s^{(k)}, X_t^{(k)}\right] = 2\lambda_k \mathbb{E}\left[\int_0^s e^{-\lambda_k(s-u)}\mathrm{d}W_u^{(k)} \int_0^t e^{-\lambda_k(t-u)}\mathrm{d}W_u^{(k)}\right] + e^{-\lambda_k(t+s)}$$

$$= 2\lambda_k e^{-\lambda_k(s+t)}\mathbb{E}\left[\left(\int_0^{s\wedge t} e^{\lambda_k u}\mathrm{d}W_u^{(k)}\right)^2 \Big| \mathcal{F}_0\right] + e^{-\lambda_k(t+s)}$$

$$= 2\lambda_k e^{-\lambda_k(s+t)}\frac{1}{2\lambda_k}\left(e^{2\lambda_k(s\wedge t)} - 1\right) + e^{-\lambda_k(t+s)}$$

$$= \left(e^{-\lambda_k|s-t|}\right).$$

$\square$

Notice that in this chapter, we set $X_0^{(k)} = 0$ and define the Gaussian field by its conditional covariance to $\mathcal{F}_0$. The covariance between $Z_t^{(k)} = X_t^{(k)}|X_0^{(k)} = 0$ and $Z_s^{(k)} = X_s^{(k)}|X_0^{(k)} = 0$ is then

$$\mathbb{C}\left[Z_t^{(k)}, Z_{t+h}^{(k)}\right] = g_k(s-t) - g_k(s)g_k(t)$$

$$= e^{-\lambda_k|s-t|} - e^{-\lambda_k(s+t)}.$$

Later, we consider d exponential autocovariance functions for $Z_t^{(k)}$ parameterized by $\lambda = \{\lambda_1, \ldots, \lambda_d\} \in \mathbb{R}^{d,+}$ and denoted by

$$g_k^C(h) := e^{-\lambda_k|h|} \quad k = 0, 1, \ldots, d. \tag{7.8}$$

(3) The last category of covariance functions is generated with a *Gaussian spectral measure*:

$$\nu(\mathrm{d}u) = \sqrt{\frac{\lambda}{\pi}}e^{-\lambda u^2}\mathrm{d}u.$$

Given that $\int_{\mathbb{R}+} e^{-\lambda u^2}\cos(hu)\,\mathrm{d}u = \frac{1}{2}\sqrt{\frac{\pi}{\lambda}}e^{-\frac{h^2}{4\lambda}}$, the autocovariance function is in this case equal to

$$g(h) = 2\sqrt{\frac{\lambda}{\pi}}\int_{\mathbb{R}+} e^{-\lambda u^2}\cos(|h|u)\,\mathrm{d}u$$

$$= e^{-\frac{h^2}{4\lambda}}.$$

We abusively refer to this covariance function as *Gaussian covariance*. This function decays at a faster pace than an exponential one and is in this sense hyper-exponential decreasing. In the following sections, we consider d Gaussian covariance functions for $Z_t^{(k)}$ parameterized by $\boldsymbol{\lambda} = \{\lambda_1, \ldots, \lambda_d\} \in \mathbb{R}^{d,+}$ and denoted as

$$g_k^G(h) = \mathrm{e}^{-\frac{h^2}{4\lambda_k}} \quad k = 0, 1, \ldots, d \,. \tag{7.9}$$

We will price calendar spread exchange options using these three autocovariance functions $g_k^L(h)$, $g_k^C(h)$, and $g_k^G(h)$. As the autocovariance functions $g_k^L(h)$ and $g_k^C(h)$ decrease, respectively, at a lower and at a faster pace than $g_k^G(h)$, our approach covers a wide spectrum of autocorrelation structures.

Options are priced under the risk neutral measure and autocovariance functions should therefore duplicate option market prices. However, if the option market is not liquid enough, an alternative solution to calibrate $g_k(\cdot)$ consists in assuming that covariances under the pricing and real measures are similar (eventually adjusted by a risk premium). Under this assumption, the functions $g_k(\cdot) \in \{g_k^L(\cdot), g_k^C(\cdot), g_k^G(\cdot)\}$ can easily be calibrated with the method of moments matching. In this case, we minimize the quadratic spread between marginal and empirical covariances denoted by $\hat{\mathbb{C}}[X_{\cdot}, X_{\cdot+\Delta h}]$:

$$\{\Sigma, \boldsymbol{\lambda}\} = \arg \min \sum_{h=0}^{H} \left(\sum_{k=1}^{d} \sigma_{1k}^2 g_k(h\Delta) - \hat{\mathbb{C}}\left[y_{\cdot}^{(1)}, y_{\cdot+h\Delta}^{(1)}\right] \right)^2 \tag{7.10}$$

$$+ \sum_{h=0}^{H} \left(\sum_{k=1}^{d} \sigma_{2k}^2 g_k(h\Delta) - \hat{\mathbb{C}}\left[y_{\cdot}^{(2)}, y_{\cdot+h\Delta}^{(2)}\right] \right)^2$$

$$+ \sum_{h=0}^{H} \left(\sum_{k=1}^{d} \sigma_{1k}\sigma_{2k} g_k(h\Delta) - \hat{\mathbb{C}}\left[y_{\cdot}^{(1)}, y_{\cdot+h\Delta}^{(2)}\right] \right)^2 \,,$$

where Δ is the time interval between two successive observations. This approach is applied in numerical illustrations in order to determine which covariance function is the most suitable for some commodity prices (silver, gold, MSE Brent, and WTI crude oil).

7.4 Simulation of a Conditional Field by Spectral Decomposition

We propose a method to simulate sample paths of asset prices. As increments of Gaussian fields involved in the dynamics of $S_t^{(1)}$ and $S_t^{(2)}$ are not independent, the simulation of these processes requires particular care. A natural approach consists in simulating the asset dynamics with a multivariate normal distribution conditioned by $X_0 = 0$, on a discrete time grid. However, there exists an elegant alternative based on a discretization in the space of frequencies. With this approach, the sample path of $S_t^{(1)}$ and $S_t^{(2)}$ is known at all times t and not only at discrete times. This feature is particularly interesting for the pricing of path-dependent options. Let us first recall that the covariance function of a homogeneous Gaussian field is defined by a measure $\nu(\cdot)$. In Sect. 7.3 we considered a (probability) measure[5] $\nu(\cdot)$ of $\mathbb{R}$. For any Borel subset $A \subset \mathbb{R}$, $\nu(A) = \int_A \nu(du)$. We defined a complex noise W based on the measure ν, or ν-*noise*, as a random process defined on Borel subsets of $\mathbb{R}$ such that for all $A, B \in \mathcal{B}$ with $\nu(A)$ and $\nu(B)$ finite, we have

$$\mathbb{E}\left[W(A)\right] = 0, \quad \mathbb{E}\left[W(A)\overline{W(A)}\right] = \nu(A),$$

$$A \cap B = \emptyset \Rightarrow W(A \cup B) = W(A) + W(B) \text{ a.s.},$$

$$A \cap B = \emptyset, \quad \mathbb{E}\left[W(A)\overline{W(B)}\right] = 0.$$

When the measure $\nu(\cdot)$ is the Lebesgue measure and $W(A) \overset{d}{\sim} N(0, \nu(A))$, W is a white noise and more precisely a Brownian motion. Having defined complex ν-noises, we build the integral, with respect to W, of a function $\varphi(t) : \mathbb{R} \to \mathbb{C}$ such that $\int \varphi(x)^2 \nu(dx) < \infty$. By analogy with the Riemann integral, the integral of $\varphi(t)$ over $\mathbb{R}$ with respect to W is defined as the limit of a sum over a partition $\Pi^n = ([t_k, t_{k+1}])_{k=0,\dots,n-1}$ of $[-T_n, T_n]$:

$$W(\varphi) := \int_{\mathbb{R}} \varphi(u)W(du) := \lim_{T_n, n \to \infty} \sum_{k=0}^{n-1} W([t_k, t_{k+1}])\varphi(t_k).$$

By construction, $\int_{\mathbb{R}} \varphi(u)W(du)$ is a random variable that has zero mean and variance given by $\int_{\mathbb{R}} \varphi(t)^2 \nu(dt)$. On the other hand, given that $\mathbb{E}\left[W(A)\overline{W(B)}\right] = 0$ if $A \cap B = \emptyset$, we have that

$$\mathbb{E}\left[W(\varphi)\overline{W(\psi)}\right] = \int_{\mathbb{R}} \varphi(t)\overline{\psi(t)}\nu(dt)$$

for any function $\psi(t) : \mathbb{R} \to \mathbb{C}$ such that $\int \psi(x)^2 \nu(dx) < \infty$. This construction allows us to state a useful representation theorem:

---

[5] Note that $\nu(\cdot)$ can be any measure on $\mathbb{R}$, not necessarily a probability measure (e.g., Lebesgue measure).

Theorem 7.3 (Spectral Representation Theorem) *Let ν be a finite measure on $\mathbb{R}$ and W a complex ν-noise. Then the complex-valued random field*

$$X_t = \int_{\mathbb{R}} e^{i(tu)} W(\mathrm{d}u) \tag{7.11}$$

has a covariance function

$$C(t, s) = g(s - t) = \int_{\mathbb{R}} e^{i((s-t)u)} \nu(\mathrm{d}u) . \tag{7.12}$$

If W is Gaussian, then so is X_t. Furthermore, to every mean-square centered (Gaussian) stationary random field on $\mathbb{R}$, with covariance function C and spectral measure, there corresponds a complex (Gaussian) ν-noise W on $\mathbb{R}$ such that Eq. (7.11) holds in mean square for each $t \in \mathbb{R}^N$. In both cases, W is called the spectral process corresponding to X_t.

In one direction, the proof of this theorem is immediate. It is a consequence of the construction of the stochastic integral $W(\varphi)$ that X_t defined by Eq. (7.11) has covariance function (7.12). The other direction is less direct, and we refer to the book of Adler and Taylor [2, Theorem 5.4.2] for a proof.

In this chapter, we consider a real Gaussian field X_t with a symmetric spectral measure. Then its covariance function can be rewritten as follows:

$$C(t, s) = \int_{\mathbb{R}} \cos((s - t)u) \, \nu(\mathrm{d}u) \tag{7.13}$$

$$= \int_{\mathbb{R}^+} \cos((s - t)u) \, \mu(\mathrm{d}u) ,$$

where we have defined a positive measure $\mu(A) = 2\nu(A)$ for all $A \subset \mathbb{R}^+$. In Sect. 7.3, we introduced three symmetric measures leading to three types of covariance. The measure $\mu(\cdot)$ for each of these cases is given by

$$\mu^L(\mathrm{d}u) = \lambda e^{-\lambda u} \mathrm{d}u ,$$

$$\mu^C(\mathrm{d}u) = \frac{2}{\pi} \frac{\lambda}{\lambda^2 + u^2} \mathrm{d}u ,$$

$$\mu^G(\mathrm{d}u) = 2\sqrt{\frac{\lambda}{\pi}} e^{-\lambda u^2} \mathrm{d}u .$$

The next proposition shows that we can reformulate the Gaussian field as integrals with respect to μ-noises.

Proposition 7.4 *A Gaussian field X_t with a covariance function (7.13) can be rewritten as the sum of two integrals with respect to independent real (Gaussian)* *valued μ-noises W_1 and W_2 such that*

$$X_t = \int_{\mathbb{R}^+} \cos(ut)\, W_1(du) + \int_{\mathbb{R}^+} \sin(ut)\, W_2(du)\,. \tag{7.14}$$

Proof Given that W_1 and W_2 are independent and that the expectation of the integral with respect to a μ-noise is null, we have the following relation

$$
\begin{aligned}
\mathbb{E}\left[X_s X_t\right] &= \mathbb{E}\left[\left(\int_{\mathbb{R}^+} \cos(us)\, W_1(du) + \int_{\mathbb{R}^+} \sin(us)\, W_2(du)\right)\right. \\
&\quad \times \left. \left(\int_{\mathbb{R}^+} \cos(ut)\, W_1(du) + \int_{\mathbb{R}^+} \sin(ut)\, W_2(du)\right)\right] \\
&= \mathbb{E}\left[\left(\int_{\mathbb{R}^+} \cos(us)\, W_1(du)\right)\left(\int_{\mathbb{R}^+} \cos(ut)\, W_1(du)\right)\right] \\
&\quad + \mathbb{E}\left[\left(\int_{\mathbb{R}^+} \sin(us)\, W_2(du)\right)\left(\int_{\mathbb{R}^+} \sin(ut)\, W_2(du)\right)\right].
\end{aligned}
$$

As $\mathbb{E}\left[W_k(\varphi)\overline{W_k(\psi)}\right] = \int_{\mathbb{R}} \varphi(t)\overline{\psi(t)}\mu(dt)$ for $k = 1, 2$, we infer that

$$
\begin{aligned}
\mathbb{E}\left[X_s X_t\right] &= \int_{\mathbb{R}^+} (\cos(us)\cos(ut) + \sin(us)\sin(ut))\,\mu(du) \\
&= \int_{\mathbb{R}^+} (\cos(us)\cos(-ut) - \sin(us)\sin(-ut))\,\mu(du) \\
&= \int_{\mathbb{R}^+} \cos(u(s-t))\,\mu(du)\,.
\end{aligned}
$$

$\square$

In practice, we use Eq. (7.14) for approximating the sample path of $X_t^{(k)}$ over a partition $\{u_0, u_1, \ldots, u_n\}$ of $\mathbb{R}^+$ (with $u_0 = 0$) as follows:

$$X_t^{(k)} \approx \sum_{j=0}^{n} \left(\cos\left(t\, u_j\right) W_1^{(k)}([u_j, u_{j+1})) + \sin\left(t\, u_j\right) W_2^{(k)}([u_j, u_{j+1}))\right), \tag{7.15}$$

where $W_1^{(k)}(\cdot)$ and $W_2^{(k)}(\cdot)$ are normal random variables:

$$W_1^{(k)}([u_j, u_{j+1})) \stackrel{d}{\sim} N\left(0\,, \mu^{(k)}\left([u_j, u_{j+1})\right)\right)\,,$$

$$W_2^{(k)}([u_j, u_{j+1})) \stackrel{d}{\sim} N\left(0\,, \mu^{(k)}\left([u_j, u_{j+1})\right)\right)\,.$$

If the measure is the Laplace measure (and the function g is power decreasing), then the measure of a time interval is equal to

$$\mu^{(k)}\left([u_j, u_{j+1})\right) = \int_{u_j}^{u_{j+1}} \lambda_k e^{-\lambda_k u} du$$

$$= e^{-\lambda_k u_j} - e^{-\lambda_k u_{j+1}}.$$

If the measure is Cauchy (and the function g is exponential) then the measure of a time interval is equal to

$$\mu^{(k)}\left([u_j, u_{j+1})\right) = \frac{2}{\pi} \int_{u_j}^{u_{j+1}} \frac{\lambda}{\lambda^2 + u^2} du$$

$$= \frac{2}{\pi}\left(\arctan\left(\frac{u_{j+1}}{\lambda}\right) - \arctan\left(\frac{u_j}{\lambda}\right)\right).$$

If the function g is associated with the Gaussian measure, then

$$\mu^{(k)}\left([u_j, u_{j+1})\right) = \int_{u_j}^{u_{j+1}} 2\sqrt{\frac{\lambda}{\pi}} e^{-\lambda u^2} du$$

$$= \mathrm{erf}\left(\sqrt{\lambda}\, u_{j+1}\right) - \mathrm{erf}\left(\sqrt{\lambda}\, u_j\right).$$

The number of intervals n and u_n is chosen such that the variance $e^{-\lambda_k u_j} - e^{-\lambda_k u_{j+1}}$ of $W_1^{(k)}$ and $W_2^{(k)}$ is close to zero. Equation (7.15) allows us to simulate a Gaussian field X_t that is not conditioned by its initial value X_0. In order to simulate $Z_t^{(k)} := X_t^{(k)}|X_0^{(k)} = 0$ for $k = 1, \ldots, d$, we need to take into account the constraint $X_0^{(k)} = 0$. With regard to Eq. (7.15), $X_0^{(k)} = 0$ if and only if the following random variable is null:

$$H := \sum_{j=0}^{n} W_1^{(k)}([u_j, u_{j+1})) = 0.$$

By construction, the random vector $\bar{W}_1^{(k)} = \left(W_1^{(k)}([u_j, u_{j+1}))\right)_{j=0,\ldots,n}$ is a multivariate normal $N\left(0_{n+1};\ \Sigma_k\right)$, where 0_{n+1} is a null vector of dimension $n + 1$ and Σ_k is an $(n + 1) \times (n + 1)$ matrix of covariance:

$$\Sigma_k := \mathrm{diag}\left(\mu^{(k)}\left([u_0, u_1)\right), \ldots, \mu^{(k)}\left([u_j, u_{j+1})\right), \ldots, \mu^{(k)}\left([u_{n-1}, u_n)\right)\right).$$

Therefore, the covariance between $W_1^{(k)}([u_j, u_{j+1}))$ and H is equal to $\mu^{(k)}\left([u_j, u_{j+1})\right)$. The variance of H is equal to $\mathbf{1}_{n+1}^{\top} \Sigma_k \mathbf{1}_{n+1}$, where $\mathbf{1}_{n+1}$ is a vector of $n + 1$ ones. The vector $\left(\bar{W}_1^{(k)}, H \right)^{\top}$ is a multivariate normal distribution with the following mean and covariance:

$$
\begin{pmatrix} \bar{W}_1^{(k)} \\ H \end{pmatrix} \overset{d}{\sim} N \left(0_{n+2} \; ; \; \begin{pmatrix} \Sigma_k & \Sigma_k 1_{n+1} \\ 1_{n+1}^{\top} \Sigma_k^{\top} & 1_{n+1}^{\top} \Sigma_k 1_{n+1} \end{pmatrix} \right).
$$

Using the properties of the conditional Gaussian multivariate distribution, we have that $\bar{B}^{(k)} := \left(\bar{W}_1^{(k)} | H = 0 \right)$ is also a multivariate normal

$$
N \left(0_{n+1} \; ; \; \Sigma_{\bar{W}_1^{(k)}|H=0} \right)
$$

with covariance matrix

$$
\Sigma_{\bar{W}_1^{(k)}|H=0} := \Sigma_k - \frac{(\Sigma_k 1_{n+1}) \left(1_{n+1}^{\top} \Sigma_k^{\top} \right)}{1_{n+1}^{\top} \Sigma_k 1_{n+1}},
$$

and where $1_{n+1}^{\top} \Sigma_k^{\top} \Sigma_k 1_{n+1}$ is an $n + 1 \times n + 1$ matrix:

$$
(\Sigma_k 1_{n+1}) \left(1_{n+1}^{\top} \Sigma_k^{\top} \right) =
$$

$$
\begin{pmatrix} \mu^{(k)} \left([u_0, u_1]\right)^2 & \cdots & \mu^{(k)} \left([u_0, u_1]\right) \mu^{(k)} \left([u_{n-1}, u_n]\right) \\ \vdots & \ddots & \vdots \\ \mu^{(k)} \left([u_0, u_1]\right) \mu^{(k)} \left([u_{n-1}, u_n]\right) \cdots & & \mu^{(k)} \left([u_{n-1}, u_n]\right)^2 \end{pmatrix}.
$$

The sample path of $Z_t^{(k)}$ over a partition $\{u_0, u_1, \ldots, u_n\}$ of $\mathbb{R}^+$ (with $u_0 = 0$) is then simulated by the following:

$$
Z_t^{(k)} \approx \sum_{j=0}^{n} \left(\cos \left(t \, u_j \right) B^{(k)} ([u_j, u_{j+1}]) + \sin \left(t \, u_j \right) W_2^{(k)} ([u_j, u_{j+1}]) \right),
$$

$$
(7.16)
$$

where $\bar{B}^{(k)} \overset{d}{\sim} N \left(0_{n+1} \; ; \; \Sigma_{\bar{W}_1^{(k)}|H=0} \right)$ and $W_2^{(k)}(\cdot)$ are normal random variables:

$$
W_2^{(k)} ([u_j, u_{j+1}]) \overset{d}{\sim} N \left(0 \, , \, \mu^{(k)} \left([u_j, u_{j+1}] \right) \right).
$$

This technique allows us to simulate sample paths of Z_t that are known for all times $t \in \mathbb{R}^+$. This method is particularly helpful for the pricing of path-dependent options.

7.5   Calendar Spread Exchange Options Pricing

One objective of this chapter is to emphasize the importance of autocorrelation in the valuation of exotic derivatives. For this purpose, we derive closed form expressions for calendar spread exchange options and, in the next section, for Asian options of the same type. Next, we will assess numerically the bias induced by a misspecified covariance structure. A *calendar spread exchange European option* is a financial derivative delivering a payoff equal to the positive difference between $\beta_1 S_T^{(1)}$ and $\beta_2 S_t^{(2)}$, at expiry T. The price of this option is equal to the expected discounted payoff under the risk neutral measure:

$$\mathbb{E}\left[e^{-rT}(\beta_1 S_T^{(1)} - \beta_2 S_t^{(2)})_+ \,|\, \mathcal{F}_0\right].$$

Gaussian fields $Z_t^{(k)}$ have Laplace, Cauchy, or Gaussian autocovariance functions: $g_k \in \{g_k^L(\cdot), g_k^C(\cdot), g_k^G(\cdot)\}$ for $k = 1, \ldots, d$. In order to obtain a closed form expression for the option price, we introduce a new probability measure, denoted by $\mathbb{Q}^{S_2}$. This measure uses $S_t^{(2)}$ as numeraire. The change of measure from $\mathbb{Q}$ to $\mathbb{Q}^{S_2}$ is defined by the following random variable:

$$\frac{d\mathbb{Q}^{S_2}}{d\mathbb{Q}} = \frac{S_t^{(2)}}{S_0^{(2)}} \frac{1}{e^{(r+c_2)t}} \tag{7.17}$$

$$= \exp\left(-\alpha_2(t) + \Sigma_{2,.} Z_t\right),$$

where $\alpha_2(t) = \frac{1}{2}\sum_{k=1}^{d}(\sigma_{2k})^2\left(1 - g_k(t)^2\right)$ and $\Sigma_{2,.}$ is the second line of the matrix Σ. This approach was used by Margrabe [18] to evaluate exchange options in a Brownian setting. By definition, $\frac{d\mathbb{Q}^{S_2}}{d\mathbb{Q}}$ is a strictly positive, $\mathcal{F}_t$-measurable random variable such that $\mathbb{E}\left[\frac{d\mathbb{Q}^{S_2}}{d\mathbb{Q}}|\mathcal{F}_0\right] = 1$. Conditionally to $\mathcal{F}_0$, it defines then an equivalent measure $\mathbb{Q}^{S_2}$ to $\mathbb{Q}$. For any $\mathcal{F}_T$-adapted random variable N_T, we have the following relation:

$$\mathbb{E}^{\mathbb{Q}^{S_2}}\left[\frac{N_T}{S_t^{(2)}} \,|\, \mathcal{F}_0\right] = \frac{\mathbb{E}\left[\frac{d\mathbb{Q}^{S_2}}{d\mathbb{Q}} \frac{N_T}{S_t^{(2)}} \,|\, \mathcal{F}_0\right]}{\mathbb{E}\left[\frac{d\mathbb{Q}^{S_2}}{d\mathbb{Q}} \,|\, \mathcal{F}_0\right]}$$

$$= \left(S_0^{(2)} e^{(r+c_2)t}\right)^{-1} \mathbb{E}\left[N_T \,|\, \mathcal{F}_0\right].$$

We immediately infer from this last expression that the expectation of N_T under $\mathbb{Q}$ is also equal to

$$\mathbb{E}\left[N_T \mid \mathcal{F}_0\right] = S_0^{(2)} e^{(r+c_2)t} \mathbb{E}^{\mathbb{Q}^{S_2}}\left[\frac{N_T}{S_t^{(2)}} \mid \mathcal{F}_0\right].$$

This result allows us to rewrite the calendar spread exchange option as follows:

$$e^{-rT}\mathbb{E}\left[(\beta_1 S_T^{(1)} - \beta_2 S_t^{(2)})_+ \mid \mathcal{F}_0\right] = \qquad (7.18)$$

$$\beta_2 S_0^{(2)} e^{-r(T-t)+c_2 t} \mathbb{E}^{\mathbb{Q}^{S_2}}\left[\left(\frac{\beta_1}{\beta_2}\frac{S_T^{(1)}}{S_t^{(2)}} - 1\right)_+ \mid \mathcal{F}_0\right].$$

The next step consists in determining the statistical distribution of $\frac{S_T^{(1)}}{S_t^{(2)}}$ under the measure $\mathbb{Q}^{S_2}$. By construction, the ratio of asset prices is given by

$$\frac{S_T^{(1)}}{S_t^{(2)}} = \frac{S_0^{(1)}}{S_0^{(2)}} \exp\left(r\,(T-t) + c_1 T - c_2 t - (\alpha_1(T) - \alpha_2(t))\right)$$

$$\times \exp\left(\Sigma_{1,.}\mathbf{Z}_T - \Sigma_{2,.}\mathbf{Z}_t\right),$$

where $\Sigma_{1,.}$ and $\Sigma_{2,.}$ indicate the first and second lines of the matrix Σ. Let us denote the logarithm of this ratio by

$$R_{t,T} = \ln\frac{S_T^{(1)} S_0^{(2)}}{S_t^{(2)} S_0^{(1)}} = r\,(T-t) + c_1 T - c_2 t - (\alpha_1(T) - \alpha_2(t))$$

$$+ \Sigma_{1,.}\mathbf{Z}_T - \Sigma_{2,.}\mathbf{Z}_t.$$

The statistical distribution of $R_{t,T}$ under the measure Q^{S_2} is Gaussian, as detailed in the next proposition:

Proposition 7.5 $R_{t,T}$ *under the measure* $\mathbb{Q}^{S_2}$ *is a normal random variable with mean* $\mu_R(t,T)$ *and variance* $\sigma_R(t,T)^2$ *equal to*

$$\mu_R(t,T) = r\,(T-t) + c_1 T - c_2 t - \alpha_1(T) - \alpha_2(t)$$

$$+ \sum_{k=1}^{d} \sigma_{1k}\sigma_{2k}\,(g_k(T-t) - g_k(T)g_k(t)),$$

$$\sigma_R^2(t,T) = 2\left(\alpha_1(T) + \alpha_2(t) - \sum_{k=1}^{d} \sigma_{1k}\sigma_{2k}\,(g_k(T-t) - g_k(T)g_k(t))\right).$$

Proof For all $\omega \in \mathbb{C}^-$, if we remember the definition (7.17) of the change of measure from $\mathbb{Q}$ to $\mathbb{Q}^{S_2}$, we can develop the moment generating function as follows:

$$\mathbb{E}^{\mathbb{Q}^{S_2}}\left[e^{\omega R_{t,T}} \mid \mathcal{F}_0\right] = e^{(\omega(r(T-t)+c_1 T - c_2 t) - \omega\alpha_1(T) + (\omega-1)\alpha_2(t))} \tag{7.19}$$

$$\times \frac{\mathbb{E}\left[e^{(\omega\Sigma_{1,.}\mathbf{Z}_T - (\omega-1)\Sigma_{2,.}\mathbf{Z}_t)} \mid \mathcal{F}_0\right]}{\mathbb{E}\left[e^{(-\alpha_2(t)+\Sigma_{2,.}\mathbf{Z}_t)} \mid \mathcal{F}_0\right]}.$$

By construction, the expectation in the denominator is equal to 1. The variances of $\Sigma_{1,.}\mathbf{Z}_T$ and $\Sigma_{2,.}\mathbf{Z}_t$ are equal to $\sum_{k=1}^{d} \sigma_{1k}^2 \left(1 - g_k(T)^2\right)$ and $\sum_{k=1}^{d} \sigma_{2k}^2 \left(1 - g_k(t)^2\right)$, respectively. Their covariance is equal to

$$\mathbb{C}\left[\Sigma_{1,.}\mathbf{Z}_T, \ \Sigma_{2,.}\mathbf{Z}_t \mid \mathcal{F}_0\right] = \sum_{k=1}^{d} \sigma_{1k}\sigma_{2k} \left(g_k(T-t) - g_k(T)g_k(t)\right).$$

Then the difference $\omega\Sigma_{1,.}\mathbf{Z}_T - (\omega - 1)\Sigma_{2,.}\mathbf{Z}_t$ is a normal random variable with null mean and variance equal to

$$(v(\omega, t, T))^2 = \omega^2 \sum_{k=1}^{d} \sigma_{1k}^2 \left(1 - g_k(T)^2\right) + (\omega - 1)^2 \sum_{k=1}^{d} \sigma_{2k}^2 \left(1 - g_k(t)^2\right)$$

$$- 2\omega (\omega - 1) \sum_{k=1}^{d} \sigma_{1k}\sigma_{2k} \left(g_k(T-t) - g_k(T)g_k(t)\right).$$

After simplification, we get that

$$(v(\omega, t, T))^2 = 2\omega^2\alpha_1(T) + 2(\omega - 1)^2 \alpha_2(t)$$

$$- 2\omega (\omega - 1) \sum_{k=1}^{d} \sigma_{1k}\sigma_{2k} \left(g_k(T-t) - g_k(T)g_k(t)\right).$$

The expectation in the numerator of Eq. (7.19) is the expectation of a log-normal random variable and the moment generating function (mgf) of $R_{t,T}$ is then equal to

$$\mathbb{E}^{\mathbb{Q}^{S_2}}\left[e^{\omega R_{t,T}} \mid \mathcal{F}_0\right] \tag{7.20}$$

$$= \exp\left((\omega (r (T - t) + c_1 T - c_2 t) - \omega\alpha_1(T) + (\omega - 1)\alpha_2(t))\right)$$

$$\times \exp\left(\frac{1}{2} (v(\omega, t, T))^2\right).$$

From this last equation, we infer that $R_{t,T}$ is a normal random variable with parameters $\mu_R(t, T)$ and $\sigma_R(t, T)$. $\qquad \square$

The exchange option price as reformulated in Eq. (7.18) is a call option on a log-normal underlying asset. As stated in the next proposition, the option price then admits a closed form expression similar to the Black & Scholes equation, and this result holds whichever the covariance function $g_k \in \{g_k^L(\cdot), g_k^C(\cdot), g_k^G(\cdot)\}$.

Proposition 7.6 *Let $\Phi(x)$ be the cumulative distribution function of a standard normal random variable. The price of a calendar spread exchange option price is given by the following expression:*

$$e^{-rT} \mathbb{E}\left[(\beta_1 S_T^{(1)} - \beta_2 S_t^{(2)})_+ \mid \mathcal{F}_0 \right] = \beta_1 S_0^{(1)} e^{c_1 T} \Phi\left(-d_1(t, T)\right) \tag{7.21}$$

$$- \beta_2 S_0^{(2)} e^{-r(T-t)+c_2 t} \Phi\left(-d_2(t, T)\right),$$

where $d_1(t, T)$ and $d_2(t, T)$ are defined as follows:

$$d_2(t, T) = \frac{\ln\left(\frac{\beta_2}{\beta_1} \frac{S_0^{(2)}}{S_0^{(1)}}\right) - \mu_R(t, T)}{\sigma_R(t, T)} \tag{7.22}$$

and

$$d_1(t, T) = d_2(t, T) - \sigma_R(t, T). \tag{7.23}$$

Proof If we rewrite the option price as an expectation under the measure Q^{S_2}, we obtain that

$$\mathbb{E}^{\mathbb{Q}^{S_2}}\left[\left(\frac{\beta_1}{\beta_2} \frac{S_0^{(1)}}{S_0^{(2)}} e^{R_{t,T}} - 1 \right)_+ \mid \mathcal{F}_0 \right] = \tag{7.24}$$

$$\frac{\beta_1}{\beta_2} \frac{S_0^{(1)}}{S_0^{(2)}} \int_{u_{\min}}^{+\infty} e^u f_{R_{t,T}}(u) du - \int_{u_{\min}}^{+\infty} f_{R_{t,T}}(u) du,$$

where $u_{\min} = \ln\left(\frac{\beta_2}{\beta_1} \frac{S_0^{(2)}}{S_0^{(1)}}\right)$. According to the definition (7.22) of d_2, the integral in the first term is equal to

$$\int_{u_{\min}}^{+\infty} f_{R_{t,T}}(u) du = \Phi\left(-d_2(t, T)\right), \tag{7.25}$$

whereas the second integral of Eq. (7.24) is developed as follows:

$$\int_{u_{\min}}^{+\infty} e^u f_{R_{t,T}}(u) du = \frac{1}{\sqrt{2\pi}\sigma_R} \int_{u_{\min}}^{+\infty} \exp\left(-\frac{1}{2} \frac{u^2 - 2u\left(\mu_R + \sigma_R^2\right) + \mu_R^2}{\sigma_R^2} \right) du.$$

If we define a new variable $v = \mu_R + \sigma_R^2$, this last equation becomes

$$\int_{u_{\min}}^{+\infty} e^u f_{R_{t,T}}(u)\,du = \frac{\exp\left(\mu_R + \frac{1}{2}\sigma_R^2\right)}{\sqrt{2\pi}\,\sigma_R} \int_{u_{\min}}^{+\infty} \exp\left(-\frac{1}{2}\frac{(u-v)^2}{\sigma_R^2}\right) du$$

$$= \exp\left(\mu_R + \frac{1}{2}\sigma_R^2\right) \Phi\left(-d_1(t,T)\right). \qquad (7.26)$$

Finally, combining equations (7.24), (7.25), and (7.26) leads to the following result:

$$\mathbb{E}^{\mathbb{Q}^{S_2}}\left[\left(\frac{\beta_1}{\beta_2}\frac{S_T^{(1)}}{S_t^{(2)}} - 1\right)_+ \Big| \mathcal{F}_0\right] =$$

$$\frac{\beta_1}{\beta_2}\frac{S_0^{(1)}}{S_0^{(2)}}\exp\left(\mu_R + \frac{1}{2}\sigma_R^2\right)\Phi\left(-d_1(t,T)\right) - \Phi\left(-d_2(t,T)\right).$$

Given that $\mu_R(t,T) - r(T-t) + c_2 t + \frac{1}{2}(\sigma_R(t,T))^2 = c_1 T$, the proof is complete.

□

In Sect. 7.7, we evaluate calendar spread exchange options for Brent against WTI crude oil and for silver against gold.

7.6 Asian Calendar Spread Exchange Options, with Geometric Average

Our approach presents the same level of tractability as a Brownian motion. To illustrate this point and to emphasize the role of the covariance structure on Asian options, we price another type of exotic derivative that depends upon the whole sample path of asset prices. We focus on Asian calendar spread options with a payoff related to geometric average returns of assets $S_t^{(1)}$ and $S_t^{(2)}$. Let us denote by $G_T^{(1)}$ and $G_t^{(2)}$ the geometric average returns of $\left(S_s^{(1)}\right)_{s=0:T}$ and $\left(S_s^{(2)}\right)_{s=0:t}$ with $t \leq T$ and defined as

$$G_T^{(1)} := e^{\frac{1}{T}\int_0^T \ln\left(\frac{S_s^{(1)}}{S_0^{(1)}}\right)ds},$$

$$G_t^{(2)} := e^{\frac{1}{t}\int_0^t \ln\left(\frac{S_s^{(2)}}{S_0^{(2)}}\right)ds}.$$

An Asian calendar spread exchange option pays the positive difference between $\beta_1 G_T^{(1)}$ and $\beta_2 G_t^{(2)}$, at expiry T. The price of this option is the expected discounted payoff under the risk neutral measure:

$$\mathbb{E}\left[e^{-rT}(\beta_1 G_T^{(1)} - \beta_2 G_t^{(2)})_+ \mid \mathcal{F}_0\right].$$

By definition, the log-return of assets is equal to

$$\ln\left(\frac{S_s^{(j)}}{S_0^{(j)}}\right) = (r + c_j)s - \alpha_j(s) + \Sigma_{j,.}\mathbf{Z}_s \quad j = 1, 2.$$

Therefore, the integral of this log-return over the interval $[0, t]$, which is equal to $\ln\left(G_t^{(j)}\right)$, is

$$\ln\left(G_t^{(j)}\right) = \frac{1}{t}\int_0^t \ln\left(\frac{S_s^{(j)}}{S_0^{(j)}}\right)\,ds \tag{7.27}$$

$$= \frac{1}{2}(r + c_j)\,t - \frac{1}{t}\int_0^t \alpha_j(s)\,ds + \frac{1}{t}\Sigma_{j,.}\int_0^t \mathbf{Z}_s\,ds \quad j = 1, 2.$$

Before any further developments, we present the properties of the integral of $Z_s^{(k)}$. This integral over $[0, T]$ is defined as the limit of a sum over a partition $\Pi^n = \left([t_j, t_{j+1}]\right)_{j=0,\ldots,n-1}$ of $[0, T]$:

$$\int_0^T Z_u^{(k)}\,du := \lim_{n\to\infty}\sum_{j=0}^{n-1} Z_{t_j}^{(k)}\left(t_{j+1} - t_j\right).$$

Given that $Z_{t_j}^{(k)}$ is a Gaussian random variable, this integral is distributed as a normal random variable of null mean under the condition that its variance exists. The variance of $\int_0^T Z_u^{(k)}\,du$ is obtained by considering the limit of the product of integrals over the partition Π^n:

$$\int_0^T Z_u^{(k)}\,du \int_0^T Z_u^{(k)}\,du = \lim_{n\to\infty}\sum_{i=0}^{n-1}\sum_{j=0}^{n-1} Z_{t_i}^{(k)} Z_{t_j}^{(k)}\left(t_{i+1} - t_i\right)\left(t_{j+1} - t_j\right).$$

As $\mathbb{E}\left[Z_t^{(k)}|\mathcal{F}_0\right] = 0$ and $Z_t^{(k)}$ for $k = 1, \ldots, d$ are independent processes, the expectation of this product with respect to $\mathcal{F}_0$ is equal to the variance of $\int_0^T Z_u^{(k)} du$:

$$\mathbb{V}\left[\int_0^T Z_u^{(k)} du \int_0^T Z_u^{(k)} du |\mathcal{F}_0\right] \tag{7.28}$$

$$= \lim_{n \to \infty} \sum_{i=0}^{n-1} \sum_{j=0}^{n-1} \mathbb{C}\left[Z_{t_i}^{(k)} Z_{t_j}^{(k)} |\mathcal{F}_0\right] (t_{i+1} - t_i)(t_{j+1} - t_j).$$

$$= \int_0^T \int_0^T (g_k(u-v) - g_k(u)g_k(v)) \, du dv.$$

This last double integral is bounded by $T^2 g(0)$ for the three covariance functions that are considered in this chapter, and the integral $\int_0^T Z_u^{(k)} du$ is a normal random variable. The next propositions detail the integrals involved in the expression (7.28) of the variance of $\int_0^T Z_u^{(k)} du$ for the three autocovariance functions $\{g_k^L(\cdot), g_k^C(\cdot), g_k^G(\cdot)\}$.

Proposition 7.7 *Let $\omega \in \mathbb{C}^-$. If $g_k(h) = g_k^L(h)$, then the double integral*

$$\int_0^t \int_0^T g_k(u-v) dv du$$

is equal to

$$\int_0^t \int_0^T g_k(u-v) dv du = t\lambda_k \tan^{-1}\left(\frac{t}{\lambda_k}\right) - \frac{1}{2}\lambda_k^2 \ln\left(1 + \frac{t^2}{(\lambda_k)^2}\right) \tag{7.29}$$

$$+ \left(\lambda_k (T-t) \tan^{-1}\left(\frac{t-T}{\lambda_k}\right) + \frac{1}{2}\lambda_k^2 \ln\left(1 + \left(\frac{t-T}{\lambda_k}\right)^2\right)\right)$$

$$- \left(\lambda_k T \tan^{-1}\left(\frac{-T}{\lambda_k}\right) + \frac{1}{2}\lambda_k^2 \ln\left(1 + \left(\frac{T}{\lambda_k}\right)^2\right)\right),$$

whereas the double integral

$$\int_0^t \int_0^T g_k(u)g_k(v) dv du$$

is given by

$$\int_0^t \int_0^T g_k(u)g_k(v) dv du = \lambda_k^2 \tan^{-1}\left(\frac{t}{\lambda_k}\right) \tan^{-1}\left(\frac{T}{\lambda_k}\right). \tag{7.30}$$

Proof For a Laplace covariance function, the double integral

$$\int_0^t \int_0^T g_k(u - v)\,dv\,du$$

is equal to

$$\int_0^t \int_0^T \frac{\lambda_k^2}{\lambda_k^2 + (u - v)^2}\,du\,dv.$$

On the other hand, the first integral is equal to

$$\int_0^T \frac{\lambda_k^2}{\lambda_k^2 + (u - v)^2}\,du = \lambda_k \tan^{-1}\left(\frac{v}{\lambda_k}\right) - \lambda_k \tan^{-1}\left(\frac{v - T}{\lambda_k}\right).$$

The integral of the arctangent function is given by

$$\int \tan^{-1}(x)\,dx = x\,\tan^{-1}(x) - \frac{1}{2}\ln(1 + x^2) + C.$$

Therefore, we infer the following result:

$$\int_0^t \int_0^T g_r(u - v)\,du\,dv$$

$$= \lambda_k \int_0^t \tan^{-1}\left(\frac{v}{\lambda_k}\right)dv - \lambda_k \int_0^t \tan^{-1}\left(\frac{v - T}{\lambda_k}\right)dv$$

$$= \lambda_k^2 \int_0^{\frac{t}{\lambda_k}} \tan^{-1}(s)\,ds - \lambda_k^2 \int_{\frac{-T}{\lambda_k}}^{\frac{t-T}{\lambda_k}} \tan^{-1}(s)\,ds.$$

Furthermore, we also have that

$$\lambda_k^2 \int_0^{\frac{t}{\lambda_k}} \tan^{-1}(s)\,ds = t\lambda_k \tan^{-1}\left(\frac{t}{\lambda_k}\right) - \frac{1}{2}\lambda_k^2 \ln\left(1 + \frac{t^2}{(\lambda_k)^2}\right)$$

and

$$\lambda_k^2 \int_{\frac{-T}{\lambda_k}}^{\frac{t-T}{\lambda_k}} \tan^{-1}(s)\,ds = \lambda_k^2 \left[x\,\tan^{-1}(x) - \frac{1}{2}\ln(1 + x^2)\right]_{x = \frac{-T}{\lambda_k}}^{x = \frac{t-T}{\lambda_k}},$$

from which we obtain the expression (7.29). On the other hand, by direct integration, we can show that

$$\int_0^t \int_0^T g_k(u)g_k(v)\mathrm{d}v\mathrm{d}u = \int_0^t \frac{\lambda_k^2}{\lambda_k^2 + v^2}\mathrm{d}v \int_0^T \frac{\lambda_k^2}{\lambda_k^2 + u^2}\mathrm{d}u$$

$$= \lambda_k^2 \arctan\left(\frac{t}{\lambda_k}\right) \arctan\left(\frac{T}{\lambda_k}\right).$$

□

From the definition (7.27) of $\ln G_t^{(j)}$, the variance of $G_t^{(j)}$ is equal to

$$\left(\sigma^{\ln G_j}(t)\right)^2 := \frac{1}{t^2}\mathbb{V}\left[\Sigma_{j,\cdot} \int_0^t \mathbf{Z}_s \mathrm{d}s \,|\, \mathcal{F}_0\right].$$

From Eq. (7.28) and Proposition 7.7, we immediately infer the next corollary:

Proposition 7.8 *If the covariance functions are Laplace,* $g_k(h) = g_k^L(h)$*, the variance of* $\ln G_t^{(j)}$*, denoted by* $\left(\sigma^{\ln G_j}(t)\right)^2$*, is equal to*

$$\left(\sigma^{\ln G_j}(t)\right)^2 := \tag{7.31}$$

$$\frac{1}{t^2}\sum_{k=1}^d \sigma_{jk}^2 \left(2t\lambda_k \tan^{-1}\left(\frac{t}{\lambda_k}\right) - (\lambda_k)^2 \ln\left(1 + \frac{t^2}{(\lambda_k)^2}\right)\right.$$

$$\left. - \left(\lambda_k \tan^{-1}\left(\frac{t}{\lambda_k}\right)\right)^2\right),$$

for $j = 1, 2$.

The drift of $\ln G_t^{(j)}$ depends upon the integral of $\alpha_j(\cdot)$. The next proposition provides the analytical integral of this function when the covariance is sub-exponential.

Proposition 7.9 *If covariance functions are Laplace,* $g_k(h) = g_k^L(h)$*, then the integral of function* $\alpha_j(\cdot)$ *is given by*

$$\frac{1}{t}\int_0^t \alpha_j(s)\mathrm{d}s = \frac{1}{2}\sum_{k=1}^d (\sigma_{jk})^2 \left(1 - \frac{\lambda_k}{2t}\tan^{-1}\left(\frac{t}{\lambda_k}\right) - \frac{\lambda_k^2}{2\left(t^2 + \lambda_k^2\right)}\right),$$

for $j = 1, 2$.

Proof By definition of $\alpha_j(t)$, we have that

$$\frac{1}{t}\int_0^t \alpha_j(s)\mathrm{d}s = \frac{1}{2}\frac{1}{t}\sum_{k=1}^{d}(\sigma_{jk})^2\left(t - \int_0^t g_k(s)^2\mathrm{d}s\right)$$

and

$$\int_0^t g_k(s)^2\mathrm{d}s = \int_0^t \frac{(\lambda_k)^4}{\left((\lambda_k)^2 + s^2\right)^2}\mathrm{d}s$$

$$= \frac{\lambda_k}{2}\tan^{-1}\left(\frac{t}{\lambda_k}\right) + \frac{\lambda_k^2 t}{2\left(t^2 + \lambda_k^2\right)}.$$

This completes the proof.   □

If the covariance function is Cauchy, then the double integrals needed to evaluate the variance of $\int_0^t \mathbf{Z}_s \mathrm{d}s$ in Eq. (7.28) are given by the following proposition.

Proposition 7.10 *Let $\omega \in \mathbb{C}^-$. If $g_k(h) = g_k^C(h)$, then the double integral $\int_0^t \int_0^T g_k(u - v)\mathrm{d}v\mathrm{d}u$ is equal to*

$$\int_0^t \int_0^T g_k(u - v)\mathrm{d}v\mathrm{d}u = \tag{7.32}$$

$$\frac{2t}{\lambda_k} - \frac{1}{\lambda_k^2}\left(1 - e^{-\lambda_k t}\right) + \frac{1}{\lambda_k^2}\left(e^{-\lambda_k T} - e^{\lambda_k(t-T)}\right),$$

whereas the double integral $\int_0^t \int_0^T g_k(u)g_k(v)\mathrm{d}v\mathrm{d}u$ is given by

$$\int_0^t \int_0^T g_k(u)g_k(v)\mathrm{d}v\mathrm{d}u = \frac{1}{\lambda_k^2}\left(1 - e^{-\lambda_k t}\right)\left(1 - e^{-\lambda_k T}\right). \tag{7.33}$$

Proof The double integral $\int_0^t \int_0^T g_k(u - v)\mathrm{d}v\mathrm{d}u$ is developed as follows:

$$\int_0^t \int_0^T g_k(u - v)\mathrm{d}v\mathrm{d}u = \int_0^t \int_0^T e^{-\lambda_k|u-v|}\mathrm{d}v\mathrm{d}u \tag{7.34}$$

$$= \int_0^t \int_0^t e^{-\lambda_k|u-v|}\mathrm{d}v\mathrm{d}u + \int_0^t \int_t^T e^{-\lambda_k|u-v|}\mathrm{d}v\mathrm{d}u.$$

On the other hand, the first integral is equal to

$$\int_0^t \int_0^t e^{-\lambda_k|u-v|}\mathrm{d}v\mathrm{d}u = 2\left(\frac{t}{\lambda} - \frac{1}{\lambda^2}\left(1 - e^{-\lambda t}\right)\right), \tag{7.35}$$

and the second integral is equal to

$$\int_0^t \int_t^T e^{-\lambda_k |u-v|} dv du = \int_0^t \int_t^T e^{\lambda_k (u-v)} dv du \tag{7.36}$$

$$= \frac{1}{\lambda_k^2} \left(1 - e^{-\lambda_k t} - e^{\lambda_k (t-T)} + e^{-\lambda_k T} \right).$$

Combining equations (7.34), (7.35), and (7.36) allows us to obtain Eq. (7.32). Furthermore, we deduce by direct integration that

$$\int_0^t \int_0^T g_k(u) g_k(v) dv du = \int_0^t e^{-\lambda_k v} dv \int_0^T e^{-\lambda_k u} du$$

$$= \frac{1}{\lambda_k^2} \left(1 - e^{-\lambda_k t} \right) \left(1 - e^{-\lambda_k T} \right).$$

$$\square$$

A direct corollary of this proposition is that the variances of $\ln G_t^{(j)}$ for $j = 1, 2$ admit a closed form expression.

Proposition 7.11 *If the autocovariance functions are Cauchy, $g_k(h) = g^C(h)$, the variance of $\ln G_t^{(j)}$, denoted by $\left(\sigma^{\ln G_j}(t) \right)^2$, is equal to*

$$\left(\sigma^{\ln G_j}(t) \right)^2 := \tag{7.37}$$

$$\frac{1}{t^2} \sum_{k=1}^d \sigma_{jk}^2 \left[2 \left(\frac{t}{\lambda_k} - \frac{1}{\lambda_k^2} \left(1 - e^{-\lambda_k t} \right) \right) - \frac{1}{\lambda_k^2} \left(1 - e^{-\lambda_k t} \right)^2 \right],$$

for $j = 1, 2$.

The drift of $\ln G_t^{(j)}$ depends upon the integral of $\alpha_j(\cdot)$, which is provided in the next proposition.

Proposition 7.12 *If the autocovariance functions are Cauchy, $g_k(h) = e^{-\lambda_k h}$, then*

$$\frac{1}{t} \int_0^t \alpha_j(s) ds = \frac{1}{2} \frac{1}{t} \sum_{k=1}^d (\sigma_{jk})^2 \left(t - \frac{1}{2\lambda_k} \left(1 - e^{-2\lambda_k t} \right) \right),$$

for $j = 1, 2$.

When the covariance functions are Gaussian, the integrals involved in the variance of $\ln G_t^{(j)}$ are given in the following proposition:

Proposition 7.13 *Let $\omega \in \mathbb{C}^-$ and $\mathrm{erf}(x) = \frac{2}{\sqrt{\pi}} \int_0^x e^{-t^2} dt$ be the error function. If $g_k(h) = g_k^G(h)$, then*

$$\int_0^t \int_0^T g_k(u-v)dvdu = \sqrt{\pi\lambda_k}\, t\, \mathrm{erf}\left(\frac{t}{2\sqrt{\lambda_k}}\right) - 2\lambda_k\left(1 - e^{-\frac{t^2}{4\lambda_k}}\right) \tag{7.38}$$

$$-\sqrt{\pi\lambda_k}(T-t)\mathrm{erf}\left(\frac{T-t}{2\sqrt{\lambda_k}}\right) + \sqrt{\pi\lambda_k}\,T\,\mathrm{erf}\left(\frac{T}{2\sqrt{\lambda_k}}\right)$$

$$-2\lambda_k\left(e^{-\frac{(t-T)^2}{4\lambda_k}} - e^{-\frac{T^2}{4\lambda_k}}\right),$$

and

$$\int_0^t \int_0^T g_k(u)g_k(v)dvdu = \pi\lambda_k\mathrm{erf}\left(\frac{t}{2\sqrt{\lambda_k}}\right)\mathrm{erf}\left(\frac{T}{2\sqrt{\lambda_k}}\right). \tag{7.39}$$

Proof The double integral $\int_0^t \int_0^T g_k(u-v)dvdu$ is equal to

$$\int_0^t \int_0^T g_k(u-v)dvdu = \int_0^t \int_0^T e^{-\frac{1}{4\lambda_k}(u-v)^2}dvdu.$$

We may check that the inner integral is given by

$$\int_0^T e^{-\frac{1}{4\lambda_k}(u-v)^2}dv = \sqrt{\pi\lambda_k}\left(\mathrm{erf}\left(\frac{u}{2\sqrt{\lambda_k}}\right) - \mathrm{erf}\left(\frac{u-T}{2\sqrt{\lambda_k}}\right)\right). \tag{7.40}$$

Furthermore, the integral of the first error function in this last equation is equal to

$$\int_0^t \mathrm{erf}\left(\frac{u}{2\sqrt{\lambda_k}}\right)du = t\,\mathrm{erf}\left(\frac{t}{2\sqrt{\lambda_k}}\right) - 2\frac{\sqrt{\lambda_k}}{\sqrt{\pi}}\left(1 - e^{-\frac{t^2}{4\lambda_k}}\right). \tag{7.41}$$

The second integral present in Eq. (7.40) is equal to

$$\int_0^t \mathrm{erf}\left(\frac{u-T}{2\sqrt{\lambda_k}}\right)dx = (T-t)\mathrm{erf}\left(\frac{T-t}{2\sqrt{\lambda_k}}\right) - T\,\mathrm{erf}\left(\frac{T}{2\sqrt{\lambda_k}}\right) \tag{7.42}$$

$$+\frac{\sqrt{\lambda_k}}{\sqrt{\pi}}2\left(e^{-\frac{(t-T)^2}{4\lambda_k}} - e^{-\frac{T^2}{4\lambda_k}}\right).$$

Combining equations (7.40), (7.41), and (7.42) allows us to obtain Eq. (7.38). As we have that

$$
\int_0^t \int_0^T g_k(u)g_k(v)\mathrm{d}v\mathrm{d}u = \int_0^t \mathrm{e}^{-\frac{v^2}{4\lambda_k}}\,\mathrm{d}v \int_0^T \mathrm{e}^{-\frac{u^2}{4\lambda_k}}\,\mathrm{d}u
$$

$$
= \pi\lambda_k \mathrm{erf}\left(\frac{t}{2\sqrt{\lambda_k}}\right)\mathrm{erf}\left(\frac{T}{2\sqrt{\lambda_k}}\right),
$$

the proof is complete.   □

When the covariance functions are generated by a Gaussian measure, the variances of $\ln G_t^{(j)}$ for $j = 1, 2$ are given by the next proposition:

Proposition 7.14 *If the autocovariance functions $g_k(h)$ are Gaussian and defined by Eq. (7.9), then the variance of $\ln G_t^{(j)}$, denoted by $\left(\sigma^{\ln G_j}(t)\right)^2$, is equal to*

$$
\left(\sigma^{\ln G_j}(t)\right)^2 := \frac{1}{t^2}\sum_{k=1}^{d}\sigma_{jk}^2\left[2\sqrt{\lambda_k\pi}\,t\,\mathrm{erf}\left(\frac{t}{2\sqrt{\lambda_k}}\right)\right. \tag{7.43}
$$

$$
\left.-4\lambda_k\left(1-\mathrm{e}^{-\frac{t^2}{4\lambda_k}}\right)-\left(\sqrt{\lambda_k\pi}\,\mathrm{erf}\left(\frac{t}{2\sqrt{\lambda_k}}\right)\right)^2\right] for \; j = 1, 2.
$$

The drift of $\ln G_t^{(j)}$ depends upon the integral of $\alpha_j(\cdot)$. The next proposition provides the analytical integral of this function when the covariance is Gaussian.

Proposition 7.15 *If the autocovariance functions are Gaussian and defined by Eq. (7.9), $g_k(h) = \mathrm{e}^{-\frac{h^2}{4\lambda_k}}$, then*

$$
\frac{1}{t}\int_0^t \alpha_j(s)\mathrm{d}s = \frac{1}{2}\frac{1}{t}\sum_{k=1}^{d}(\sigma_{jk})^2\left(t - \frac{\sqrt{\pi\lambda_k}}{\sqrt{2}}\mathrm{erf}\left(\frac{t}{\sqrt{2\lambda_k}}\right)\right),
$$

for $j = 1, 2$.

Proof By definition of $\alpha_j(s)$, we have that

$$
\frac{1}{t}\int_0^t \alpha_j(s)\mathrm{d}s = \frac{1}{2}\frac{1}{t}\sum_{k=1}^{d}(\sigma_{jk})^2\left(t - \int_0^t \mathrm{e}^{-\frac{h^2}{2\lambda_k}}\mathrm{d}s\right).
$$

Given that

$$
\int_0^t \mathrm{e}^{-\frac{h^2}{2\lambda_k}}\,\mathrm{d}s = \frac{\sqrt{\pi\lambda_k}}{\sqrt{2}}\mathrm{erf}\left(\frac{t}{\sqrt{2\lambda_k}}\right),
$$

the proof is complete.   □

In order to price the Asian calendar spread exchange option, we introduce a new probability measure, denoted by $\mathbb{Q}^{G_2}$. The change of measure from $\mathbb{Q}$ to $\mathbb{Q}^{G_2}$ is defined by the following random variable:

$$\frac{\mathrm{d}\mathbb{Q}^{G_2}}{\mathrm{d}\mathbb{Q}} = \frac{G_t^{(2)}}{\mathbb{E}\left[G_t^{(2)}|\mathcal{F}_0\right]} \tag{7.44}$$

$$= \exp\left(-\frac{1}{2}\left(\sigma^{\ln G_2}(t)\right)^2 + \frac{1}{t}\Sigma_{2,.}\int_0^t Z_s \mathrm{d}s\right).$$

By definition, $\frac{\mathrm{d}\mathbb{Q}^{G_2}}{\mathrm{d}\mathbb{Q}}$ is a strictly positive random variable, $\mathcal{F}_t$-adapted, and such that $\mathbb{E}\left(\frac{\mathrm{d}\mathbb{Q}^{G_2}}{\mathrm{d}\mathbb{Q}}|\mathcal{F}_0\right) = 1$. It defines then an equivalent measure $\mathbb{Q}^{G_2}$ to $\mathbb{Q}$. For any $\mathcal{F}_T$-adapted random variable N_T, the following relation holds:

$$\mathbb{E}^{\mathbb{Q}^{G_2}}\left[\frac{N_T}{G_t^{(2)}}|\mathcal{F}_0\right] = \mathbb{E}\left(\frac{G_t^{(2)}}{\mathbb{E}\left[G_t^{(2)}|\mathcal{F}_0\right]}\frac{N_T}{G_t^{(2)}}|\mathcal{F}_0\right)$$

$$= \left(\mathbb{E}\left[G_t^{(2)}|\mathcal{F}_0\right]\right)^{-1}\mathbb{E}\left[N_T|\mathcal{F}_0\right].$$

The expectation of N_T under the risk neutral measure is then equal to

$$\mathbb{E}\left[N_T|\mathcal{F}_0\right] = \mathbb{E}\left[G_t^{(2)}|\mathcal{F}_0\right]\mathbb{E}^{\mathbb{Q}^{G_2}}\left[\frac{N_T}{G_t^{(2)}}|\mathcal{F}_0\right],$$

where $\mathbb{E}\left[G_t^{(2)}|\mathcal{F}_0\right] = \exp\left(\frac{1}{2}(r + c_2)t - \frac{1}{t}\int_0^t \alpha_2(s)\mathrm{d}s + \frac{1}{2}\left(\sigma^{\ln G_2}(t)\right)^2\right)$ given that $G_t^{(2)}$ is a log-normal random variable. Using this change of measure allows us to write the option price as follows:

$$e^{-rT}\mathbb{E}\left[(\beta_1 G_T^{(1)} - \beta_2 G_t^{(2)})_+|\mathcal{F}_0\right]$$

$$= \beta_2 e^{-rT}\exp\left(\frac{1}{2}(r + c_2)t - \frac{1}{t}\int_0^t \alpha_2(s)\mathrm{d}s + \frac{1}{2}\left(\sigma^{\ln G_2}(t)\right)^2\right)$$

$$\times \mathbb{E}^{\mathbb{Q}^{G_2}}\left[\left(\frac{\beta_1}{\beta_2}\frac{G_T^{(1)}}{G_t^{(2)}} - 1\right)_+|\mathcal{F}_0\right].$$

Next, we study the moment generating function of the ratio $\frac{G_T^{(1)}}{G_t^{(2)}}$. If we remember the definitions of $G_T^{(1)}$ and $G_t^{(2)}$, this ratio may be written as follows:

$$\frac{G_T^{(1)}}{G_t^{(2)}} = \exp\left(\frac{1}{T}\int_0^T \ln\left(\frac{S_s^{(1)}}{S_0^{(1)}}\right)ds - \frac{1}{t}\int_0^t \ln\left(\frac{S_s^{(2)}}{S_0^{(2)}}\right)ds\right)$$

$$= \exp\left(A_{t,T}\right),$$

where the exponent in this last expression is a random variable denoted by $A_{t,T}$ and defined as follows:

$$A_{t,T} := \frac{1}{2}r\,(T-t) + \frac{1}{2}(c_1 T - c_2 t) - \frac{1}{T}\int_0^T \alpha_1(s)ds \tag{7.45}$$

$$+\frac{1}{t}\int_0^t \alpha_2(s)ds + \frac{1}{T}\Sigma_{1,.}\int_0^T \mathbf{Z}_s ds - \frac{1}{t}\Sigma_{2,.}\int_0^t \mathbf{Z}_s ds\,.$$

The next result states that $A_{t,T}$ is a normal random variable under $\mathbb{Q}^{G_2}$.

Proposition 7.16 *The random variable $A_{t,T}$ defined in Eq. (7.45) is a normal random variable with mean and variance, respectively, equal to*

$$\mu_A(t,T) = \frac{1}{2}r\,(T-t) + \frac{1}{2}(c_1 T - c_2 t) - \frac{1}{T}\int_0^T \alpha_1(s)ds \tag{7.46}$$

$$+\frac{1}{t}\int_0^t \alpha_2(s)ds - \left(\sigma^{\ln G_2}(t)\right)^2$$

$$+\frac{1}{T\,t}\sum_{k=1}^d \sigma_{1k}\sigma_{2k}\int_0^t\int_0^T g_k(u-v) - g_k(u)g_k(v)\,dvdu\,,$$

and

$$\sigma_A^2(t,T) = \left(\sigma^{\ln G_1}(T)\right)^2 + \left(\sigma^{\ln G_2}(t)\right)^2 \tag{7.47}$$

$$-\frac{2}{T\,t}\times\sum_{k=1}^d \sigma_{1k}\sigma_{2k}\int_0^t\int_0^T g_k(u-v) - g_k(u)g_k(v)\,dvdu\,.$$

Proof Let $\omega \in \mathbb{C}^-$. By definition of the change of measure (7.17), the moment generating function (mgf) of $\frac{G_T^{(1)}}{G_t^{(2)}}$ is developed as follows:

$$\mathbb{E}^{\mathbb{Q}^{G_2}}\left[e^{\omega A_{t,T}} \mid \mathcal{F}_0\right] \tag{7.48}$$

$$= e^{\left(\omega \frac{1}{2} r (T-t) + \omega \frac{1}{2}(c_1 T - c_2 t) - \frac{\omega}{T}\int_0^T \alpha_1(s)\mathrm{d}s + \frac{\omega}{t}\int_0^t \alpha_2(s)\mathrm{d}s\right)}$$

$$\times \mathbb{E}^{\mathbb{Q}^{G_2}}\left[e^{\omega\left(\frac{1}{T}\Sigma_{1,\cdot}\int_0^T \mathbf{Z}_s \mathrm{d}s - \frac{1}{t}\Sigma_{2,\cdot}\int_0^t \mathbf{Z}_s \mathrm{d}s\right)} \mid \mathcal{F}_0\right]$$

$$= e^{\left(\omega \frac{1}{2} r (T-t) + \omega \frac{1}{2}(c_1 T - c_2 t) - \frac{\omega}{T}\int_0^T \alpha_1(s)\mathrm{d}s + \frac{\omega}{t}\int_0^t \alpha_2(s)\mathrm{d}s - \frac{1}{2}\left(\sigma^{\ln G_2}(t)\right)^2\right)}$$

$$\times \mathbb{E}\left[e^{\left(\omega \frac{1}{T}\Sigma_{1,\cdot}\int_0^T \mathbf{Z}_s \mathrm{d}s - (\omega-1)\frac{1}{t}\Sigma_{2,\cdot}\int_0^t \mathbf{Z}_s \mathrm{d}s\right)} \mid \mathcal{F}_0\right].$$

The variances of $\frac{1}{T}\Sigma_{1,\cdot}\int_0^T \mathbf{Z}_s \mathrm{d}s$ and $\frac{1}{t}\Sigma_{2,\cdot}\int_0^t \mathbf{Z}_s \mathrm{d}s$ are $\left(\sigma^{\ln G_1}(T)\right)^2$ and $\left(\sigma^{\ln G_2}(t)\right)^2$, respectively, as defined by Eq. (7.31). The weighted sum of random field integrals has a covariance equal to

$$\mathbb{C}\left[\frac{1}{T}\Sigma_{1,\cdot}\int_0^T \mathbf{Z}_s \mathrm{d}s \,,\, \frac{1}{t}\Sigma_{2,\cdot}\int_0^t \mathbf{Z}_s \mathrm{d}s \mid \mathcal{F}_0\right]$$

$$= \frac{1}{T\,t}\mathbb{C}\left[\Sigma_{1,\cdot}\int_0^T \mathbf{Z}_s \mathrm{d}s \,,\, \Sigma_{2,\cdot}\int_0^t \mathbf{Z}_s \mathrm{d}s \mid \mathcal{F}_0\right]$$

$$= \frac{1}{T\,t}\sum_{k=1}^d \sigma_{1k}\sigma_{2k}\int_0^t \int_0^T \left(g_k(u-v) - g_k(u)g_k(v)\right)\mathrm{d}v\mathrm{d}u \,.$$

The random variable $\omega \frac{1}{T}\Sigma_{1,\cdot}\int_0^T \mathbf{Z}_s \mathrm{d}s - (\omega-1)\frac{1}{t}\Sigma_{2,\cdot}\int_0^t \mathbf{Z}_s \mathrm{d}s$ is then normal, with null mean and variance equal to

$$(v(\omega))^2 = \omega^2 \left(\sigma^{\ln G_1}(T)\right)^2 + (\omega-1)^2 \left(\sigma^{\ln G_2}(t)\right)^2 - \frac{2\omega(\omega-1)}{T\,t}$$

$$\times \sum_{k=1}^d \sigma_{1k}\sigma_{2k}\int_0^t \int_0^T g_k(u-v) - g_k(u)g_k(v)\,\mathrm{d}v\mathrm{d}u \,,$$

and we can conclude that the mgf of $A_{t,T}$ with $t \leq T$ under the measure $\mathbb{Q}^{G_2}$ is

$$\mathbb{E}^{\mathbb{Q}^{G_2}}\left[e^{\omega A_{t,T}} \mid \mathcal{F}_0\right] = \exp\left(\omega\frac{1}{2}r\ (T-t) + \omega\frac{1}{2}\ (c_1T - c_2t)\right) \qquad (7.49)$$

$$\exp\left(-\frac{\omega}{T}\int_0^T \alpha_1(s)ds + \frac{\omega}{t}\int_0^t \alpha_2(s)ds\right)$$

$$\times \exp\left(-\frac{1}{2}\left(\sigma^{\ln G_2}(t)\right)^2 + \frac{1}{2}(v(\omega))^2\right).$$

After simplification, we obtain Eqs. (7.46) and (7.47). $\qquad\qquad\qquad\qquad\square$

Finally, we infer a closed form expression for the Asian exchange calendar option.

Proposition 7.17 *Let $\Phi(x)$ be the cumulative distribution function of a standard normal random variable. The value of an Asian exchange calendar option is given by the following expression:*

$$e^{-rT}\mathbb{E}\left[(\beta_1 G_T^{(1)} - \beta_2 G_t^{(2)})_+ \mid \mathcal{F}_0\right] \qquad (7.50)$$

$$= \beta_2 e^{-rT}\exp\left(\frac{1}{2}\ (r+c_2)\,t - \frac{1}{t}\int_0^t \alpha_2(s)ds + \frac{1}{2}\left(\sigma^{\ln G_2}(t)\right)^2\right) \times$$

$$\left(\frac{\beta_1}{\beta_2}\exp\left(\mu_A(t,T) + \frac{1}{2}\sigma_A^2(t,T)\right)\Phi\left(-d_1(t,T)\right) - \Phi\left(-d_2(t,T)\right)\right),$$

where

$$d_2(t,T) = \frac{\ln\left(\frac{\beta_2}{\beta_1}\right) - \mu_A(t,T)}{\sigma_A(t,T)}, \qquad (7.51)$$

and

$$d_1(t,T) = d_2(t,T) - \sigma_A(t,T). \qquad (7.52)$$

Proof Using a change of numeraire allows us to rewrite the option price as follows:

$$e^{-rT}\mathbb{E}\left[(\beta_1 G_T^{(1)} - \beta_2 G_t^{(2)})_+ \mid \mathcal{F}_0\right]$$

$$= \beta_2 e^{-rT}\exp\left(\frac{1}{2}\ (r+c_2)\,t - \frac{1}{t}\int_0^t \alpha_2(s)ds + \frac{1}{2}\left(\sigma^{\ln G_2}(t)\right)^2\right)$$

$$\times \mathbb{E}^{\mathbb{Q}^{G_2}}\left[\left(\frac{\beta_1}{\beta_2}\frac{G_T^{(1)}}{G_t^{(2)}} - 1\right)_+ \mid \mathcal{F}_0\right].$$

If $f_A(u)$ denotes the density of $A_{t,T}$ under the measure $\mathbb{Q}^{G_2}$, the expectation in this last equation is given by

$$\mathbb{E}^{\mathbb{Q}^{G_2}} \left[\left(\frac{\beta_1}{\beta_2} \frac{G_T^{(1)}}{G_t^{(2)}} - 1 \right)_+ \mid \mathcal{F}_0 \right] = \int_{\ln\left(\frac{\beta_2}{\beta_1}\right)}^{\infty} \left(\frac{\beta_1}{\beta_2} e^u - 1 \right) f_A(u) du . \quad (7.53)$$

As $A_{t,T}$ is normal, and by definition of $d_2(t, T)$, the second integral is equal to

$$\int_{\ln\left(\frac{\beta_2}{\beta_1}\right)}^{\infty} f_A(u) du = \Phi\left(-d_2(t, T)\right) , \quad (7.54)$$

whereas the first integral is rewritten after a change of variable as follows:

$$\frac{\beta_1}{\beta_2} \int_{\ln\left(\frac{\beta_2}{\beta_1}\right)}^{\infty} e^u f_A(u) du . = \frac{\beta_1}{\beta_2} \exp\left(\mu_A + \frac{1}{2} \sigma_A^2 \right) \Phi\left(-d_1(t, T)\right) . \quad (7.55)$$

Finally,

$$\mathbb{E}^{\mathbb{Q}^{G_2}} \left[\left(\frac{\beta_1}{\beta_2} e^{A_{t,T}} - 1 \right)_+ \mid \mathcal{F}_0 \right] =$$
$$\frac{\beta_1}{\beta_2} \exp\left(\mu_A + \frac{1}{2} \sigma_A^2 \right) \Phi\left(-d_1(t, T)\right) - \Phi\left(-d_2(t, T)\right) .$$

Combining equations (7.54), (7.55), and (7.53) leads to the following expression for the Asian exchange calendar option:

$$e^{-rT} \mathbb{E} \left[(\beta_1 G_T^{(1)} - \beta_2 G_t^{(2)})_+ \mid \mathcal{F}_0 \right]$$
$$= \beta_2 e^{-rT} \exp\left(\frac{1}{2}(r + c_2) t - \frac{1}{t} \int_0^t \alpha_2(s) ds + \frac{1}{2} \left(\sigma^{\ln G_2}(t) \right)^2 \right)$$
$$\times \left(\frac{\beta_1}{\beta_2} \exp\left(\mu_A + \frac{1}{2} \sigma_A^2 \right) \Phi\left(-d_1(t, T)\right) - \Phi\left(-d_2(t, T)\right) \right) .$$

□

In the next section, we compare Asian option values obtained with the different covariance structures fitted to some metals and oil prices.

7.7   Numerical Illustration

The market of exotic options lacks liquidity and the absence of data prevents us from estimating models by replicating market prices. An alternative solution for calibrating functions $g_k(\cdot)$ consists in assuming that the covariance structures under the pricing and real measures are similar (eventually adjusted by a risk premium). We adopt this approach and calibrate models by minimizing the quadratic spread between marginal and empirical covariances. The model is fitted to two pairs of commodities:[6] MSE Brent vs WTI oil and Silver vs Gold. The dataset contains daily log-returns from 11/4/08 to 11/4/18 (2522 observations). The number of lags considered in the optimized objective (7.10) is set to 150 days. The square roots of the quadratic error between empirical and model covariances are reported in Tables 7.1 and 7.2 for different numbers of homogeneous fields. We draw two conclusions from these figures. Firstly, increasing the number of fields reduces the error whatever the covariance structure. Secondly, the best fit is obtained with Laplace covariance functions. Tables 7.3 and 7.4 contain parameter estimates with $d = 3$ fields.

Recall that the field with an exponential covariance (Cauchy) corresponds to an Ornstein–Uhlenbeck (OU) process. Modeling commodity prices with this mean reverting process as in Cortazar and Schwartz [10] is mathematically convenient, mainly because we can rely on stochastic calculus to price options. However our results clearly suggest that these processes do not replicate the empirical covariance. This is confirmed by Fig. 7.1, which compares the covariance structures of fitted models with $d = 3$ homogeneous fields to empirical covariances. The observed autocovariances are concave decreasing functions and the exponential (Cauchy) model fails to replicate this concavity, particularly at short term. Tables 7.5 and 7.6 report calendar spread option prices[7] written on pairs Brent–WTI crude oil and silver–gold. These figures emphasize the impact of a misspecified autocovariance function on the value of calendar spread options. Assuming an OU dynamics for Brent and WTI oil leads to an overestimate of option values by 5–37% compared to those computed with a Laplace covariance. The gap between option prices computed with Gaussian and Laplace covariances is much smaller and ranges from -1.55 to 2.95%. For silver against gold, the conclusions are similar: the bias caused by a wrong choice of covariance structure can be significant (up to 6.2% with an OU model and -19.6% with a Gaussian covariance). Table 7.7 presents calendar spread option prices computed by Monte Carlo simulations with different levels of discretization in the space of frequencies. As expected, increasing u_n and decreasing the size of the step of discretization reduce the gap between values obtained analytically and numerical estimates.

---

[6] As mentioned in Sect. 7.2, homogeneous fields have constant asymptotic variance. This feature makes them more suitable for the modeling of commodities or interest rates.

[7] The discount rate is set to $r = 5\%$ and $c_1 = c_2 = 0$.

Table 7.1 Square root errors: Brent vs WTI oil. Number of lags: 150 days

d	$g_k(h) = \frac{\lambda_k^2}{\lambda_k^2 + h^2}$	$g_k(h) = e^{-\lambda_k |h|}$	$g_k(h) = e^{-\lambda_k h^2}$
1	0.0669	0.0567	0.0943
2	0.0094	0.0455	0.0137
3	0.0085	0.0451	0.0121
4	0.0051	0.0450	0.0101

Table 7.2 Square root errors: Silver vs Gold. Number of lags: 150 days

d	$g_k(h) = \frac{\lambda_k^2}{\lambda_k^2 + h^2}$	$g_k(h) = e^{-\lambda_k h}$	$g_k(h) = e^{-\lambda_k h^2}$
1	0.0537	0.0483	0.0540
2	0.0189	0.0154	0.0268
3	0.0067	0.0134	0.0211
4	0.0046	0.0133	0.0205

Table 7.3 Model parameters: Brent vs WTI oil. Number of lags: 150 days and $d = 3$

k	λ_k	$\sigma_{1,k}$	$\sigma_{2,k}$
Laplace			
1	0.1887	0.1546	0.1778
2	0.3614	0.2502	0.2477
3	2.2482	0.2132	0.1626
Cauchy			
1	0.100	0.1428	0.08124
2	1.454	0.2130	0.18387
3	2.146	0.2749	0.29315
Gaussian			
1	1.844	0.31737	0.2776
2	17.102	0.09077	0.1352
3	30.638	0.15033	0.1542

Table 7.4 Model parameters: Silver vs Gold. Number of lags: 150 days and $d = 3$

k	λ_k	$\sigma_{1,k}$	$\sigma_{2,k}$
Laplace			
1	0.2091	0.1486	0.0374
2	0.2636	0.0021	0.0784
3	0.9922	0.2819	0.1705
Cauchy			
1	0.2131	0.0804	0.0560
2	0.7088	0.2772	0.1840
3	1.8922	0.1475	0.0005
Gaussian			
1	0.8151	0.2569	0.1793
2	4.0357	0.1654	0.0198
3	40.6157	0.0855	0.0615

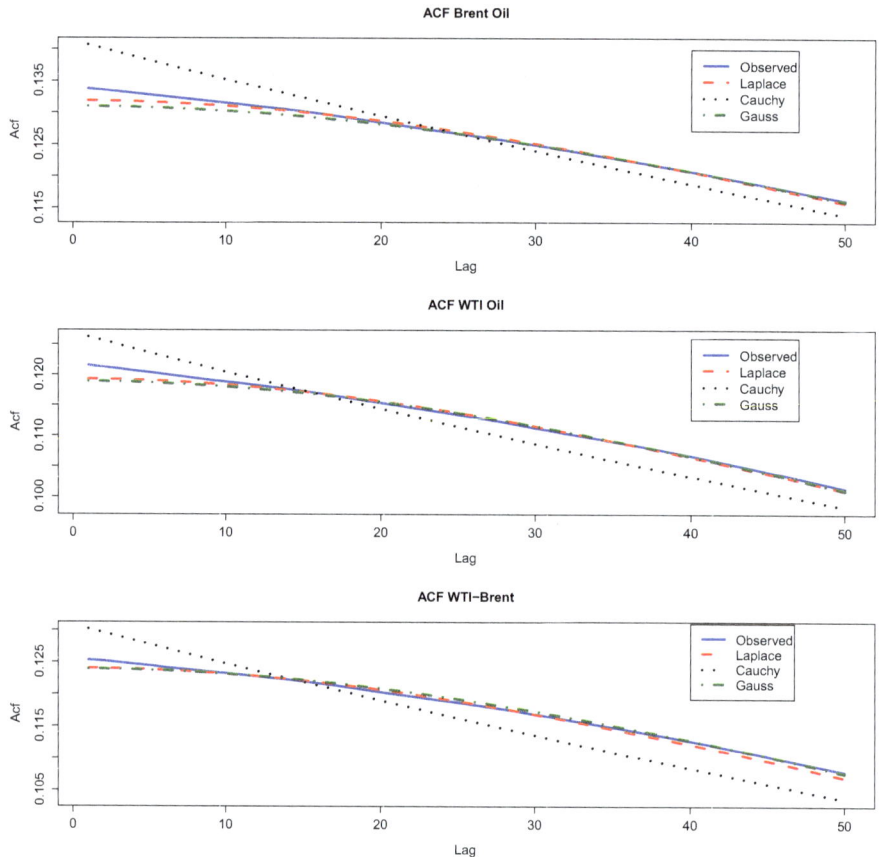

Fig. 7.1 Plot of autocovariances of Brent and WTI oil log-returns. Time lags are expressed in days. "Observed": empirical covariances. "Laplace," "Cauchy," and "Gauss": modeled covariances with corresponding measure

To understand the influence of autocorrelation parameters λ_k on option values, we perform a sensitivity analysis for the Laplace model driven by $d = 2$ conditional Gaussian fields. Table 7.8 reports the parameters of this model fitted to Brent and WTI crude oil. We appraise calendar spread exchange options with three sets of autocovariance parameters: (λ_1, λ_2), $\left(\frac{\lambda_1}{2}, \lambda_2\right)$ and $\left(\lambda_1, \frac{\lambda_2}{2}\right)$. Figure 7.2 emphasizes the impact of these modifications on autocovariance functions $g_1(t)$ and $g_2(t)$. We observe that reducing λ_1 or λ_2 decreases the level of autocorrelation of $Z_t^{(1)}$ and $Z_t^{(2)}$. Table 7.9 underlines the important role played by λ_1 or λ_2 on option prices. The relative spread between prices computed with (λ_1, λ_2) and $\left(\frac{\lambda_1}{2}, \lambda_2\right)$ ranges from 5% up to 55%, whereas the spread between prices obtained with (λ_1, λ_2) and $\left(\lambda_1, \frac{\lambda_2}{2}\right)$ varies from 7.5% up to 11%. These results confirm the importance of

Table 7.5 $S_t^{(1)}$: Brent, $S_t^{(2)}$: WTI. $d = 3$. Option payoff: $(\beta_1 S_T^{(1)} - \beta_2 S_t^{(2)})_+$ where $T = 1$ year and t ranges from 0.3 to 0.9 years. β_1 and β_2 are such that $\beta_1 S_0^{(1)} = \beta_2 S_0^{(2)} = 1$

	Calendar spread exchange options				
			Relative error (%)		Relative error (%)
t	Laplace	Cauchy	Cauchy vs Laplace	Gaussian	Gauss vs Laplace
0.3	0.16476	0.17342	5.251	0.1660	0.7250
0.4	0.16184	0.16881	4.303	0.1604	−0.9173
0.5	0.15502	0.16116	3.955	0.1526	−1.5519
0.6	0.14417	0.15035	4.283	0.1422	−1.3608
0.7	0.12776	0.13569	6.212	0.1265	−0.9992
0.8	0.10243	0.11556	12.810	0.1023	−0.1369
0.9	0.06236	0.08564	37.344	0.0642	2.9498

Table 7.6 $S_t^{(1)}$: Silver, $S_t^{(2)}$: Gold. $d = 3$. Option payoff: $(\beta_1 S_T^{(1)} - \beta_2 S_t^{(2)})_+$ where $T = 1$ year and t ranges from 0.3 to 0.9 years. β_1 and β_2 are such that $\beta_1 S_0^{(1)} = \beta_2 S_0^{(2)} = 1$

	Calendar spread exchange options				
			Relative error (%)		Relative error (%)
t	Laplace	Cauchy	Cauchy vs Laplace	Gaussian	Gauss vs Laplace
0.3	0.12271	0.12251	−0.163	0.12494	1.821
0.4	0.11613	0.11670	0.492	0.11633	0.170
0.5	0.10874	0.11035	1.483	0.10661	−1.957
0.6	0.10067	0.10339	2.693	0.09570	−4.942
0.7	0.09202	0.09571	4.011	0.08357	−9.178
0.8	0.08274	0.08718	5.368	0.07065	−14.614
0.9	0.07305	0.07758	6.200	0.05875	−19.574

Table 7.7 $S_t^{(1)}$: Brent, $S_t^{(2)}$: WTI. $d = 3$. Option payoff: $(\beta_1 S_T^{(1)} - \beta_2 S_t^{(2)})_+$ where $T = 1$ year and t equals to 0.4 years. β_1 and β_2 are such that $\beta_1 S_0^{(1)} = \beta_2 S_0^{(2)} = 1$. Prices are obtained with 10,000 simulations

	Monte Carlo simulations			
u_n	$u_{j+1} - u_j$	Laplace	Cauchy	Gaussian
40	0.30	0.1554	0.1563	0.1585
	0.25	0.1619	0.1606	0.1582
	0.20	0.1598	0.1630	0.1594
	0.15	0.1617	0.1621	0.1616
	0.10	0.1597	0.1654	0.1596
	0.05	0.1603	0.1651	0.1618
50	0.30	0.1537	0.1643	0.1610
	0.25	0.1588	0.1628	0.1577
	0.20	0.1638	0.1679	0.1584
	0.15	0.1590	0.1627	0.1569
	0.10	0.1631	0.1673	0.1556
	0.05	0.1629	0.1683	0.1578
Analytical price		0.1618	0.1688	0.1604

Table 7.8 Model parameters: Brent vs WTI oil. Number of lags: 150 days and $d = 2$. The last columns contain values of λ_k used to test the sensitivity of option prices to autocovariance parameters

k	Laplace parameters for the sensitivity analysis				
	$\sigma_{1,k}$	$\sigma_{2,k}$	(λ_1, λ_2)	$\left(\frac{\lambda_1}{2}, \lambda_2\right)$	$\left(\lambda_1, \frac{\lambda_2}{2}\right)$
1	0.2475	0.2655	0.2451	0.1226	0.2451
2	0.2701	0.2149	0.9393	0.9393	0.4696

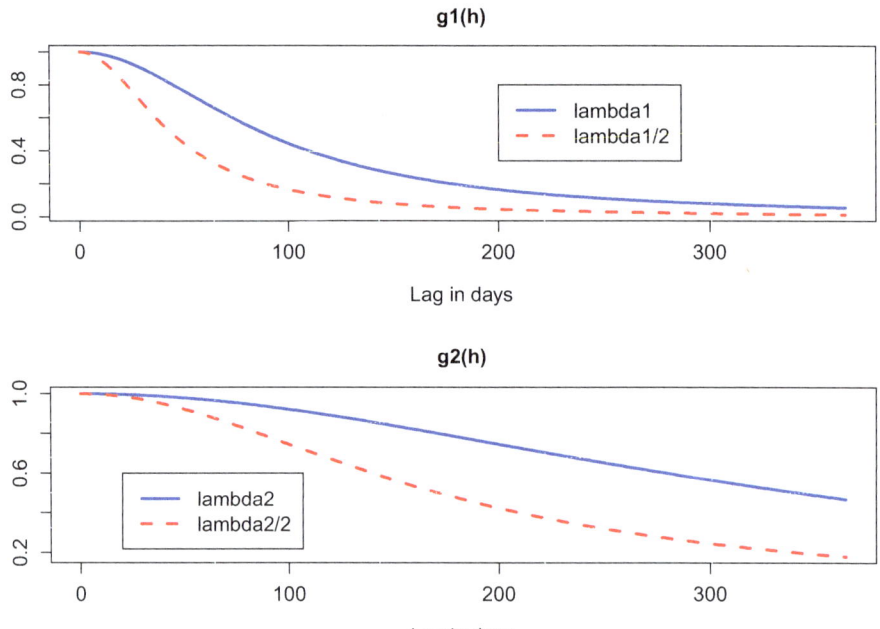

Fig. 7.2 Plot of Laplace covariance functions $g_1(t)$ and $g_2(t)$ for the model fitted to Brent and WTI oil log-returns with $d = 2$. Parameters λ_1 and λ_2 are reported in Table 7.8

autocorrelation in the valuation of exotic derivatives like calendar spread exchange options.

Tables 7.10 and 7.11 report Asian calendar spread option prices on Brent vs WTI crude oil and silver vs gold. The spreads between prices obtained with Laplace and Cauchy covariance structures is smaller than these of calendar options. However, they remain significant: from -1.84 to 3.48% for Brent vs WTI and from -1.74 to 1.54%. Prices obtained with a Gaussian covariance model deviate more widely from these computed with a Laplace covariance (from 7.16 to 20.6%).

To understand the influence of the autocovariance parameters λ_k on Asian options, we perform a sensitivity analysis for the Laplace model driven by $d = 2$ conditional Gaussian fields. We appraise Asian calendar spread exchange options

Table 7.9 $S_t^{(1)}$: Brent, $S_t^{(2)}$: WTI. $d = 2$. Option payoff: $(\beta_1 S_T^{(1)} - \beta_2 S_t^{(2)})_+$ where $T = 1$ year and t ranges from 0.3 to 0.9 years. β_1 and β_2 are such that $\beta_1 S_0^{(1)} = \beta_2 S_0^{(2)} = 1$

	Sensitivity of calendar spread exchange options to autocovariance parameters				
	Price	Price	relative spread (%)	Price	Relative spread (%)
t	(λ_1, λ_2)	$\left(\frac{\lambda_1}{2}, \lambda_2\right)$	$\left(\frac{\lambda_1}{2}, \lambda_2\right)$ vs (λ_1, λ_2)	$\left(\lambda_1, \frac{\lambda_2}{2}\right)$	$\left(\lambda_1, \frac{\lambda_2}{2}\right)$ vs (λ_1, λ_2)
0.3	0.16828	0.17708	5.227	0.18335	8.956
0.5	0.15547	0.16553	6.473	0.17268	11.072
0.7	0.12814	0.14851	15.895	0.14165	10.540
0.9	0.06316	0.09783	54.905	0.06793	7.564

Table 7.10 $S_t^{(1)}$: Brent, $S_t^{(2)}$: WTI. $d = 3$. Option payoff: $(\beta_1 G_T^{(1)} - G_2 S_t^{(2)})_+$ where $T = 1$ year and t ranges from 0.3 to 0.9 years. $\beta_1 = \beta_2 = 1$

	Asian calendar spread exchange options				
			Relative error (%)		Relative error (%)
t	Laplace	Cauchy	Cauchy vs Laplace	Gaussian	Gauss vs Laplace
0.3	0.06406	0.06629	3.477	0.06865	7.162
0.4	0.05797	0.05863	1.143	0.06224	7.368
0.5	0.05059	0.05039	−0.392	0.05440	7.530
0.6	0.04228	0.04170	−1.377	0.04565	7.960
0.7	0.03331	0.03270	−1.845	0.03637	9.188
0.8	0.02393	0.02360	−1.384	0.02687	12.314
0.9	0.01470	0.01504	2.306	0.01773	20.600

Table 7.11 $S_t^{(1)}$: Silver, $S_t^{(2)}$: Gold. $d = 3$. Option payoff: $(\beta_1 G_T^{(1)} - G_2 S_t^{(2)})_+$ where $T = 1$ year and t ranges from 0.3 to 0.9 years. $\beta_1 = \beta_2 = 1$

	Asian calendar spread exchange options				
			Relative error (%)		Relative error (%)
t	Laplace	Cauchy	Cauchy vs Laplace	Gaussian	Gauss vs Laplace
0.3	0.06044	0.05984	−0.984	0.06408	6.024
0.4	0.05628	0.05537	−1.605	0.05787	2.828
0.5	0.05207	0.05116	−1.743	0.05175	−0.607
0.6	0.04796	0.04726	−1.447	0.04586	−4.376
0.7	0.04409	0.04376	−0.761	0.04036	−8.474
0.8	0.04062	0.04073	0.264	0.03547	−12.687
0.9	0.03768	0.03826	1.540	0.03146	−16.505

with the sets of autocovariance parameters, (λ_1, λ_2), $\left(\frac{\lambda_1}{2}, \lambda_2\right)$ and $\left(\lambda_1, \frac{\lambda_2}{2}\right)$, presented in Table 7.9. Table 7.12 reveals that the relative spread between prices computed with (λ_1, λ_2) and $\left(\frac{\lambda_1}{2}, \lambda_2\right)$ ranges from 0.75% up to 4.3%. Whereas the spread between prices obtained with (λ_1, λ_2) and $\left(\lambda_1, \frac{\lambda_2}{2}\right)$ varies from 0.6% up to

Table 7.12 $S_t^{(1)}$: Brent, $S_t^{(2)}$: WTI. $d = 2$. Option payoff: $(\beta_1 G_T^{(1)} - G_2 S_t^{(2)})_+$ where $T = 1$ year and t ranges from 0.3 to 0.9 years. $\beta_1 = \beta_2 = 1$

	Sensitivity of Asian calendar spread exchange options to autocovariance parameters				
	Price	Price	Relative spread (%)	Price	Relative spread (%)
t	(λ_1, λ_2)	$\left(\frac{\lambda_1}{2}, \lambda_2\right)$	$\left(\frac{\lambda_1}{2}, \lambda_2\right)$ vs (λ_1, λ_2)	$\left(\lambda_1, \frac{\lambda_2}{2}\right)$	$\left(\lambda_1, \frac{\lambda_2}{2}\right)$ vs (λ_1, λ_2)
0.3	0.06781	0.06995	3.147	0.07124	5.0538
0.5	0.05369	0.05410	0.754	0.05613	4.5379
0.7	0.03607	0.03669	1.714	0.03712	2.9144
0.9	0.01798	0.01876	4.375	0.01809	0.6626

5.5%. The impact of the autocorrelation on Asian option prices is still significant but less important than for exchange options due to the smoothing induced by the definition of the payoff.

7.8 Further Reading

Gaussian fields have been used for several decades in the analysis of spatial statistics. One of the key features in spatial statistics is the autocorrelation of data. Observations made at locations in close spatial proximity often tend to be more similar than observations made at locations far apart. We refer the interested reader to books of Cressie [11], Adler [1] and Matern [19] for a detailed presentation of spatial statistics and the theory of Gaussian fields. An introduction to Gaussian processes on general parameter spaces may be found in Adler and Taylor [3]. We find applications of Gaussian fields in the financial literature. For example, Goldstein [13] generalizes the work of Kennedy [16] and proposes a model for interest rates based on a two-dimensional random field. In this model, increments along time are independent but the correlation structure between bond yields of different maturities can be arbitrarily chosen. Kimmel [17] introduces a state-dependent volatility in this model. Albeverio et al. [4] extend previous models for yield curves with Lévy fields. Özkan and Schmidt [20] use similar fields to model the yield curve in the presence of credit risk.

Gaussian fields are also used for actuarial modeling. Biffis and Millossovich [8] model the intensity of mortality as a random field, in order to capture cross-generation (risk class) effects induced by the on-going management of portfolios of policies. Based on this framework, Biagini et al. [7] study the pricing and hedging of life insurance liabilities. Wu [23] develops Gaussian process regression methods for mortality forecasting. Finally, this chapter was inspired by the article of Hainaut [15] about option pricing on commodities.

References

1. Adler, R.J.: The Geometry of Random Fields. Wiley, Chichester (1981)
2. Adler, R., Taylor, J.E.: Random Fields and Geometry. Springer, New York (2007)
3. Adler, R., Taylor, J.E.: Topological Complexity of Smooth Random Fields Functions: Ecole d'été de probabilité de Saint-Flour XXXIX, pp. 13–35. Springer, Berlin (2009)
4. Albeverio, S., Lytvynov, E., Mahnig, A.: A model of the term structure of interest rates based on Lévy fields. Stoch. Process. Appl. **114**(2), 251–263 (2004)
5. Bender, C., Sottinen, T., Valkeil, E.: Arbitrage with fractional Brownian motion? Theory Stoch. Process. **12**(28), 23–34 (2006)
6. Bessembinder, H., Coughenour, J.F., Seguin, P.J., Smoller, M.M.: Mean reversion in equilibrium asset prices: evidence from the futures term structure. J Finance **50**(1), 361–75 (1995)
7. Biagini, F., Botero, C., Schreiber, I.: Risk-minimization for life insurance liabilities with dependent mortality risk. Math. Finance **27**(2), 505–533 (2017)
8. Biffis, E., Millossovich, P.: A bidimensional approach to mortality risk. Decis. Econ. Finance **29**, 71–94 (2006)
9. Cheridito, P.: Arbitrage in fractional Brownian motion models. Finance Stoch. **7**(4), 533–553 (2003)
10. Cortazar, G., Schwartz, E.: The valuation of commodity contingent claims. J. Derivatives **1**, 27–29 (1994)
11. Cressie, N.A.: Statistics for Spatial Data, 2nd edn. Wiley, New York (1993)
12. Gibson, R., Schwartz, E.S.: Stochastic convenience yield and the pricing of oil contingent claims. J. Finance **45**(3), 959–976 (1990)
13. Goldstein, R.: The term structure of interest rates as a random field. Rev. Financial Stud. **13**(2), 365–384 (2000)
14. Hainaut, D.: Continuous mixed-laplace jump diffusion models for stocks and commodities. Quant. Finance Econ. **1**(2), 145–173 (2017)
15. Hainaut, D.: Calendar spread exchange options pricing with gaussian random fields. Risks **6**(3), 77, 1–33 (2018)
16. Kennedy, D.P.: The term structure of interest rates as a Gaussian random field. Math. Finance **4**, 257–258 (1994)
17. Kimmel, R.L.: Modeling the term structure of interest rates: a new approach. J. Financial Econ. **72**, 143–183 (2004)
18. Margrabe, W.: The value of an option to exchange one asset for another. J. Finance **33**(1), 177–186 (1978)
19. Matern, B.: Spatial Variation. Lecture Notes in Statistics, vol. 36. Springer, Berlin (1986)
20. Özkan, F., Schmidt, T.: Credit risk with infinite dimensional Lévy processes. Statist. Decisions **23**, 281–299 (2009)
21. Schwartz, E.: The stochastic behavior of commodity prices: implications for valuation and hedging. J. Finance **52**(3), 923–973 (1997)
22. Willinger, W., Taqqu, M.S., Teverovsky V.: Stock market prices and long-range dependence. Finance Stoch. **3**, 1–13 (1999)
23. Wu, R.: Gaussian Process and Functional Data Methods for Mortality Modelling, PhD Thesis, Department of Mathematics, University of Leicester (2016)

Chapter 8
Lévy Interest Rate Models with a Long Memory

From the 1980s to the present, many interest rate models have been proposed in the literature. They all aim to explain changes in bond or swap quotes and to replicate risks within the interest rates market. Three dominating frameworks coexist: short-term rate, forward rate, and the Libor market models. In this last approach, proposed by Brace et al. [3], interest rates are driven by geometric diffusions. In the forward rate models, pioneered by Heath et al. [18], the term structure of rates is specified through instantaneous forward rates. Mercurio and Moraleda [22] and Falini [10] propose a forward model with a humped structure of volatilities. Li et al. [20] develop a forward rate model unifying most of the existing Gaussian models for interest rates. Short-term rate models, such as promoted by Hull and White [19], specify a mean reverting dynamic for the instantaneous risk-free rate. The framework developed in this chapter belongs to this third category.

A large majority of short-term rate models are driven by Markov mean reverting processes. In this setting, the interest rate depends on its path through the integral of past occurrences weighted by an exponential decaying function, called the memory kernel. In this framework, the influence of previous variations of interest rates on today's value decreases exponentially with time. In this chapter, we consider an alternative that consists of replacing this memory kernel by a Mittag-Leffler decaying function. This function is sub-exponential and may be seen as a generalization of the exponential. In this setting, the interest rate remembers its previous occurrences for a longer period than in the exponential framework. We will meet the Mittag-Leffler function again in Chap. 10 about sub-diffusions.

A consequence of this change of memory kernel is that the interest rate process is not Markov anymore. Bond prices depend on the sample path of past rates. Nevertheless, we take advantage of the representation of the memory kernel as a Laplace–Stieltjes integral. This allows us to rewrite the short-term rate as an infinite-dimensional Markov process. Approximating this process allows us to infer the dynamics of bond prices and many of their features using standard Itô calculus.

D. Hainaut, *Continuous Time Processes for Finance*,
Bocconi & Springer Series 12, https://doi.org/10.1007/978-3-031-06361-9_8

We can draw a parallel with Chap. 5 in which a non-Markov self-exciting process is seen as a function of an infinity of pure jump processes.

This chapter starts with an introduction to the standard Lévy mean reverting model for interest rates. We next present two Lévy models with a Mittag-Leffler memory and some empirical arguments in favor of such an approach. After a reformulation as an infinite-dimensional Markov process, we find bond prices and their dynamics under the real and forward measures. We conclude with a discussion of methods to evaluate options on zero-coupon bonds.

8.1   A Lévy Model with an Exponential Memory Kernel

In the proposed model, the short-term rate is driven by $d \in \mathbb{N}$ independent Lévy processes, denoted by $(L_t^{(j)})_{t \geq 0}$ for $j = 1, \ldots, d$. These processes are defined on a probability space Ω, endowed with the natural filtration $(\mathcal{F}_t)_{t \geq 0}$ and the risk measure $\mathbb{Q}$. Each process $(L_t^{(j)})_{t \geq 0}$ has independent and stationary increments. It is fully described by a triplet $\left(\mu_j, \sigma_j^2, \nu_j \right)$, where $\mu_j \in \mathbb{R}$, $\sigma_j \in \mathbb{R}^+$, and $\nu_j(\cdot)$ is a Lévy measure. Note that most expectations in this chapter are calculated under the risk neutral measure $\mathbb{Q}$. For this reason, to lighten the notation, we shall write $\mathbb{E}^{\mathbb{Q}}(\cdot)$ simply as $\mathbb{E}(\cdot)$. The moment generating function (mgf) is denoted by $\phi_t^{(j)}(\omega)$ for $j = 1, \ldots, d$ and is equal to

$$\phi_t^{(j)}(\omega) := \mathbb{E}\left(\exp\left(\omega (L_t^{(j)} - L_0^{(j)}) \right) \mid \mathcal{F}_0 \right) \tag{8.1}$$

$$= \exp\left(t \left(\mu_j \omega + \frac{1}{2} \omega^2 \sigma_j^2 + \int_{\mathbb{R}} \left(e^{\omega z} - 1 - \omega z \mathbf{1}_{|z| < \epsilon} \right) \nu_j(\mathrm{d}z) \right) \right)$$

$$= \exp\left(t \psi_j(\omega) \right),$$

where $\epsilon > 0$. Notice that if the Lévy measure, $\nu_j(\cdot)$, has no singularity at $z = 0$, we can set ϵ to zero. The function $\psi_j(\omega)$ is the characteristic exponent of this Lévy process. Without loss of generality, we assume that $L_0^{(j)} = 0$. According to the Lévy–Itô decomposition, each $L_t^{(j)}$ is the sum of three components: a deterministic drift $\mu_j t$, a Brownian motion with variance σ_j^2, and a jump process, $J_j(t, z)$, of intensity $\nu_j(\mathrm{d}z)$ well defined on $[-\infty, 0) \cup (0, +\infty]$. This Lévy measure is such that the probability of observing k jumps between $[\tau_1, \tau_2]$ of a size included in a set $B \subset \mathbb{R}_{\backslash\{0\}}$ is given by

$$\mathbb{Q}\left(J_j([\tau_1, \tau_2] \times B) = k \right) = e^{-\int_{\tau_1}^{\tau_2} \int_B \nu_j(\mathrm{d}z)\mathrm{d}t} \frac{\left(\int_{\tau_1}^{\tau_2} \int_B \nu_j(\mathrm{d}z)\mathrm{d}t \right)^k}{k!}, \tag{8.2}$$

for $j = 1, \ldots, d$. If $\left(W_t^{(j)} \right)_{t \geq 0}$, for $j = 1, \ldots, d$, are independent Brownian motions on $(\Omega, \mathcal{F}, \mathbb{Q})$, $L_t^{(j)}$ may be split as the sum of a drift, a Brownian motion, and a jump process

$$\mathrm{d}L_t^{(j)} = \mu_j \mathrm{d}t + \sigma_j \mathrm{d}W_t^{(j)} + \int_{\mathbb{R}} z \, \tilde{J}_j(\mathrm{d}t \, , \, \mathrm{d}z), \tag{8.3}$$

where

$$\tilde{J}_j(\mathrm{d}t \, , \, \mathrm{d}z) = J_j(\mathrm{d}t \, , \, \mathrm{d}z) - \mathbf{1}_{|z| < \epsilon} \nu_j(\mathrm{d}z)\mathrm{d}t \, .$$

Without loss of generality, we assume that $\mathbb{E}\left(L_t^{(j)} \right) = 0$. This constraint implies that the drift is equal to

$$\mu_j = -\int_{|z| \geq \epsilon} z \, \nu_j(\mathrm{d}z) \quad , \, j = 1, \ldots, d.$$

Notice that if the Lévy measure, $\nu_j(\cdot)$, has no singularity at $z = 0$, we can set ϵ to zero. From Cont and Tankov [5, Lemma 15.1, p. 482], for any integrable function $f : \mathbb{R} \to \mathbb{C}$, the following relation holds:

$$\mathbb{E}\left(e^{\omega \int_s^t f(u)\mathrm{d}L_u^{(j)}} |\mathcal{F}_s \right) = \exp\left(\int_s^t \psi_j\left(\omega f(u) \right) \mathrm{d}u \right) \quad , \, j = 1, \ldots, d. \tag{8.4}$$

This can be proved by approaching $f(\cdot)$ with a stepwise function and by using the property of independence of increments. This property will be useful in later sections. We will also need mean reverting Lévy processes of the form

$$\mathrm{d}Y_t^{(j)} = -\kappa_j Y_t^{(j)} \mathrm{d}t + \mathrm{d}L_t^{(j)} \quad , \, j = 1, \ldots, d, \tag{8.5}$$

which are also called *Ornstein–Uhlenbeck (OU) processes*. We assume that $Y_0^{(j)} = 0$. The solution of the previous stochastic differential equation is given by

$$Y_t^{(j)} = e^{-\kappa_j(t-s)} Y_s^{(j)} + \int_s^t e^{-\kappa_j(t-u)} \mathrm{d}L_u^{(j)} \quad , \, j = 1, \ldots, d. \tag{8.6}$$

In this OU process, the influence on $Y_t^{(j)}$ of the past sample paths of $L_t^{(j)}$ decays in an exponential manner. We call the function $e^{-\kappa_j(t-u)}$ the *memory kernel* of $Y_t^{(j)}$. A classical approach to modeling the short-term rate, denoted $(r_t)_{t \geq 0}$, consists in postulating the following dynamic:

$$r_t = \varphi(t) + \sum_{j=1}^d Y_t^{(j)}, \tag{8.7}$$

where $\varphi(t)$ is a differentiable function. Such an approach was developed, for example, in Hainaut and Macgilchrist [16] and approached by a pentanomial tree in order to price interest rate derivatives. As $\mathbb{E}\left(L_u^{(j)}\right) = 0$, the expectation of r_t conditionally to the information up to time s is given by

$$\mathbb{E}\left(r_t|\mathcal{F}_s\right) = \varphi(t) + \sum_{j=1}^{d} e^{-\kappa_j(t-s)} Y_s^{(j)}. \tag{8.8}$$

This last equation emphasizes that the impact of current values of $Y_s^{(j)}$ on the expected interest rate decays exponentially with time. The conditional variance of the short-term rate is a function of the time horizon:

$$\mathbb{V}\left(r_t|\mathcal{F}_s\right) = \sum_{j=1}^{d}\left(\sigma_j^2 + \int_{\mathbb{R}} z^2 \nu_j(\mathrm{d}z)\right) \frac{\left(1 - e^{-2\kappa_j(t-s)}\right)}{2\kappa_j}, \tag{8.9}$$

whereas the autocovariance of r_t and r_u for $t \geq u \geq s$ is given by

$$\mathbb{C}\left(r_t \, r_u|\mathcal{F}_s\right) = \sum_{j=1}^{d}\left(\sigma_j^2 + \int_{\mathbb{R}} z^2 \nu_j(\mathrm{d}z)\right) \frac{e^{-\kappa_j(t-u)} - e^{-\kappa_j(t+u-2s)}}{2\kappa_j}.$$

This last equation reveals that the autocovariance between the current short rate and the future rate decays exponentially. In the next section, we propose a model in which this covariance decreases at a slower pace. Notice that the covariance between r_t and $r_{t-\Delta}$ for any $\Delta > 0$ converges to a function of Δ:

$$\lim_{t \to \infty} \mathbb{C}\left(r_t \, r_{t-\Delta}|\mathcal{F}_0\right) = \sum_{j=1}^{d}\left(\sigma_j^2 + \int_{\mathbb{R}} z^2 \nu_j(\mathrm{d}z)\right) \frac{e^{-\kappa_j \Delta}}{2\kappa_j}. \tag{8.10}$$

8.2 A Lévy Model with a Mittag-Leffler Kernel

We have seen that the influence on $Y_t^{(j)}$ of the sample path of $L_t^{(j)}$ decays in an exponential manner. This section proposes an alternative model in which this decay is sub-exponential. For this purpose, we postulate that the risk-free rate, $(r_t)_{t \geq 0}$, is the sum of a deterministic function $\varphi(t) : \mathbb{R} \to \mathbb{R}$ and of d processes $\left(X_t^{(d)}\right)_{t \geq 0}$:

$$r_t = \varphi(t) + \sum_{j=1}^{d} X_t^{(j)}, \tag{8.11}$$

where the $X_t^{(j)}$ are defined in a similar manner to $Y_t^{(j)}$, except that the exponential memory kernel is replaced by another decreasing function $g_j(\cdot)$:

$$X_t^{(j)} = g_j(t) X_0^{(j)} + \int_0^t g_j(t-u)\, dL_u^{(j)}\,. \tag{8.12}$$

The function $g_j(\cdot) : \mathbb{R}^+ \to \mathbb{R}$ is a continuously decreasing kernel, with an initial value $g_j(0) = 1$ and that admits a representation as a Laplace–Stieltjes integral. Contrary to the OU process, $X_t^{(j)}$ is in general not Markov and $X_t^{(j)}|\mathcal{F}_s$ may not be reformulated as a function of $X_s^{(j)}$. The interest rate process $(r_t)_{t \geq 0}$ is therefore not Markov. This feature makes difficult the evaluation of bond prices at a given time $t > 0$ since their value depends on the whole sample path of interest rates up to t. Nevertheless, for some well-chosen kernel functions, we will see that the interest rate may be represented as an infinite-dimensional Markov process. In particular, we consider two types of kernel functions, both based on the Mittag-Leffler function, denoted by $E_\alpha(\cdot)$ where $\alpha \in [0, 1]$:

$$g_j(t) = E_{\alpha_j}(-\beta_j t) \quad \text{or} \quad g_j(t) = E_{\alpha_j}(-\beta_j t^{\alpha_j})\,. \tag{8.13}$$

To understand the motivation for working with such functions, we need to review the main properties of the Mittag-Leffler function. As we will see in Chap. 10, the Mittag-Leffler function of order $\alpha > 0$ plays a fundamental role in the fractional calculus and can be considered as an extension of the exponential function. It is defined as an infinite sum

$$E_\alpha(t) = \sum_{n=0}^{\infty} \frac{t^n}{\Gamma(n\alpha + 1)}\,.$$

In this chapter, we assume that $\alpha \in [0, 1]$. For this range of values, $E_\alpha(-t)$ is an intermediary between the power and the exponentially decreasing functions:

$$E_0(-t) = \frac{1}{1+t} \ \forall |t| < 1 \ \text{ and } E_1(-t) = e^{-t}\,.$$

We refer the reader to the book by Gorenflo et al. [13] for a detailed presentation of this function. Let us recall that a function $f : (0, \infty) \to \infty$ is called *completely monotonic* if it possesses derivatives $f^{(n)}(t)$ of any order $n = 0, 1, 2, \ldots$ and the derivatives are alternating in sign, i.e.,

$$(-1)^n f^{(n)}(t) \geq 0 \ \forall t \in ((0, \infty))\,.$$

The above property is equivalent to the existence of a representation of the function f in the form of a Laplace–Stieltjes integral with non-decreasing density and non-negative measure $d\gamma(\cdot)$ such that

$$f(t) = \int_0^\infty e^{-ut} d\gamma(u).$$

The Mittag-Leffler function of negative argument $E_\alpha(-\beta t)$ is completely mono-tonic for all $0 \leq \alpha \leq 1$ and $\beta \in \mathbb{R}^+$. A proof of this result can be found in Gorenflo et al. [13, p. 47]. The authors show that

$$E_\alpha(-\beta t) = \int_0^\infty e^{-tu} d\gamma_{\alpha,\beta}(u), \tag{8.14}$$

where the derivative of $\gamma_{\alpha,\beta}(u)$ is equal to

$$\frac{d\gamma_{\alpha,\beta}(u)}{du} = \frac{1}{\pi\alpha} \sum_{k=1}^\infty \frac{(-1)^{k-1}}{k!} \sin(\pi\alpha k)\, \Gamma(\alpha k + 1) \frac{u^{k-1}}{\beta^k}. \tag{8.15}$$

Given that $E_\alpha(0) = 1$, we have that $\int_0^\infty d\gamma_{\alpha,\beta}(u) = 1$ and $\gamma_{\alpha,\beta}(\cdot)$ is then a probability measure on $\mathbb{R}^+$. In the remainder of this chapter, $E_\alpha(-\beta t)$ is called the *decreasing Mittag-Leffler (ML) kernel*. When $t \to 0$, the ML behaves at short term as

$$E_\alpha(-\beta t) = 1 - \frac{\beta t}{\Gamma(1+\alpha)} + \cdots \sim \exp\left(-\frac{\beta t}{\Gamma(1+\alpha)}\right).$$

From Haubold et al. [17], we known that when $t \to \infty$, the ML converges to

$$E_\alpha(-\beta t) \sim \frac{(\beta t)^{-1}}{\Gamma(1-\alpha)}.$$

For intermediate times t, the function $E_\alpha(-\beta t)$ interpolates between the decreasing exponential and the inverse power law. The exponential models decay fast for small time t, whereas the asymptotic inverse power law entails a slow decrease at long term. This point is illustrated in the left plot of Fig. 8.1, which compares the ML kernel to the exponential and inverse power functions.

The Mittag-Leffler function is also related to the fractional calculus. To explain this link, we recall that *Caputo's fractional derivative* of order $\alpha \in (0, 1)$ for a function $h(t) : \mathbb{R}^+ \to \mathbb{R}, C^1$ with respect to t is defined by

$$\frac{\partial^\alpha}{\partial t^\alpha} h(t) = \frac{1}{\Gamma(1-\alpha)} \int_0^t (t-s)^{-\alpha} \frac{\partial}{\partial s} h(s) ds. \tag{8.16}$$

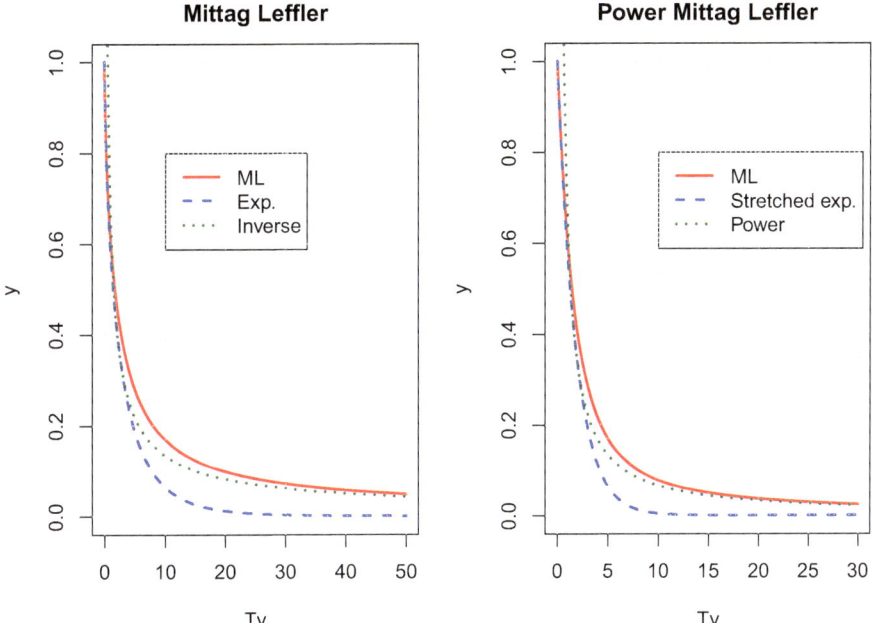

Fig. 8.1 Left and right plots: comparison of the Mittag-Leffler (ML) and power Mittag-Leffler (PML) kernels with exponential and power decreasing functions ($\alpha = 0.7$ and $\beta = 0.5$)

When $\alpha = 1$, this derivative corresponds to the first-order derivative. The solution of the fractional differential equation

$$\frac{\partial^\alpha}{\partial t^\alpha} y(t) = -\beta y(t) \qquad 0 < \alpha < 1$$

with the initial condition $y(0) = b_0$ is $y(t) = E_\alpha(-\beta t^\alpha)$. We abusively call this function the *power decreasing Mittag-Leffler kernel (PML)*. The Laplace transform of $E_\alpha(-\beta t^\alpha)$ is given by $\frac{s^{\alpha-1}}{s^\alpha+\beta}$, where $\text{Re}(s) > |\beta|^{\frac{1}{\alpha}}$. By inverting this transform, we can prove as in Gorenflo and Mainardi [12] that the PML is the Laplace transform

$$E_\alpha(-\beta t^\alpha) = \int_0^\infty \mathrm{e}^{-ut} \, \mathrm{d}\gamma_{\alpha,\beta}^p(u) , \tag{8.17}$$

where the derivative of $\gamma_{\alpha,\beta}^p(u)$ is given by

$$\frac{\mathrm{d}\gamma_{\alpha,\beta}^p(u)}{\mathrm{d}u} = \frac{1}{\pi} \frac{\beta \, u^{\alpha-1} \sin(\alpha\pi)}{u^{2\alpha} + 2\beta u^\alpha \cos(\alpha\pi) + \beta^2} . \tag{8.18}$$

Since $E_\alpha(-\beta t^\alpha) = 1$, we have $\int_0^\infty \mathrm{d}\gamma_{\alpha,\beta}^P(u) = 1$ and $\gamma_{\alpha,\beta}^P(u)$ is a probability measure on $\mathbb{R}^+$. As underlined by Mainardi [21], when $t \to 0$ the PML behaves at short term as

$$E_\alpha(-\beta t^\alpha) = 1 - \frac{\beta t^\alpha}{\Gamma(1+\alpha)} + \cdots \sim \exp\left(-\frac{\beta t^\alpha}{\Gamma(1+\alpha)}\right).$$

From Erdélyi et al. [9], we know that when $t \to \infty$, the PML converges to

$$E_\alpha(-\beta t^\alpha) \sim \frac{\left(\beta^{1/\alpha} t\right)^{-\alpha}}{\Gamma(1-\alpha)}.$$

As a consequence, for intermediate times t, the function $E_\alpha(-\beta t^\alpha)$ interpolates between the stretched exponential and the negative power law. The stretched exponential models the very fast decay at short-term, whereas the asymptotic power law is due to the very slow decay for large time t. The right plot of Fig. 8.1 illustrates this convergence.

Figure 8.2 presents simulated sample paths of the short-term rate with a one-dimensional ($d = 1$) model ruled by a Brownian motion. These paths are computed for various α with the same random occurrences in order to make comparison feasible. For the ML kernel, the trajectories are nearly similar. An analysis of the figures reveals that the sample path is smoother for $\alpha = 0.50$ than for $\alpha = 0.90$. This trend becomes visible if we choose a highest value for β. For PML sample paths, the difference is clearly visible. Decreasing α reduces the volatility of rates and smoothes the sample path.

To conclude this section, we present the first two moments of the short-term rate and its autocovariance function. The ML and PML kernels defined in Eq. (8.13) are continuous decreasing and integrable functions on any bounded interval of $\mathbb{R}$. From Eq. (8.4), the mgf of $X_t^{(j)}$ for $j = 1, \ldots, d$ is then equal to

$$\mathbb{E}\left(e^{\omega X_t^{(j)}} | \mathcal{F}_0\right) = \exp\left(\omega\, g_j(t)\, X_0^{(j)} + \int_0^t \psi_j\left(\omega\, g_j(t-u)\right)\, \mathrm{d}u\right).$$

Differentiating the mgf allows us to find the first moments of $X_t^{(j)}$ conditionally to the initial information. On the other hand, from the representation

$$r_t = \varphi(t) + \sum_{j=1}^d g_j(t) X_0^{(j)} + \sum_{j=1}^d \int_0^s g_j(t-u)\, \mathrm{d}L_u^{(j)} \qquad (8.19)$$

$$+ \sum_{j=1}^d \int_s^t g_j(t-u)\, \mathrm{d}L_u^{(j)}$$

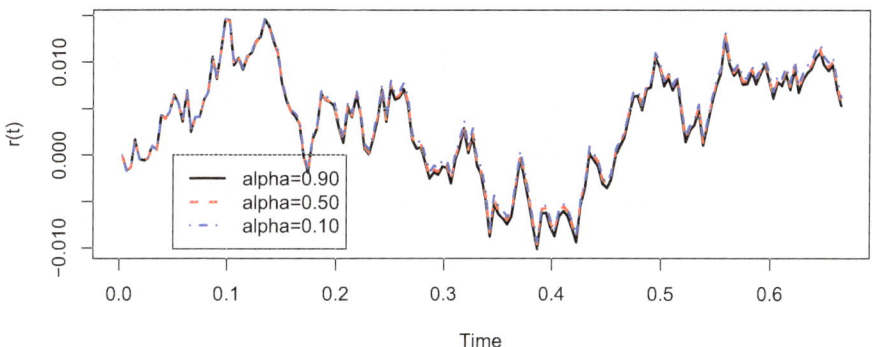

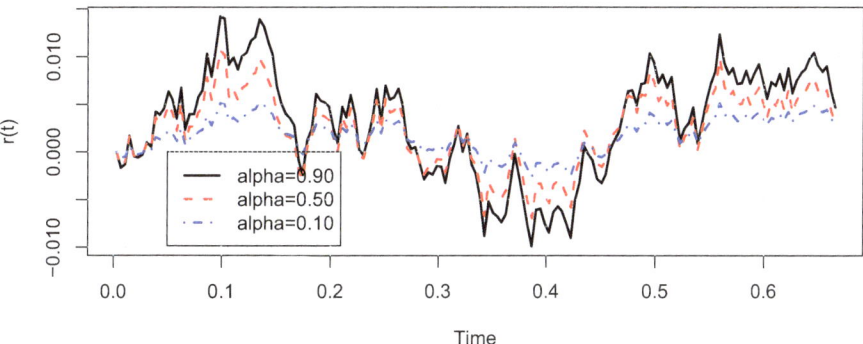

Fig. 8.2 Upper and lower plots: comparison of sample paths of $(r_t)_{t \geq 0}$ with Mittag-Leffler (ML) and power Mittag-Leffler (PML) kernels. $d = 1$, $\beta = 3$, $X_0^{(1)} = 0$, $\varphi(t) = 0$, and L_t is a Brownian motion with $\sigma_1 = 0.05$

of r_t, we infer that the conditional expectation of the short-term rate is given by

$$\mathbb{E}\left(r_t \mid \mathcal{F}_s\right) = \varphi(t) + \sum_{j=1}^{d} g_j(t) X_0^{(j)} + \sum_{j=1}^{d} \int_0^s g_j(t - u)\, \mathrm{d}L_u^{(j)}, \qquad (8.20)$$

whereas the conditional variance of r_t is equal to the sum of the products of the variances of Lévy processes and the integral of the squared kernels

$$\mathbb{V}\left(r_t \mid \mathcal{F}_s\right) = \sum_{k=1}^{d} \left(\sigma_j^2 + \int_{\mathbb{R}} z^2 \nu_j(\mathrm{d}z) \right) \int_s^t g_j(t - u)^2 \mathrm{d}u. \qquad (8.21)$$

Unfortunately, the integral of $g_j(\cdot)^2$ does not admit a closed form expression for the ML and PML kernels. Nevertheless, they can be numerically estimated. A direct calculation allows us to infer that the autocovariance is given by

$$\mathbb{C}\left(r_t r_u | \mathcal{F}_s\right) = \sum_{k=1}^{d} \mathbb{E}\left(\int_s^t g_j(t-v)\, dL_v^{(j)} \int_s^u g_j(u-v)\, dL_v^{(j)}\right) \qquad (8.22)$$

$$= \sum_{k=1}^{d}\left(\sigma_j^2 + \int_{\mathbb{R}} z^2 v_j(dz)\right)\int_s^u g_j(t-v)g_j(u-v)\, dv\,.$$

In contrast to the exponential case, this covariance function between r_t and $r_{t-\Delta}$ does not admit a closed form expression when $t \to \infty$.

8.3 Empirical Motivation

This short section provides some empirical arguments that motivate the developments of this chapter, in particular the choice of an ML or PML memory kernel. We fit univariate Gaussian models ($d = 1$) with exponential, ML, and PML kernels to the Eonia time series from the 13th of March 2016 to the 1st of August 2019. The dataset counts $n_{obs} = 866$ daily observations. We select this time window mainly because the Eonia was relatively stable during this period, as illustrated in the right plot of Fig. 8.3. This allows us to assume that the trend function, $\varphi(t)$, is constant. Considering a larger dataset would raise the question of modeling a nonlinear trend in view of the left plot of Fig. 8.3.

The dates of observations are denoted by $t_0, t_1, \ldots, t_{n_{obs}-1}$, whereas the step of time between two successive observations is Δ_t. We first consider a one-dimensional exponential kernel model, as detailed in Sect. 8.1. Under the assumption that $\varphi(t_k) = \varphi(t_k + \Delta_t)$ and $X_0 = 0$, we find that

$$\Delta L_{k+1} = L_{t_{k+1}} - L_{t_k} \approx r_{t_{k+1}} - r_{t_k} - \sum_{j=0}^{k}\left(e^{-\kappa(t_{k+1}-t_j)} - e^{-\kappa(t_k-t_j)}\right)\Delta L_j\,,$$

for $k = 1, \ldots, n_{obs} - 1$. Under the assumption that L_t is a Brownian motion without drift and of variance σ^2, parameters are estimated by log-likelihood maximization. We next consider models with ML and PML kernel functions. In these cases, the variations of the underlying Lévy process are approached by

$$\Delta L_{k+1} = L_{t_{k+1}} - L_{t_k} \approx r_{t_{k+1}} - r_{t_k} - \sum_{j=0}^{k}\left(g(t_{k+1}-t_j) - g(t_k-t_j)\right)\Delta L_j\,,$$

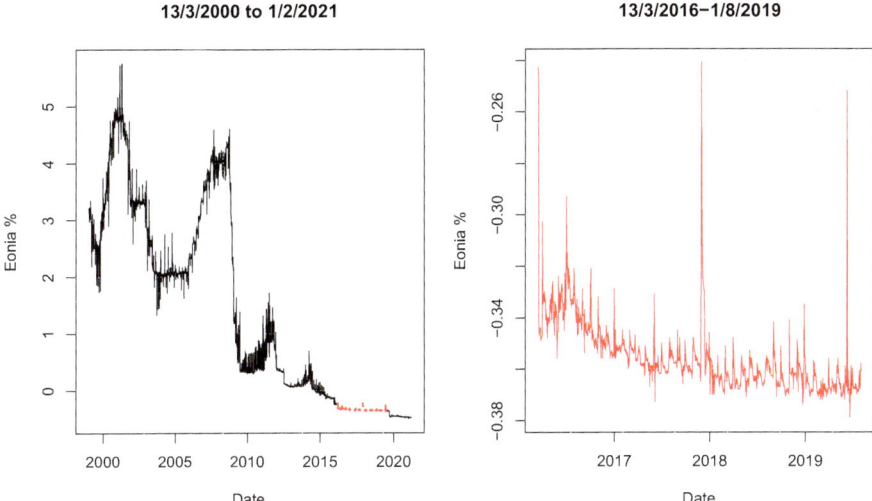

Fig. 8.3 Left plot: Eonia time series from 13/3/2000 to 1/2/2021. Right plot: dataset used to estimate parameters

for $k = 1, \ldots, n_{\text{obs}} - 1$, where $g(\cdot)$ is either the ML or the PML function. Under the assumption of normality, parameters are estimated by log-likelihood maximization, whereas the goodness of fit is measured with the Akaike Information Criterion (AIC). The results of this procedure are reported in Table 8.1. In terms of log-likelihood, the exponential and ML models achieve the same goodness of fit but the exponential model is more parsimonious and therefore preferred. They share the same estimated volatilities, but the parameters of reversion κ and β slightly differ. The best fit, based on the AIC, is obtained with a power Mittag-Leffler kernel. The parameter α that drives the decay of the memory is small. This means that the process has a longer-term memory than the exponential model. On the other hand, the mean reversion speed is significantly higher than those of other models. We remind the reader that estimated parameters reflect the dynamics of r_t under the real measure $\mathbb{P}$ and not the risk neutral measure.

Table 8.1 Parameter estimates by log-likelihood maximization of one-dimensional models with exponential, ML, and PML kernels

	Exponential		ML	PML
σ	1.4763e−3	σ	1.4767e−3	1.3927e−3
κ	0.6926	α	0.9954	0.1924
		β	0.7218	1.2171
Log-likelihood	6 809.75	Log-likelihood	6 809.93	6 861.15
AIC	−13615.50		−13613.86	−13716.30

8.4 Alternative Formulation

If the memory is not exponential, the interest rate process $(r_t)_{t \geq 0}$ is no longer Markov. This feature makes difficult the evaluation of bond prices or interest rate derivatives because their values depend on the whole history of the rate process. Nevertheless, if the memory kernel admits a representation as a Laplace–Stieltjes integral, we can convert the short-term rate into an infinite-dimensional Markov process. Approximating this process allows us to infer the dynamics of bond prices. As the details are similar in both cases, we denote by $\gamma_j(\cdot)$ the measures $\gamma_{\alpha_j,\beta_j}^{(j)}(\cdot)$ and $\gamma_{\alpha_j,\beta_j}^{p(j)}(\cdot)$ of $E_{\alpha_j}(-\beta_j t)$ and $E_{\alpha_j}(-\beta_j t^{\alpha_j})$, respectively. Let us recall that $g_j(t) = \int_0^\infty e^{-\xi t} d\gamma_j(\xi)$. Using Fubini's theorem, the processes $X_t^{(j)}$ are then rewritten as

$$
\begin{aligned}
X_t^{(j)} &= X_0^{(j)} \int_0^\infty e^{-t\xi} d\gamma_j(\xi) + \int_0^t \int_0^\infty e^{-\xi(t-u)} d\gamma_j(\xi) \, dL_u^{(j)} \\
&= \int_0^\infty \underbrace{\left(e^{-t\xi} Y_0^{(j,\xi)} + \int_0^t e^{-\xi(t-u)} \, dL_u^{(j)} \right)}_{Y_t^{(j,\xi)}} d\gamma_j(\xi)
\end{aligned}
$$

for $j = 1, \ldots, d$. The $Y_t^{(j,\xi)}$ are Ornstein–Uhlenbeck processes for $j = 1, \ldots, d$ and for all $\xi \in \mathbb{R}^+$. Their initial values are $X_0^{(j)} = Y_0^{(j,\xi)}$, and they satisfy the following stochastic differential equation (SDE):

$$
dY_t^{(j,\xi)} = -\xi Y_t^{(j,\xi)} dt + dL_t^{(j)} . \tag{8.23}
$$

As this SDE admits the solution

$$
Y_t^{(j,\xi)} = e^{-\xi(t-s)} Y_s^{(j,\xi)} + \int_s^t e^{-\xi(t-u)} dL_u^{(j)} , \tag{8.24}
$$

the short-term rate is then a sum of integrals of $Y_t^{(j,\xi)}$:

$$
r_t = \varphi(t) + \sum_{j=1}^d \int_0^\infty Y_t^{(j,\xi)} d\gamma_j(\xi) . \tag{8.25}
$$

Given that $\gamma_{\alpha_j,\beta_j}^{(j)}(\cdot)$ and $\gamma_{\alpha_j,\beta_j}^{p(j)}(\cdot)$ are probability measures, the differential of r_t is on the other hand equal to

$$
dr_t = d\varphi(t) - \sum_{j=1}^d \int_0^\infty \xi Y_t^{(j,\xi)} d\gamma_j(\xi) + \sum_{j=1}^d \underbrace{\int_0^\infty d\gamma_j(\xi)}_{=1} dL_t^{(j)} .
$$

Combining Eqs. (8.24) and (8.25) allows us to rewrite the short-term rate as the following sum:

$$r_t = \varphi(t) + \sum_{j=1}^{d} \int_0^\infty \left(e^{-\xi(t-s)} Y_s^{(j,\xi)} + \int_s^t e^{-\xi(t-u)} dL_u^{(j)} \right) d\gamma_j(\xi) \qquad (8.26)$$

$$= \varphi(t) + \sum_{j=1}^{d} \int_0^\infty e^{-\xi(t-s)} Y_s^{(j,\xi)} d\gamma_j(\xi) + \sum_{j=1}^{d} \int_s^t g_j(t-u) dL_u^{(j)},$$

which no longer depends upon information prior to s. This relation emphasizes that r_t and $\left(Y_t^{(j,\xi)} \right)_{\xi \in \mathbb{R}^+, j \in \{1,...,d\}}$ form an infinite-dimensional Markov process. If we remember that $\mathbb{E}(L_t^{(j)}) = 0$, we infer an equivalent representation of Eq. (8.20) for the conditional expectation of r_t in terms of $Y_t^{(j,\xi)}$:

$$\mathbb{E}(r_t | \mathcal{F}_s) = \varphi(t) + \sum_{j=1}^{d} \int_0^\infty e^{-\xi(t-s)} Y_s^{(j,\xi)} d\gamma_j(\xi).$$

We will explain in the next section how to approximate the infinity of processes by discretizing the measures $\gamma_j(\xi)$. But first we evaluate zero-coupon bonds and instantaneous forward rates in this setting.

8.5 Bond Prices and Forward Rates

A zero-coupon bond of maturity t delivers a unit cash-flow at expiry. Its price at time s prior to t is denoted by $P(s, t)$ and, in the absence of arbitrage, is the expected discount factor under the risk neutral measure, $P(s, t) = \mathbb{E}\left(e^{-\int_s^t r_u \, du} | \mathcal{F}_s \right)$. The next proposition presents a semi-closed form expression for this price.

Proposition 8.1 *Let us, respectively, define the functions $B^{(\xi)}(s, t)$ and $h_j(s, t)$ as follows:*

$$B^{(\xi)}(s, t) = \frac{1}{\xi} \left(1 - e^{-\xi(t-s)} \right), \qquad (8.27)$$

$$h_j(s, t) = \int_0^\infty B^{(\xi)}(s, t) \, d\gamma_j(\xi), \qquad (8.28)$$

for $s \leq t$ and $j = 1, \ldots, d$. The zero-coupon bond price is equal to

$$P(s, t) = \exp\Bigg(- \int_s^t \varphi(u)\mathrm{d}u + \sum_{j=1}^d \int_s^t \psi_j(-h_j(u, t))\mathrm{d}u \tag{8.29}$$

$$- \sum_{j=1}^d \Bigg(X_0^{(j)} \int_s^t g_j(u)\mathrm{d}u + \int_0^s \int_0^\infty e^{-\xi(s-u)} B^\xi(s, t)\,\mathrm{d}\gamma_j(\xi)\,\mathrm{d}L_u^{(j)} \Bigg) \Bigg) \,.$$

Proof From Eq. (8.19), we infer after a change of the integration order that

$$\int_s^t r_u\,\mathrm{d}u = \int_s^t \varphi(u)\mathrm{d}u + \sum_{j=1}^d X_0^{(j)} \int_s^t g_j(u)\mathrm{d}u$$

$$+ \sum_{j=1}^d \int_0^s \int_s^t g_j(u - v)\,\mathrm{d}u\,\mathrm{d}L_v^{(j)} + \sum_{j=1}^d \int_s^t \int_v^t g_j(u - v)\,\mathrm{d}u\,\mathrm{d}L_v^{(j)} \,.$$

A direct calculation allows us to rewrite the integrals of $g_j(\cdot)$ in this last expression as

$$\int_s^t g_j(u - v)\,\mathrm{d}u = \int_0^\infty e^{-(s-v)} B^\xi(t, s)\,\mathrm{d}\gamma_j(\xi)\,,$$

$$\int_v^t g_j(u - v)\,\mathrm{d}u = \int_0^\infty \frac{1}{\xi}\left(1 - e^{-(t-v)\xi}\right)\mathrm{d}\gamma_j(\xi)\,.$$

The zero-coupon bond price then becomes

$$\mathbb{E}\left(e^{-\int_s^t r_u\,\mathrm{d}u} \,|\, \mathcal{F}_s\right) = \mathbb{E}\Bigg(\exp\Bigg(-\sum_{j=1}^d \int_s^t h_j(v, t)\,\mathrm{d}L_v^{(j)}\Bigg) \,|\, \mathcal{F}_s\Bigg)$$

$$\times \exp\Bigg(-\int_s^t \varphi(u)\mathrm{d}u - \sum_{j=1}^d \Bigg(X_0^{(j)} \int_s^t g_j(u)\mathrm{d}u\Bigg)\Bigg)$$

$$\times \exp\Bigg(-\sum_{j=1}^d \Bigg(\int_0^s \int_0^\infty e^{-(s-v)} B^\xi(s, t)\,\mathrm{d}\gamma_j(\xi)\,\mathrm{d}L_v^{(j)}\Bigg)\Bigg)\,.$$

Finally, the expectation in this last expression is calculated with Eq. (8.4). □

Observe that $\lim_{\xi \to \infty} B^{(\xi)}(v, t) = 0$ and that $\lim_{\xi \to 0} B^{(\xi)}(v, t) = t - v$. Furthermore, the function $B^{(\xi)}(v, t)$ admits a single maximum $\xi^* < \infty$ such that $B^{(\xi^*)}(v,t) \geq B^{(\xi)}(v,t)$ for all $\xi \in \mathbb{R}^+$. Therefore the integral $h_j(v,t) = \int_0^\infty B^{(\xi)}(v,t)\,\mathrm{d}\gamma_j(\xi)$ is bounded by $B^{(\xi^*)}(v,t)$ and is well defined. Equation (8.29) emphasizes the dependence of the bond price on the history of the Lévy processes. A more convenient formulation in terms of $\left(Y_s^{(j,\xi)}\right)_{\xi \geq 0}$, for $j = 1,\ldots,d$, is presented in the next proposition. In this alternative valuation formula, the bond price is exclusively calculated with information available at the time of valuation.

Proposition 8.2 *The value at time $s > 0$ of a zero-coupon bond price expiring at $t > s$ is also equal to*

$$P(s,t) = \exp\left(-\int_s^t \varphi(u)\mathrm{d}u - \sum_{j=1}^d \int_0^\infty Y_s^{(j,\xi)} B^{(\xi)}(s,t)\,\mathrm{d}\gamma_j(\xi)\right) \quad (8.30)$$

$$\times \prod_{j=1}^d \exp\left(\int_s^t \psi_j\left(-h_j(u,t)\right)\mathrm{d}u\right),$$

where $B^{(\xi)}(s,t)$ and $h_j(v,t)$ for $j = 1,\ldots,d$ are defined by Eqs. (8.27) and (8.28).

Proof Starting from Eq. (8.29), we rewrite the last terms as follows:

$$X_0^{(j)}\int_s^t g_j(u)\mathrm{d}u + \int_0^s \int_0^\infty e^{-\xi(s-u)} B^\xi(s,t)\,\mathrm{d}\gamma_j(\xi)\,\mathrm{d}L_u^{(j)}$$

$$= \int_0^\infty B^\xi(s,t)\left(X_0^{(j)}e^{-s\xi} + \int_0^s e^{-\xi(s-v)}\,\mathrm{d}L_v^{(j)}\right)\mathrm{d}\gamma_j(\xi)$$

$$= \int_0^\infty B^\xi(s,t)Y_s^{(j,\xi)}\,\mathrm{d}\gamma_j(\xi).$$

We conclude that the bond price is given by Eq. (8.30).   □

The integrals in the bond price formula do not admit closed form expressions. Nevertheless, the integrals $\int_s^t \psi_j\left(-h_j(u,t)\right)\mathrm{d}u$ can be numerically computed without any particular difficulty, whereas $\int_0^\infty Y_s^{(j,\xi)} B^{(\xi)}(s,t)\,\mathrm{d}\gamma_j(\xi)$ is approached by a discretization scheme developed in the following section.

Let us recall that the initial values of the processes $\left(Y_t^{(j,\xi)}\right)_{t \geq 0,\, j \in \{1,\ldots,d\}}$ are null, that is, $Y_0^{(j,\xi)} = 0$, for all $\xi \in \mathbb{R}^+$ and $j = 1,\ldots,d$. Therefore, Eq. (8.30) provides us a way to estimate the function $\varphi(\cdot)$ such that the model perfectly matches the initial term structure of bond prices. More precisely, the integral of $\varphi(\cdot)$ is such that

$$\int_0^t \varphi(u)\mathrm{d}u = -\ln\left(P(0,t)\right) + \sum_{j=1}^d \int_0^t \psi_j\left(-h_j(v,t)\right)\mathrm{d}v.$$

Differentiating this last expression leads to the following function:

$$\varphi(t) = -\frac{\partial}{\partial t}\ln\left(P(0,t)\right) - \sum_{j=1}^{d}\psi_j\left(-h_j(0,t)\right). \tag{8.31}$$

The dynamic of interest rates can also be reformulated in terms of instantaneous forward rates. Let us recall that the instantaneous forward rate, denoted by $f(s,t)$, is such that $P(s,t) = \exp\left(-\int_s^t f(s,u)\,du\right)$ and is therefore equal to

$$f(s,t) = -\frac{\partial}{\partial t}\ln P(s,t) \quad 0 \le s \le t.$$

The next proposition states that the forward rate is the sum of the expected interest rate under the risk neutral measure and of an adjustment that directly depends upon the memory kernel.

Proposition 8.3 *The instantaneous forward rate at time s and of maturity t is given by*

$$f(s,t) = \mathbb{E}\left(r_t|\mathcal{F}_s\right) + \sum_{j=1}^{d}\int_s^t g_j(t-v)\left(\mu_j - h_j(v,t)\sigma_j^2\right)dv \tag{8.32}$$

$$+ \sum_{j=1}^{d}\int_s^t g_j(t-v)\int_{\mathbb{R}} z\left(e^{-h_j(v,t)z} - \mathbf{1}_{|z|<\epsilon}\right)v_j(dz)\,dv,$$

where $\mathbb{E}\left(r_t|\mathcal{F}_s\right) = \varphi(t) + \sum_{j=1}^{d}\int_0^\infty e^{-\xi(t-s)}Y_s^{(j,\xi)}d\gamma_j(\xi)$. Its dynamic is independent from the processes $Y_s^{(j,\xi)}$ and is equal to

$$df(s,t) = \sum_{j=1}^{d}g_j(t-s)\,dL_s^{(j)} - \sum_{j=1}^{d}g_j(t-s)\left(\mu_j - h_j(v,t)\sigma_j^2\right)ds$$

$$- \sum_{j=1}^{d}g_j(t-s)\left(\int_{\mathbb{R}} z\left(e^{-h_j(v,t)z} - \mathbf{1}_{|z|<\epsilon}\right)v_j(dz)\right)ds.$$

Proof From Eq. (8.30) and $\frac{\partial B^{(\xi)}(s,t)}{\partial t} = e^{-\xi(t-s)}$, we directly infer that

$$-\frac{\partial}{\partial t}\ln P(s,t)$$

$$= \mathbb{E}\left(r_t|\mathcal{F}_s\right) - \sum_{j=1}^{d}\int_s^t\frac{\partial}{\partial t}\psi_j\left(-\int_0^\infty B^{(\xi)}(v,t)\,d\gamma_j(\xi)\right)dv.$$

If we remember the expression (8.30) of the characteristic exponent, we have that

$$\frac{\partial}{\partial t} \psi_j \left(-\int_0^\infty B^{(\xi)}(v,t)\, d\gamma_j(\xi) \right) = -\mu_j \int_0^\infty e^{-\xi(t-v)}\, d\gamma_j(\xi)$$

$$+ \left(\int_0^\infty B^{(\xi)}(v,t)\, d\gamma_j(\xi) \right) \int_0^\infty e^{-\xi(t-v)}\, d\gamma_j(\xi)\, \sigma_j^2$$

$$- \int_{\mathbb{R}} \left(z \int_0^\infty e^{-\xi(t-v)}\, d\gamma_j(\xi) e^{-\int_0^\infty B^{(\xi)}(v,t)\, d\gamma_j(\xi)\, z} \right) \nu_j(dz)$$

$$+ \int_{\mathbb{R}} \left(\int_0^\infty e^{-\xi(t-v)}\, d\gamma_j(\xi)\, z \mathbf{1}_{|z|<\epsilon} \right) \nu_j(dz).$$

As $\int_0^\infty e^{-\xi(t-v)}\, d\gamma_j(\xi) = g_j(t-v)$, we obtain Eq. (8.32). The differential of $f(s,t)$ is obtained by applying Itô's lemma for Lévy processes. $\qquad \square$

This last proposition emphasizes that our short-term rate model can be reformulated as a forward rate model in which the forward rate dynamic is independent from the processes $\left(Y_s^{(j,\xi)} \right)_{\xi \geq 0}$ for $j = 1, \ldots, d$.

8.6 Discretization Scheme

Instead of considering an infinity of processes $\left(Y_t^{(j,\xi)} \right)_{t \geq 0,\, j \in \{1,\ldots,d\}}$, we approach the model with a finite number of equivalent processes. This presents several advantages. Firstly, it makes it possible to implement our model. Secondly, we can rely on the Itô calculus in most of the arguments. This allows us to deduce the dynamics of bond prices in both the approximated and original models. The key step consists in approximating the $\gamma_j(\cdot)$'s by discrete measures with a finite number of atoms. For this purpose, we consider a partition $\mathcal{E}^{(n)} := \{0 < \xi_0^{(n)} < \xi_1^{(n)} < \ldots < \xi_n^{(n)} < \infty\}$. On each interval $(\xi_k^{(n)}, \xi_{k+1}^{(n)})$, we define the barycenter of $\gamma_j(\cdot)$:

$$b_{k+1}^{(j)} = \frac{\int_{\xi_k^{(n)}}^{\xi_{k+1}^{(n)}} z\, d\gamma_j(z)}{\int_{\xi_k^{(n)}}^{\xi_{k+1}^{(n)}} d\gamma_j(z)} \qquad j = 1, \ldots, d \qquad (8.33)$$

and the mass of the corresponding atoms is defined as the measure of intervals of the partition:

$$m_{k+1}^{(j)} = \int_{\xi_k^{(n)}}^{\xi_{k+1}^{(n)}} d\gamma_j(z), \qquad (8.34)$$

for $k = 0, \ldots, n - 1$. In practice, we choose $\xi_0^{(n)} = 0$ and set $\xi_n^{(n)}$ to a percentile of the density $\gamma_j(z)$ (e.g., 95%). The discrete measure for a partition of size n is defined as follows:

$$\gamma_j^{(n)}(z) = \sum_{k=1}^{n} m_k^{(j)} \delta_{b_k^{(j)}}(z) , \qquad (8.35)$$

where $\delta_{b_k^{(j)}}(z)$ is the Dirac measure located at point $b_k^{(j)}$. We assume the following assumptions hold for the partition $\mathcal{E}^{(n)}$:

- $\xi_0^{(n)} \to 0$ and $\xi_n^{(n)} \to \infty$ when $n \to \infty$.
- $\max |\xi_{i+1}^{(n)} - \xi_i^{(n)}| \to 0$ when $n \to \infty$.
- $\mathcal{E}^{(n)} \subset \mathcal{E}^{(n+1)}$.

In this case, for any function $f(\cdot)$ integrable with respect to $\gamma_j(\cdot)$, we have that $\lim_{n \to \infty} \int_0^\infty f(z) \, d\gamma_j^{(n)}(z) = \int_0^\infty f(z) \, d\gamma_j(z)$. Figure 8.4 shows the differential of measures $\gamma(\cdot)$ of the ML and PML kernels. For the ML kernels (left plot), decreasing the α clearly makes the right tail of $d\gamma(\cdot)$ fatter. The right plot also reveals that the construction of the partition requires particular care for the PML

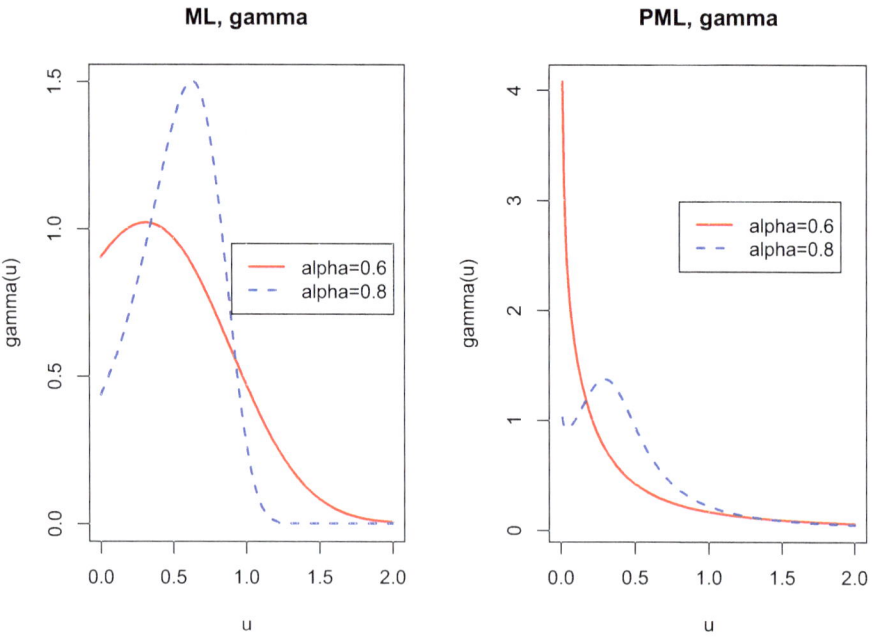

Fig. 8.4 Left and right plots: measures $\gamma(\cdot)$ of the Mittag-Leffler (ML) and power Mittag-Leffler (PML) kernels. $\beta = 0.5$

kernel as we have to numerically integrate the measure $\gamma_{\alpha,\beta}^{(p)}(z)$, which is not defined for $z = 0$. The next two propositions, respectively, present the expressions of integrals needed for calculating the discrete equivalent distributions of $\gamma_j(\cdot)$ in the ML and PML cases.

Proposition 8.4 *For the ML kernel, the discrete probability mass of atoms is given by*

$$m_{k+1}^{(j)} = \int_{\xi_k^{(n)}}^{\xi_{k+1}^{(n)}} d\gamma_j(z) \tag{8.36}$$

$$= \frac{1}{\pi \alpha_j} \sum_{l=1}^{\infty} \frac{(-1)^{l-1}}{l \, \beta_j^l \, l!} \sin\left(\pi \alpha_j l\right) \Gamma\left(\alpha_j l + 1\right) \left[\left(\xi_{k+1}^{(n)}\right)^l - \left(\xi_k^{(n)}\right)^l \right],$$

whereas the numerator in the expression of barycenters (8.33) is

$$\int_{\xi_k^{(n)}}^{\xi_{k+1}^{(n)}} z \, d\gamma_j(z) = \frac{1}{\pi \alpha_j} \sum_{l=1}^{\infty} \frac{(-1)^{l-1}}{\beta^l \, (l+1)!} \sin\left(\pi \alpha_j l\right) \Gamma(\alpha_j l + 1) \tag{8.37}$$

$$\times \left[\left(\xi_{k+1}^{(n)}\right)^{l+1} - \left(\xi_k^{(n)}\right)^{l+1} \right].$$

This result is a direct consequence of the definition of the ML function. To the best of our knowledge, these expressions do not admit any other representations. In the PML case, barycenters have a closed form expression.

Proposition 8.5 *For the PML kernel, the discrete probability mass of atoms is given by*

$$m_{k+1}^{(j)} = \int_{\xi_k^{(n)}}^{\xi_{k+1}^{(n)}} d\gamma_j(z) = \frac{1}{\pi \alpha_j} \arctan\left(\frac{\left(\xi_{k+1}^{(n)}\right)^{\alpha_j} + \beta_j \cos(\pi \alpha_j)}{\beta_j \sin(\pi \alpha_j)} \right) \tag{8.38}$$

$$- \frac{1}{\pi \alpha_j} \arctan\left(\frac{\left(\xi_k^{(n)}\right)^{\alpha_j} + \beta_j \cos(\pi \alpha_j)}{\beta_j \sin(\pi \alpha_j)} \right).$$

The numerator in the expression of barycenters (8.33) does not admit a closed form expression but may be approached by

$$
\int_{\xi_k^{(n)}}^{\xi_{k+1}^{(n)}} z \, d\gamma_j(z) \approx \left(\frac{\xi_{k+1}^{(n)} + \xi_k^{(n)}}{2\pi\alpha_j} \right) \tag{8.39}
$$

$$
\times \left[\arctan\left(\frac{\left(\xi_{k+1}^{(n)}\right)^\alpha + \beta_j \cos(\pi\alpha_j)}{\beta_j \sin(\pi\alpha_j)} \right) \right.
$$

$$
\left. - \arctan\left(\frac{\left(\xi_k^{(n)}\right)^\alpha + \beta_j \cos(\pi\alpha_j)}{\beta_j \sin(\pi\alpha_j)} \right) \right].
$$

Proof The expression of the probability mass $m_{k+1}^{(j)}$ comes from the relation

$$
\frac{\beta \, \sin(\alpha\pi)}{\pi} \int \frac{z^{\alpha-1}}{z^{2\alpha} + 2\beta z^\alpha \cos(\alpha\pi) + \beta^2} dz = \frac{1}{\pi\alpha} \arctan\left(\frac{z^\alpha + \beta \cos(\pi\alpha)}{\beta \sin(\pi\alpha)} \right).
$$

The second result is obtained by an integration by parts in which the integral is approached with the trapezoidal method. $\square$

To lighten the notation, we define $Y_s^{(j,k)} := Y_s^{(j, b_k^{(n)})}$ for $k = 1, \ldots, n$ and $j = 1, \ldots, d$. Let us first recall that $Y_0^{(j,k)} = 0$, and therefore $Y_s^{(j,k)}$ is an integral from zero to s with respect to the jth Lévy process

$$
Y_s^{(j,k)} = \int_0^s e^{-b_k^{(j)}(s-u)} dL_u^{(j)}. \tag{8.40}
$$

Replacing the continuous measures $\left(\gamma_j\right)_{j=1,\ldots,d}$ by their discrete equivalents leads to the approximation $r_s^{(n)}$ of the short-term rate r_s:

$$
r_s^{(n)} = \varphi^{(n)}(s) + \sum_{j=1}^d \sum_{k=1}^n m_k^{(j)} Y_s^{(j,k)}. \tag{8.41}
$$

We adopt the following notation: $B^{(j,k)}(s, t) = \frac{1}{b_k^{(j)}}\left(1 - e^{-b_k^{(j)}(t-s)}\right)$ for $k = 1, \ldots, n$ and $j = 1, \ldots, d$. The bond price in the discretized model with partitions of size n is denoted by $P^{(n)}(s, t)$ and is equal to

$$P^{(n)}(s, t) = \tag{8.42}$$

$$\exp\left(-\int_s^t \varphi^{(n)}(u)du - \sum_{j=1}^d \sum_{k=1}^n Y_s^{(j,k)} B^{(j,k)}(s, t) m_k^{(j)}\right)$$

$$\times \prod_{j=1}^d \exp\left(\int_s^t \psi_j\left(-\sum_{k=1}^n B^{(j,k)}(u, t) m_k^{(j)}\right) du\right),$$

where $\varphi^{(n)}(t)$ is the discretized version of $\varphi(t)$:

$$\varphi^{(n)}(t) = -\frac{\partial}{\partial t}\ln\left(P(0, t)\right) - \sum_{j=1}^d \psi_j\left(-\sum_{k=1}^n B^{(j,k)}(0, t) m_k^{(j)}\right). \tag{8.43}$$

As the number of processes in the discretized model is finite, we can apply Itô's lemma in order to establish the dynamic of $P^{(n)}(s, t)$.

Proposition 8.6 *The zero-coupon bond price $P^{(n)}(s, t)$ is a geometric Lévy process solution of the SDE:*

$$\frac{dP^{(n)}(s, t)}{P^{(n)}(s, t)} = r_s^{(n)}ds - \sum_{j=1}^d \sigma_j \sum_{k=1}^n B^{(j,k)}(s, t) m_k^{(j)}dW_s^{(j)} \tag{8.44}$$

$$+ \sum_{j=1}^d \int_{\mathbb{R}} \left(e^{-z\left(\sum_{k=1}^n m_k^{(j)} B^{(j,k)}(s,t)\right)} - 1\right)\left(J_j(ds, dz) - \nu_j(dz)ds\right),$$

with the terminal condition $P^{(n)}(t, t) = 1$.

Proof If we remember that $dY_s^{(j,k)} = dL_s^{(j)} - b_k^{(j)}Y_s^{(j,k)}ds$, Itô's lemma gives us the following differential for $P^{(n)}(s, t)$:

$$dP^{(n)}(s, t) = \frac{\partial P^{(n)}(s, t)}{\partial s}ds - \sum_{j=1}^d \sum_{k=1}^n \frac{\partial P^{(n)}(s, t)}{\partial Y^{(j,k)}}b_k^{(j)}Y_s^{(j,k)}ds \tag{8.45}$$

$$+ \sum_{j=1}^d \sum_{k=1}^n \frac{\partial P^{(n)}(s, t)}{\partial Y^{(j,k)}}\mu_j ds + \sum_{j=1}^d \sum_{k=1}^n \frac{\partial P^{(n)}(s, t)}{\partial Y^{(j,k)}}\int_{\mathbb{R}} z \tilde{J}_j(dt, dz)$$

$$+ \sum_{j=1}^d \sum_{k=1}^n \frac{\partial P^{(n)}(s, t)}{\partial Y^{(j,k)}}\sigma_j dW_s^{(j)} + \frac{1}{2}\sum_{j=1}^d \sum_{k=1}^n \sum_{l=1}^n \frac{\partial^2 P^{(n)}(s, t)}{\partial Y^{(j,k)}\partial Y^{(j,l)}}\sigma_j^2 ds$$

$$+ \sum_{j=1}^{d} \int_{\mathbb{R}} P^{(n)}(s,t) \left(e^{-z \left(\sum_{k=1}^{n} m_k^{(j)} B^{(j,k)}(s,t) \right)} - 1 \right)$$

$$- z \sum_{k=1}^{n} \frac{\partial P^{(n)}(s,t)}{\partial Y^{(j,k)}} J_j(ds, dz). \tag{8.46}$$

The first- and second-order partial derivatives of $P^{(n)}(s,t)$ are equal to

$$\frac{\partial P^{(n)}(s,t)}{\partial Y^{(j,k)}} = -P^{(n)}(s,t) B^{(j,k)}(s,t) m_k^{(j)}, \tag{8.47}$$

$$\frac{\partial^2 P^{(n)}(s,t)}{\partial Y^{(j,k)} \partial Y^{(i,l)}} = P^{(n)}(s,t) B^{(j,k)}(s,t) B^{(i,l)}(s,t) m_k^{(j)} m_l^{(i)}, \tag{8.48}$$

whereas its partial derivative with respect to time is given by

$$\frac{\partial P^{(n)}(s,t)}{\partial s} = P^{(n)}(s,t) \left(\varphi^{(n)}(s) + \sum_{j=1}^{d} \sum_{k=1}^{n} Y_s^{(j,k)} e^{-b_k^{(j)}(t-s)} m_k^{(j)} \right. \tag{8.49}$$

$$\left. - \sum_{j=1}^{d} \psi_j \left(-\sum_{k=1}^{n} B^{(j,k)}(s,t) m_k^{(j)} \right) \right),$$

where the characteristic exponent is developed as the following sum:

$$\psi_j \left(-\sum_{k=1}^{n} B^{(j,k)}(s,t) m_k^{(j)} \right) = \tag{8.50}$$

$$- \mu_j \sum_{k=1}^{n} B^{(j,k)}(s,t) m_k^{(j)} + \frac{1}{2} \left(\sum_{k=1}^{n} B^{(j,k)}(s,t) m_k^{(j)} \right)^2 \sigma_j^2$$

$$\int_{\mathbb{R}} \left(e^{-\left(\sum_{k=1}^{n} B^{(j,k)}(s,t) m_k^{(j)} \right) z} - 1 + \left(\sum_{k=1}^{n} B^{(j,k)}(s,t) m_k^{(j)} \right) z \mathbf{1}_{|z| < \epsilon} \right) \nu_j(dz).$$

Combining Eqs. (8.45), (8.47), (8.48), (8.49), and (8.50) allows us to rewrite the dynamic of bond prices as follows:

$$dP^{(n)}(s,t) = P^{(n)}(s,t) \left(\varphi^{(n)}(s) + \sum_{j=1}^{d} \sum_{k=1}^{n} Y_s^{(j,k)} e^{-b_k^{(j)}(t-s)} m_k^{(j)} \right) ds$$

$$+ P^{(n)}(s,t) \sum_{j=1}^{d} \sum_{k=1}^{n} Y_s^{(j,k)} B^{(j,k)}(s,t) m_k^{(j)} b_k^{(j)} ds$$

$$-P^{(n)}(s,t) \sum_{j=1}^{d} \sigma_j \sum_{k=1}^{n} B^{(k)}(s,t) m_k^{(n)} dW_s^{(j)}$$

$$+P^{(n)}(s,t) \sum_{j=1}^{d} \int_{\mathbb{R}} \left(e^{-z\left(\sum_{k=1}^{n} m_k^{(j)} B^{(j,k)}(s,t)\right)} - 1 \right) \left(J_j(ds,dz) - \nu_j(dz)ds \right).$$

We infer the result if we remember Eq. (8.41) and notice that

$$\sum_{k=1}^{n} Y_s^{(j,k)} e^{-b_k^{(j)}(t-s)} m_k^{(j)} + \sum_{k=1}^{n} Y_s^{(j,k)} B^{(k)}(s,t) m_k^{(j)} b_k^{(j)} = \sum_{k=1}^{n} Y_s^{(j,k)} m_k^{(j)}.$$

□

As by construction $\lim_{n\to\infty} \gamma_j^{(n)}(z) = \gamma_j(z)$, we immediately infer the dynamic of the bond price in the non-discretized model by considering the limit of Eq. (8.44).

Corollary 8.7 *If the short-term rate is driven by Eq. (8.25), the bond price is the solution of the following SDE:*

$$\frac{dP(s,t)}{P(s,t)} = r_s ds - \sum_{j=1}^{d} \sigma_j h_j(s,t) dW_s^{(j)} \tag{8.51}$$

$$+ \sum_{j=1}^{d} \int_{\mathbb{R}} \left(e^{-z h_j(s,t)} - 1 \right) \left(J_j(ds,dz) - \nu_j(dz)ds \right),$$

where $h_j(s,t) = \int_0^\infty B^{(\xi)}(s,t) d\gamma_j(\xi)$ and with the terminal condition $P(t,t) = 1$.

In order to understand the role played by the memory kernel in the evolution of the term structure of interest rates, we analyze the expectation and variance of bond yields. The bond yield of a zero-coupon bond in the discretized model is defined as

$$y^{(n)}(s,t) = -\frac{\ln P^{(n)}(s,t)}{t-s}.$$

From Eq. (8.42), we infer that this yield is proportional to a weighted sum of mean reverting processes $Y_s^{(j,k)}$. Since $\mathbb{E}\left(Y_s^{(j,k)} | \mathcal{F}_0 \right) = 0$ by construction, the expected yield conditionally to the initial filtration is equal to

$$\mathbb{E}\left(y^{(n)}(s,t)\,|\,\mathcal{F}_0\right) = \frac{1}{t-s}\int_s^t \varphi^{(n)}(u)\mathrm{d}u \tag{8.52}$$

$$-\frac{1}{t-s}\sum_{j=1}^d\int_s^t \psi_j\left(-\sum_{k=1}^n B^{(j,k)}(v,t)\,m_k^{(j)}\right)\mathrm{d}v\,.$$

The variance of the yield is the sum of d variances of a linear combination of processes $Y_s^{(j,k)}$. If we remember that the covariance of $Y_s^{(j,k)}$ and $Y_s^{(l,k)}$ is equal to

$$\mathbb{C}\left(Y_s^{(j,k)}Y_s^{(j,l)}|\mathcal{F}_0\right) = \left(\sigma_j^2 + \int_{\mathbb{R}} z^2 v_j(\mathrm{d}z)\right)\int_0^s e^{-\left(b_k^{(j)}+b_l^{(j)}\right)(s-u)}\mathrm{d}u \tag{8.53}$$

$$= \frac{\sigma_j^2 + \int_{\mathbb{R}} z^2 v_j(\mathrm{d}z)}{b_k^{(j)}+b_l^{(j)}}\left(1-e^{-\left(b_k^{(j)}+b_l^{(j)}\right)s}\right),$$

the variance of the yield, conditionally to $\mathcal{F}_0$, is the following triple sum:

$$\mathbb{V}\left(y^{(n)}(s,t)\,|\,\mathcal{F}_0\right) = \sum_{j=1}^d \mathbb{V}\left(\sum_{k=1}^n \frac{B^{(j,k)}(s,t)\,m_k^{(j)}}{t-s}Y_s^{(j,k)}|\mathcal{F}_0\right) \tag{8.54}$$

$$= \sum_{j=1}^d\left[\sum_{k=1}^n\sum_{l=1}^n \frac{B^{(j,k)}(s,t)\,m_k^{(j)}}{t-s}\frac{B^{(j,l)}(s,t)\,m_l^{(j)}}{t-s}\mathbb{C}\left(Y_s^{(j,k)}Y_s^{(j,l)}|\mathcal{F}_0\right)\right].$$

To illustrate these results, we consider a univariate jump-diffusion model ($d=1$) in which a single Lévy process is ruled by the following SDE:

$$\mathrm{d}L_t^{(1)} = -\lambda_1\eta_1\mathrm{d}t + \sigma_1\mathrm{d}W_t^{(1)} + \eta_1\,\mathrm{d}N_t^{(1)}\,. \tag{8.55}$$

$N_t^{(1)}$ is a Poisson process with constant intensity λ_1. The characteristic exponent is in this case equal to

$$\psi_1(\omega) = -\lambda_1\eta_1\omega + \frac{1}{2}\omega^2\sigma_1^2 + \lambda_1\left(e^{\omega\eta_1}-1\right), \tag{8.56}$$

whereas the Lévy measure is $v_1(z) = \lambda_1\delta_{\eta_1}(z)$, where $\delta_{\eta_1}(z)$ is the Dirac measure located at $z=\eta_1$. The instantaneous variance of $\mathrm{d}L_t^{(1)}$ is equal to

$$\sigma_1^2 + \int_{\mathbb{R}} z^2 v_1(\mathrm{d}z) = \sigma_1^2 + \lambda_1\eta_1^2\,.$$

Of course, we can consider other types of Lévy processes like the variance gamma and the normal inverse gaussian, but this does not fundamentally modify the conclusions drawn in this section. We first fit the curve $\varphi(t)$ to the term structure of zero-coupon bond yields, bootstrapped from the ICE swap rates on 26/2/21 at noon. Since this function is proportional to instantaneous forward rates, we have to interpolate bond yields. For this purpose, we use a Nelson–Siegel (NS) model. Appendix of this chapter recalls this model and reports the swap rates curve on 26/2/21 and the parameter estimates. There exist more advanced interpolation methods, such as splines or kriging techniques, as detailed in Cousin et al. [6], but the NS model is sufficiently accurate for our purposes, that is, to understand the dynamic of yields over time.

We calculate the term structure of expected yields in 5, 10, and 15 years with $n = 40$ atoms. Increasing n does not significantly modify the results. A sensitivity analysis with respect to the number of atoms is proposed in Sect. 8.7. For the parameters, we set $\beta = 1.5$, $\lambda_1 = 0.5$, $\sigma_1 = 0.01$, and $\eta_1 = -0.0002$, which are possible realistic values in view of results from Sect. 8.3. We remind the reader that α is the parameter tuning the memory of the model. If $\alpha = 1$, the memory kernel is exponential and the model forgets in an exponential manner past fluctuations of interest rates, whereas for lower values of α, the model forgets these variations according to a power decaying function. Figures 8.5 and 8.6 show expected yields computed with the ML and PML models for $\alpha = 0.5$ and $\alpha = 0.9$. We use a discretization scheme with 40 atoms that cover the ML and PML density up to their 90% percentiles. For both models, the range between short- and long-term yields is narrowing with the time horizon. In the ML model, the lower the α, the longer the memory and the quicker the convergence to a bumped (but nearly flat) yield curve. In the PML model, expected yield curves with a low α dominate those with a high α and have a more pronounced curvature.

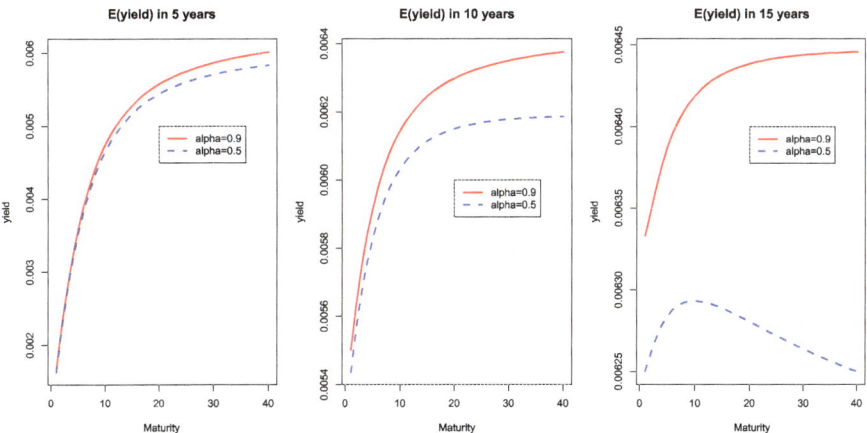

Fig. 8.5 Expected yield curves in 5, 10, and 15 years, $\alpha = 0.5$ and 0.9, $\beta = 1.5$, $\lambda_1 = 0.5$, $\eta_1 = -0.002$, and $\sigma_1 = 0.01$. ML kernel with $n = 40$ atoms

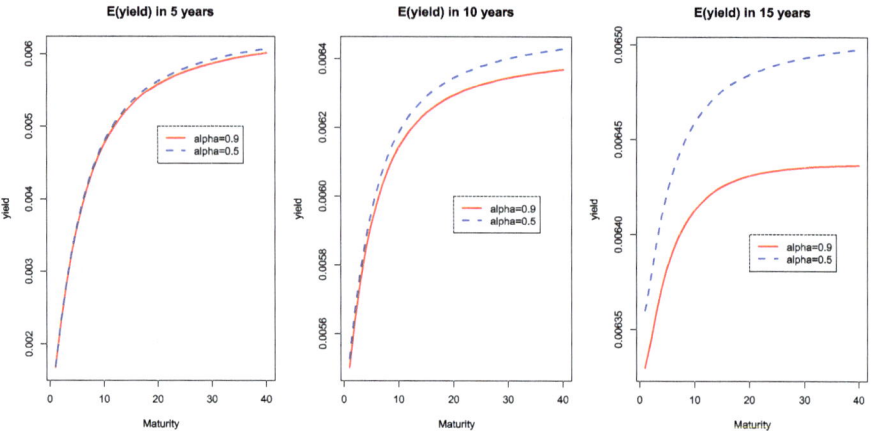

Fig. 8.6 Expected yield curves in 5, 10, and 15 years, $\alpha = 0.5$ and 0.9, $\beta = 1.5$, $\lambda_1 = 0.5$, $\eta_1 = -0.002$, and $\sigma_1 = 0.01$. PML kernel with $n = 40$ atoms

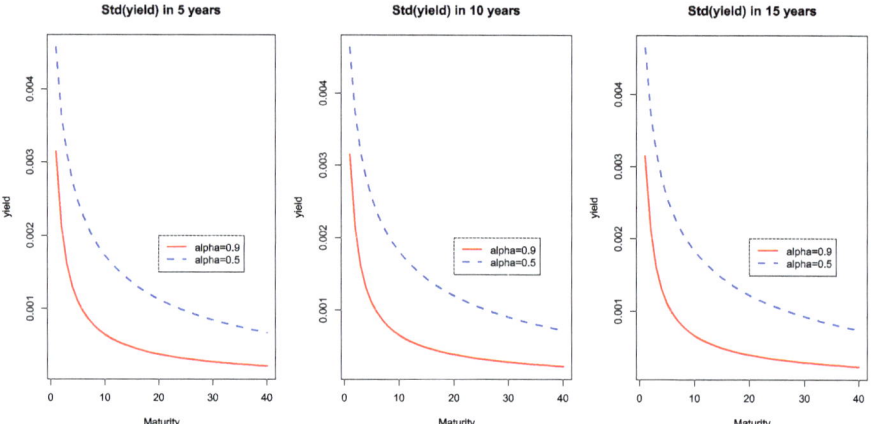

Fig. 8.7 Expected standard deviations of yields in 5, 10, and 15 years, $\alpha = 0.5$ and 0.9, $\beta = 1.5$, $\lambda_1 = 0.5$, $\eta_1 = -0.002$, and $\sigma_1 = 0.01$. ML kernel with $n = 40$ atoms

Figures 8.7 and 8.8 report the expected standard deviations of future yields calculated with the ML and PML models. In both cases, the term structures of expected standard deviations do not significantly evolve with the time horizon and are decreasing functions of the maturity. We also notice that standard deviations are, respectively, inversely and directly proportional to α in the ML and PML models.

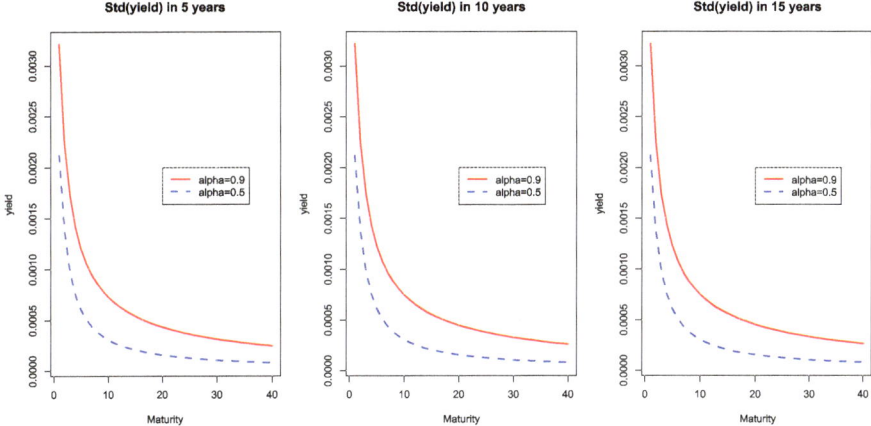

Fig. 8.8 Expected standard deviations of yields in 5, 10, and 15 years, $\alpha = 0.5$ and 0.9, $\beta = 1.5$, $\lambda_1 = 0.5$, $\eta_1 = -0.002$, and $\sigma_1 = 0.01$. PML kernel with $n = 40$ atoms

8.7 Pricing of Bond Options

The purpose of this section is to illustrate how long-term memory models can be used for the pricing of derivatives. We investigate this for European options written on a zero-coupon bond. Our choice is motivated by the fact that these are the building blocks of many structured products. The underlying bond has a maturity, denoted t_2, while the option expires at time $t_1 \le t_2$ with a strike price K. Without loss of generality, we focus on a call option, but everything remains valid for any other type of European options. The value at time $s \le t_1 \le t_2$ of the call option is denoted by $C(s)$ and is equal to the expected discounted payoff under the risk neutral measure $\mathbb{Q}$:

$$C(s) = \mathbb{E}\left(e^{-\int_s^{t_1} r_u\,du}\,(P(t_1, t_2) - K)_+ \mid \mathcal{F}_s\right).$$

This price does not admit a closed form expression. Nevertheless we can apply the standard technique of changes of numeraire combined with a discrete Fourier transform (DFT) to evaluate this call option. The first stage consists in defining a forward measure, denoted by $\mathbb{F}$, that has $P(s, t_1)$ as numeraire, i.e., every asset discounted by $P(s, t_1)$ is a martingale. If $S_t^{(0)} = e^{\int_0^t r_u\,du}$ is the cash account (numeraire of $\mathbb{Q}$), the Radon–Nikodym derivative defining this change of measure is $\left.\dfrac{d\mathbb{F}}{d\mathbb{Q}}\right|_{t_1} = \dfrac{S_0^{(0)}}{S_{t_1}^{(0)} P(0, t_1)}$. Using standard arguments, the call price is equal to the product of a discount bond and of the expected payoff under this forward measure,

$$C(s) = P(s, t_1)\mathbb{E}^{\mathbb{F}}\left((P(t_1, t_2) - K)_+ \mid \mathcal{F}_s\right). \tag{8.57}$$

The next proposition details the dynamics of underlying Lévy processes under the forward measure.

Proposition 8.8 *Under the forward measure $\mathbb{F}$ with $P(s, t_1)$ as numeraire, the process $\left(W_s^{(j)\mathbb{F}} \right)_{s \geq 0}$ such that*

$$dW_s^{(j)\mathbb{F}} = dW_t^{(j)} + h_j(u, t_1)du , \qquad (8.58)$$

for $j = 1, \ldots d$, is a Brownian motion. Furthermore, the random measure

$$\tilde{J}_j^{\mathbb{F}}(du, dz) = \tilde{J}_j(du, dz) - \mathbf{1}_{|z|<\epsilon} \left(e^{-h_j(u,t_1)z} - 1 \right) v_j(dz)du , \qquad (8.59)$$

for $j = 1, \ldots d$, defines a jump process under $\tilde{\mathbb{P}}$ with the time-varying measure $v_j^{\mathbb{F}}(dt, dz) = e^{-h_j(u,t_1)z} v_j(dz)$.

Proof From Applebaum [1, chapter 5, sections 2 and 4], the forward change of measure

$$\left. \frac{d\mathbb{F}}{d\mathbb{Q}} \right|_{t_1} = \frac{S_0^{(0)}}{S_{t_1}^{(0)} P(0, t_1)} \qquad (8.60)$$

$$= \exp \left(\sum_{j=1}^{d} \int_0^{t_1} -h_j(u, t_1)\sigma_j dW_u^{(j)} - \frac{1}{2} h_j(u, t_1)^2 \sigma_j^2 du \right.$$

$$- \sum_{j=1}^{d} \int_0^{t_1} \int_{\mathbb{R}} \left(e^{-h_j(u,t_1)z} - 1 + h_j(u, t_1)z \mathbf{1}_{|z|<\epsilon} \right) v_j(dz)du$$

$$+ \left. \sum_{j=1}^{d} \int_0^{t_1} \int_{|z|\geq\epsilon} -h_j(u, t_1)z \, \tilde{J}_j(du , dz) \right)$$

is a martingale under the measure $\mathbb{Q}$ which defines an equivalent measure $\mathbb{F}$ under which $\left(W_s^{(j)\mathbb{F}} \right)_{s \geq 0}$ and $\tilde{J}_j^{\mathbb{F}}(du, dz)$ for $j = 1, \ldots d$ are, respectively, Brownian motions and random jump measures satisfying Eqs. (8.58) and (8.59). $\square$

Combining Corollary 8.7 and Proposition 8.8 allows us to infer that a zero-coupon bond of maturity $t \geq t_1$ is ruled under the forward measure by the SDE

$$
\frac{dP(s,t)}{P(s,t)} = \left(r_s + \sum_{j=1}^{d} \sigma_j h_j(s,t_1)^2 + \sum_{j=1}^{d} \int_{\mathbb{R}} \left(e^{-zh_j(s,t_1)} - 1 \right)^2 v_j(dz) \right) ds
$$

$$
+ \sum_{j=1}^{d} \int_{\mathbb{R}} \left(e^{-zh_j(s,t_1)} - 1 \right) \left(J_j^{\mathbb{F}}(ds, dz) - e^{-h_j(s,t_1)z} v_j(dz) ds \right)
$$

$$
- \sum_{j=1}^{d} \sigma_j h_j(s,t_1) dW_s^{(j)\mathbb{F}} . \tag{8.61}
$$

This equation emphasizes that the average bond return under the forward measure is the risk-free rate plus a positive time-varying premium that converges toward zero when $s \to t_1$. The next proposition reveals that the processes $Y_t^{(j,\xi)}$ no longer revert toward zero under $\mathbb{F}$.

Proposition 8.9 *Let us define the function*

$$
\theta^{(j,\xi)}(u) = \begin{cases} -\left(h_j(u,t_1)\sigma_j + \int_{\mathbb{R}} z \left(1 - e^{-h_j(u,t_1)z} \mathbf{1}_{|z|<\epsilon} \right) v_j(dz) \right) & \xi > 0 \\ 0 & \xi = 0. \end{cases}
$$

$$
\tag{8.62}
$$

Under the forward measure $\mathbb{F}$, $Y_t^{(j,\xi)}$ *is a process reverting to* $\theta^{(j,\xi)}(t)/\xi$ *at a speed* ξ, *such that*

$$
Y_t^{(j,\xi)} = e^{-\xi(t-s)} Y_s^{(j)} + \int_s^t e^{-\xi(t-u)} \theta^{(j,\xi)}(u) du \tag{8.63}
$$

$$
+ \sigma_j \int_s^t e^{-\xi(t-u)} dW_u^{(j)\mathbb{F}} + \int_s^t e^{-\xi(t-u)} \int_{\mathbb{R}} z \tilde{J}_j^{\mathbb{F}}(du, dz),
$$

for all $j = 1, \ldots, d$ *and* $\xi \in \mathbb{R}^+$.

Proof By definition of $Y_s^{(j,\xi)}$ and from Corollary 8.7, we immediately infer its dynamic under the forward measure,

$$
dY_s^{(j,\xi)} = -\xi Y_s^{(j,\xi)} ds + \sigma_j dW_s^{(j)\mathbb{F}} + \int_{\mathbb{R}} z \tilde{J}_j^{\mathbb{F}}(ds, dz) \tag{8.64}
$$

$$
+ \left(\mu_j - h_j(s,t_1)\sigma_j + \int_{|z|<\epsilon} z \left(e^{-h_j(s,t_1)z} - 1 \right) v_j(dz) \right) ds .
$$

Since we impose that $\mu_j = -\int_{\mathbb{R}} \left(z - z1_{|z|<\epsilon} \right) \nu_j(dz)$ for $j = 1, \ldots, d$ in order to ensure that $\mathbb{E}\left(L_t^{(j)} \right) = 0$, we deduce that $Y_s^{(j,\xi)}$ reverts to $\theta^{(j,\xi)}(t)/\xi$:

$$dY_s^{(j,\xi)} = \xi \left(\frac{1}{\xi} \theta^{(j,\xi)}(s) - Y_s^{(j,\xi)} \right) ds + \sigma_j dW_s^{(j)\mathbb{F}} + \int_{\mathbb{R}} z \tilde{J}_j^{\mathbb{F}}(ds, dz) ,$$

and the solution of this SDE is Eq. (8.63).   □

Equations (8.60) and (8.64) are useful when simulating sample paths of interest rates and bond prices under the forward measure, for example, for pricing exotic options. One solution to price a call option on a zero-coupon bond consists in numerically inverting the Fourier transform of the density of its bond log-return under the forward measure. This step is performed with a standard discrete Fourier transform (DFT) algorithm that approximates the probability density function of the log-return under the forward measure $\mathbb{F}$. The expectation of the payoff under the forward measure is then computed with this distribution. The moment generating function of the bond log-return, presented in the next proposition, admits a closed form formula and is a key result to implement the DFT procedure.

Proposition 8.10 *The moment generating function (mgf) of $\ln P(t_1, t_2)$ under the measure $\mathbb{F}$, conditionally to the filtration $\mathcal{F}_s$, is given by*

$$\mathbb{E}^{\mathbb{F}}\left(e^{\omega \ln P(t_1,t_2)} | \mathcal{F}_s \right) = \exp\left(-\omega \int_{t_1}^{t_2} \varphi(u) du \right. \tag{8.65}$$

$$- \sum_{j=1}^{d} \int_{s}^{t_1} \psi_j\left(-h_j(u, t_1) \right) du + \omega \sum_{j=1}^{d} \int_{t_1}^{t_2} \psi_j\left(-h_j(u, t_2) \right) du$$

$$- \omega \sum_{j=1}^{d} \int_{0}^{\infty} Y_s^{(j,\xi)} e^{-\xi(t_1-s)} B^{(\xi)}(t_1, t_2) d\gamma_j(\xi)$$

$$+ \sum_{j=1}^{d} \int_{s}^{t_1} \psi_j\left(-\int_{0}^{\infty} \frac{1}{\xi} \left(1 - (1+\omega) e^{-\xi(t_1-u)} + \omega e^{-\xi(t_2-u)} \right) d\gamma_j(\xi) \right) du \right).$$

Proof The mgf can be rewritten as an expectation under the risk neutral measure using Bayes' rule

$$\mathbb{E}^{\mathbb{F}}\left(e^{\omega \ln P(t_1,t_2)} | \mathcal{F}_s \right) = \frac{\mathbb{E}\left(\left. \frac{d\mathbb{F}}{d\mathbb{Q}} \right|_{t_1} e^{\omega \ln P(t_1,t_2)} | \mathcal{F}_s \right)}{\mathbb{E}\left(\left. \frac{d\mathbb{F}}{d\mathbb{Q}} \right|_{t_1} | \mathcal{F}_s \right)}, \tag{8.66}$$

where $\left.\dfrac{d\mathbb{F}}{d\mathbb{Q}}\right|_{t_1}$ is given in Eq. (8.60). From Eq. (8.30) and due to the independence between the processes $Y_{t_1}^{(j,\xi)}$ and $Y_{t_1}^{(k,\xi)}$ for $j \neq k$, the expectation (8.66) becomes

$$\mathbb{E}^{\mathbb{F}}\left(e^{\omega \ln P(t_1,t_2)}|\mathcal{F}_s\right) = \exp\left(-\omega\int_{t_1}^{t_2}\varphi(u)du\right)$$

$$\times \exp\left(-\sum_{j=1}^{d}\int_{s}^{t_1}\psi_j\left(-h_j(u,t_1)\right)du + \omega\sum_{j=1}^{d}\int_{t_1}^{t_2}\psi_j\left(-h_j(u,t_2)\right)du\right)$$

$$\times \prod_{j=1}^{d}\mathbb{E}\left(\exp\left(-\int_{s}^{t_1}h_j(u,t_1)\,dL_u^{(j)} - \omega\int_{0}^{\infty}Y_{t_1}^{(j,\xi)}B^{(\xi)}(t_1,t_2)\,d\gamma_j(\xi)\right)\mid\mathcal{F}_s\right).$$

From Eq. (8.24) and after a change of integration order, the integral of the processes $Y_{t_1}^{(j,\xi)}$ is

$$\int_{0}^{\infty}Y_{t_1}^{(j,\xi)}B^{(\xi)}(t_1,t_2)\,d\gamma_j(\xi)$$

$$= \int_{0}^{\infty}Y_s^{(j,\xi)}e^{-\xi(t_1-s)}B^{(\xi)}(t_1,t_2)\,d\gamma_j(\xi)$$

$$+ \int_{0}^{\infty}\int_{s}^{t_1}e^{-\xi(t_1-u)}dL_u^{(j)}B^{(\xi)}(t_1,t_2)\,d\gamma_j(\xi)$$

$$= \int_{0}^{\infty}Y_s^{(j,\xi)}e^{-\xi(t_1-s)}B^{(\xi)}(t_1,t_2)\,d\gamma_j(\xi)$$

$$+ \int_{s}^{t_1}\int_{0}^{\infty}e^{-\xi(t_1-u)}B^{(\xi)}(t_1,t_2)\,d\gamma_j(\xi)\,dL_u^{(j)}.$$

This allows us to rewrite $\mathbb{E}^{\mathbb{F}}\left(e^{\omega \ln P(t_1,t_2)}|\mathcal{F}_s\right)$ as follows:

$$\mathbb{E}^{\mathbb{F}}\left(e^{\omega \ln P(t_1,t_2)}|\mathcal{F}_s\right) = \exp\left(-\omega\int_{t_1}^{t_2}\varphi(u)du\right)$$

$$\times \exp\left(-\sum_{j=1}^{d}\int_{s}^{t_1}\psi_j\left(-h_j(u,t_1)\right)du + \omega\sum_{j=1}^{d}\int_{t_1}^{t_2}\psi_j\left(-h_j(u,t_2)\right)du\right)$$

$$\times \exp\left(-\omega\sum_{j=1}^{d}\int_{0}^{\infty}Y_s^{(j,\xi)}e^{-\xi(t_1-s)}B^{(\xi)}(t_1,t_2)\,d\gamma_j(\xi)\right)$$

$$\times \prod_{j=1}^{d}\mathbb{E}\left(\exp\left(Z(u)\right)\mid\mathcal{F}_s\right),$$

where

$$Z(u) = -\int_s^{t_1} \left(h_j(u, t_1) - \omega \int_0^\infty e^{-\xi(t_1-u)} B^{(\xi)}(t_1, t_2) \, d\gamma_j(\xi) \right).$$

By definition of $h_j(u, t) := \int_0^\infty B^{(\xi)}(u, t) \, d\gamma_j(\xi)$, we have that

$$h_j(u, t_1) - \omega \int_0^\infty e^{-\xi(t_1-u)} B^{(\xi)}(t_1, t_2) \, d\gamma_j(\xi)$$

$$= \int_0^\infty \left(B^{(\xi)}(u, t_1) - \omega e^{-\xi(t_1-u)} B^{(\xi)}(t_1, t_2) \right) d\gamma_j(\xi),$$

wherein the integrand is equal to

$$B^{(\xi)}(u, t_1) - \omega e^{-\xi(t_1-u)} B^{(\xi)}(t_1, t_2)$$

$$= \frac{1}{\xi} \left(1 - (1 + \omega) e^{-\xi(t_1-u)} + \omega e^{-\xi(t_2-u)} \right).$$

Combining these last three equations leads to the result.   □

The expression (8.65) of the mgf involves integrals of $Y_s^{(j,\xi)}$ with respect to $\xi \in \mathbb{R}^+$. These integrals do not admit analytical formulas and are in practice approached with the discretization scheme introduced in Sect. 8.6. Let n be the number of atoms of the discrete approximation of $(\gamma_j)_{j=1,\ldots,d}$. To lighten the notation, we make the following definitions:

$$h_j^{(n)}(u, t) := \sum_{k=1}^n \frac{m_k^{(j)}}{b_k^{(j)}} \left(1 - e^{-b_k^{(j)}(t-u)} \right),$$

$$g_j^{(n)}(\omega, u, t_1, t_2) := \sum_{k=1}^n \frac{m_k^{(j)}}{b_k^{(j)}} \left(1 - (1 + \omega) e^{-b_k^{(j)}(t_1-u)} + \omega e^{-b_k^{(j)}(t_2-u)} \right),$$

for $j = 1, \ldots, d$. The discretized version of the mgf of the bond log-return under the forward measure is as follows:

$$\mathbb{E}^{\mathbb{F}} \left(e^{\omega \ln P^{(n)}(t_1, t_2)} | \mathcal{F}_s \right) =$$

$$\exp \left(\sum_{j=1}^d \int_s^{t_1} \psi_j \left(-g_j^{(n)}(\omega, u, t_1, t_2) \right) - \psi_j \left(-h_j^{(n)}(u, t_1) \right) du \right)$$

$$\times \exp\left(\int_{t_1}^{t_2} \omega \sum_{j=1}^{d} \psi_j\left(-h_j^{(n)}(u, t_2)\right) - \omega \varphi^{(n)}(u) \, du\right)$$

$$\times \exp\left(-\sum_{j=1}^{d} \sum_{k=1}^{n} m_k^{(j)} Y_s^{(j,k)} e^{-b_k^{(j)}(t_1 - s)} B^{(j,k)}(t_1, t_2)\right),$$

where $\varphi^{(n)}(t)$, the discretized version of $\varphi(t)$, is given in Eq. (8.43). The integrals with respect to times are numerically computed, for example, using Simpson's rule. Let us denote by $f^{(n)}(x)$ the probability density of $\ln P^{(n)}(t_1, t_2)$ conditionally to $\mathcal{F}_s$ and by

$$\Upsilon^{(n)}(i\omega) = \mathbb{E}^{\mathbb{F}}\left(e^{i\omega \ln P^{(n)}(t_1, t_2)} \mid \mathcal{F}_s\right) = \int e^{i\omega x} f^{(n)}(x) \, dx$$

its Fourier transform. The probability density function can therefore be expressed as the real part of the inverse Fourier transform:

$$f^{(n)}(x) = \frac{1}{2\pi} \int_{-\infty}^{+\infty} \Upsilon^{(n)}(i\omega) \, e^{-ix\omega} d\omega \qquad (8.67)$$

$$= \frac{1}{\pi} \mathrm{Re}\left(\int_0^{+\infty} \Upsilon^{(n)}(i\omega) e^{-ix\omega} d\omega\right)$$

and is numerically computed with a DFFT algorithm such as detailed in Proposition 1.8 of Chap. 1.

Figure 8.9 shows the probability density functions of $P^{(n)}(5, 10)$ under the forward measure of numeraire $P^{(n)}(0, 5)$. We consider for this illustration a one-factor Lévy model ($d = 1$) in which the driving process is a jump-diffusion such as presented in Eq. (8.55), with the same parameters as those of Sect. 8.6. The DFT parameters are $x_{\max} = 0.2$ and $M = 2^{10}$. The plots reveal that variances of bond prices under $\mathbb{F}$ display the same sensitivity to the memory parameter α as under the risk neutral measure. The PML and ML bond variances are, respectively, directly and inversely proportional to α.

After computing the log-bond density, the call price is calculated by the following sum:

$$C(s) = P(s, t_1)\mathbb{E}^{\mathbb{F}}\left((P(t_1, t_2) - K)_+ \mid \mathcal{F}_s\right)$$

$$\approx P(s, t_1) \sum_{k=1}^{M} \mathbf{1}_{\{x_k \geq \ln K\}} \left(e^{x_k} - K\right) f^{(n)}(x_k) \Delta_x,$$

where $\mathbf{1}_{\{x_k \geq \ln K\}}$ is an indicator variable equal to one on $[\ln L, +\infty)$ and zero otherwise. Figure 8.10 shows call prices on $P^{(n)}(5, 10)$ for various strike prices in the ML and PML kernels. In the PML model, increasing the memory parameter α

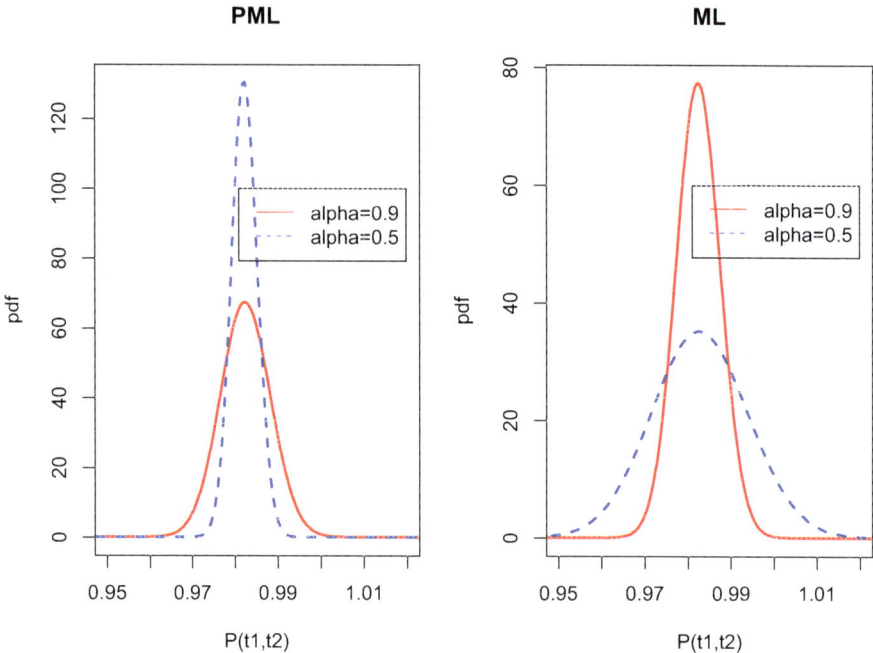

Fig. 8.9 Probability density function of $P^{(n)}(5, 10)$ under the forward measure of numeraire $P^{(n)}(0, 5)$, $\alpha = 0.5$ and 0.9, $\beta = 1.5$, $\lambda_1 = 0.5$, $\eta_1 = -0.002$, and $\sigma_1 = 0.01$. ML and PML kernels with $n = 40$ atoms

drives up the option values whatever the strike price. For the ML kernel, we observe the opposite trend. This is consistent with our previous conclusion that the variance is the main driving factor of option prices.

Figure 8.11 allows us to confirm the convergence of the call option when the number, n, of atoms in the discretization scheme grows. We recall that the partition ranges from $\xi_0 = 0$ up to ξ_n, which is a percentile of $\gamma_j(z)$. Here percentiles 90, 95, and 97% are considered. For the ML kernel (right plot), the convergence is quick and the difference between prices computed with the 90 and 97% is negligible (around 2.6e−5). For the PML model (right plot), prices are higher (and then conservative) with a 90% percentile and converge with fewer atoms than with a partition covering 97% of $\gamma_j(\cdot)$.

8.8 Further Reading

The family of short-term rate models has gathered multiple frameworks. We refer the reader, for example, to Boero and Torricelli [2] or to Schmidt [25] for a review. On the other hand, various processes have been proposed to explain the interest

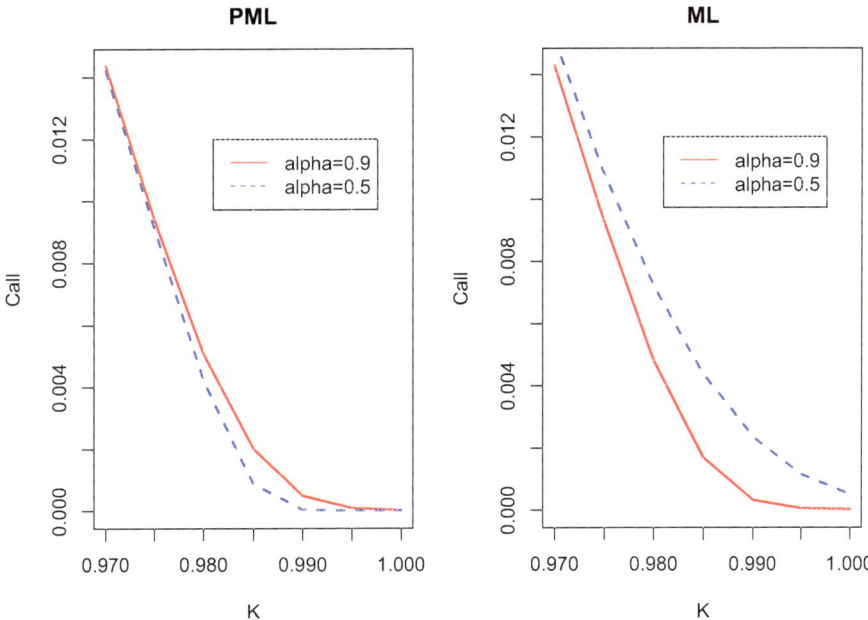

Fig. 8.10 Call prices on $P^{(n)}(5, 10)$, $\alpha = 0.5$ and 0.9, $\beta = 1.5$, $\lambda_1 = 0.5$, $\eta_1 = -0.002$, and $\sigma_1 = 0.01$. ML and PML kernels with $n = 40$ atoms

rate risk. The literature is too vast to give an exhaustive list here. We restrict our attention to the contributions that are related to this book and which do not attempt a general overview, referring instead to Brigo and Mercurio [4] for a detailed account on the topic. For instance, Eberlein and Raible [8] were among the first to propose a short-term rate model driven by Lévy processes. In their setting, the short-term rate is ruled by Ornstein–Uhlenbeck processes reverting in an exponential manner to a mean level. Within this framework, Eberlein and Kluge [7] derive analytical formulae for the prices of caps and floors using bilateral Laplace transforms. In a similar setting, Hainaut and Macgilchrist [16] propose a pentanomial tree for pricing derivatives. Hainaut [14] studies the properties of a Gaussian short-term rate with a Markov Switching Multifractal volatility. Moreno and Platania [23] study a square-root model replicating economic cycles in the dynamic of interest rates. Hainaut [15] and Njike Leunga and Hainaut [24] develop unicurve and multicurve models in which the short-term rate is exposed to self-exciting jumps. Fontana et al. [11] propose a modeling framework exploiting the self-exciting behavior of continuous-state branching processes with immigration (CBI).

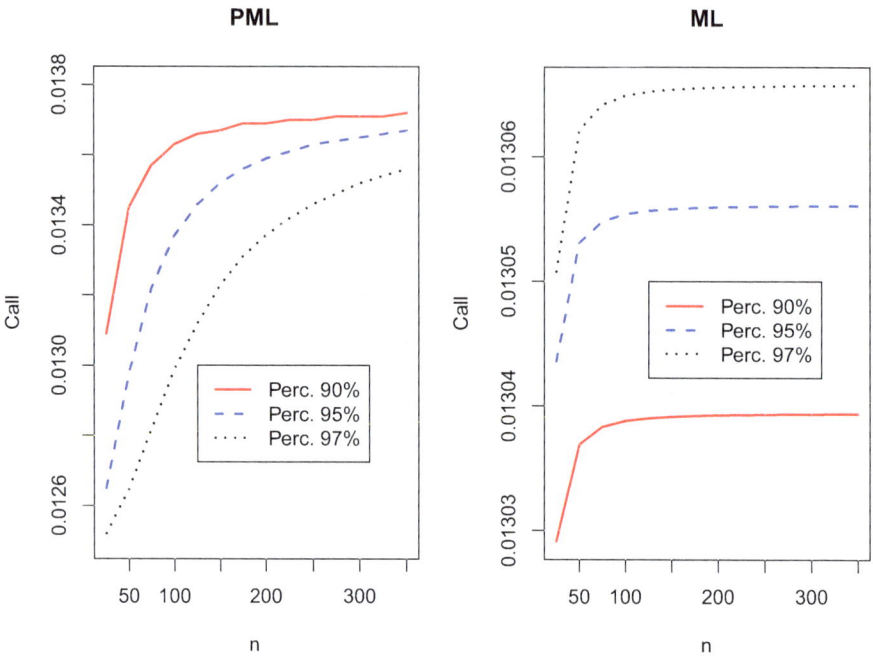

Fig. 8.11 Call prices on $P^{(n)}(5, 10)$ as a function of n, $\alpha = 0.5$ and 0.9, $\beta = 1.5$, $\lambda_1 = 0.5$, $\eta_1 = -0.002$, and $\sigma_1 = 0.01$. Strike $= 0.97$

Appendix

The Nelson–Siegel model postulates the following representation of instantaneous forward rates:

$$f(0, t) = b_0 + (b_{10} + b_{11}t) \exp(-c_1 t) , \qquad (8.68)$$

the parameters b_0, b_{10}, and b_{11}, respectively, determine the mean level, the slope, and the curvature of the forward rates curve. The zero-coupon rate in the Nelson–Siegel model (8.68) is obtained by integrating these forward rates and by dividing the result by the maturity:

$$y_m(t) = b_0 + \frac{1}{t} \frac{b_{10}}{c_1} \left(1 - e^{-c_1 t}\right) + \frac{1}{t} \frac{b_{11}}{(c_1)^2} \left(1 - (c_1 t + 1) e^{-c_1 t}\right) . \quad (8.69)$$

The parameters are estimated by minimizing the quadratic spread between market and model zero-coupon yields. In the numerical illustration, this model is fitted to ICE Euro swap rates on 26/2/21 at 12 o'clock, presented in Table 8.2. The corresponding Nelson–Siegel parameters are reported in Table 8.3.

Table 8.2 ICE Euro swap rates on 26/2/21, 12:00. Source: https://fred. stlouisfed.org/

Maturity	Swap rates (%)	Maturity	Swap rates (%)
1	−0.466	8	−0.027
2	−0.419	9	0.036
3	−0.363	10	0.148
4	−0.298	12	0.275
5	−0.229	15	0.376
6	−0.160	20	0.393
7	−0.092	25	0.373

Table 8.3 Nelson–Siegel parameters for the ICE Euro swap rates on 26/2/21

Parameters	Values
b_0	0.00504905
b_{10}	−0.00892662
b_{11}	−0.00350623
c_1	0.29428630

References

1. Applebaum, D.: Lévy processes and Stochastic Calculus. Cambridge Studies in Advanced Mathematics. Cambridge University Press, Cambridge (2004)
2. Boero, G., Torricelli, C.: A comparative evaluation of alternative models of the term structure of interest rates. Eur. J. Oper. Res. **93**(1), 205–223 (1996)
3. Brace, A., Gatarek, M., Musiela, M.: The market model of interest rate dynamics. Math. Finance **7**(2), 127–147 (1997)
4. Brigo, D., Mercurio, F.: Interest Rate Models: Theory and Practice: With Smile, Inflation. Springer, Berlin (2006)
5. Cont, R., Tankov, P.: Financial Modelling with Jump Processes. Financial Mathematics Series. Chapman & Hall, Boca Raton (2003)
6. Cousin, A., Maatouk, H., Rullière, D.: Kriging of financial term-structures. Eur. J. Oper. Res. **255**(2), 631–648 (2016)
7. Eberlein, E., Kluge, W.: Exact pricing formulae for caps and swaptions in a Lévy term structure model. J. Comput. Finance **9**(2), 99–125 (2005)
8. Eberlein, E., Raible, S.: Term structure models driven by general Lévy processes. Math. Finance **9**(1), 31–53 (1999)
9. Erdélyi, A., Magnus, W., Oberhettinger, F., Tricomi, F.: Higher Transcendental Functions, vol. 3. McGraw-Hill, New York (1955)
10. Falini, J.: Pricing caps with HJM models: the benefits of humped volatility. Eur. J. Oper. Res. **207**(3), 1358–1367 (2010)
11. Fontana, C., Gnoatto, A., Szulda, G.: Multiple yield curve modelling with CBI processes. Math. Financ. Econ. **15**, 579–610 (2020)
12. Gorenflo, R., Mainardi, F.: Fractional calculus. In: Carpinteri, A., Mainardi, F. (eds.) Fractals and Fractional Calculus in Continuum Mechanics. International Centre for Mechanical Sciences (Courses and Lectures), vol. 378, pp. 223–276. Springer, Berlin (1997)
13. Gorenflo, R., Kilbas, A.A., Mainardi, F., Rogosin, S.V.: Mittag-Leffler Functions, Related Topics and Applications. Springer, Berlin (2014)
14. Hainaut, D.: A fractal version of the Hull–White interest rate model. Econ. Model. **31**, 323–334 (2013)
15. Hainaut, D.: A model for interest rates with clustering effects. Quant. Finan. **16**(8), 1203–1218 (2016)

16. Hainaut, D., Macgilchrist, R.: An interest rate tree driven by a Lévy process. J. Deriv. **18**(2), 33–45 (2010)
17. Haubold, H.J., Mathai, A.M., Saxena, R.K.: Mittag-Leffler functions and their applications. J. Appl. Math. **2011**, 298628 (2011).
18. Heath, D., Jarrow, R., Morton, A.: Bond pricing and the term structure of interest rates: a new methodology for contingent claims valuation. Econometrica **60**(1), 77–105 (1992)
19. Hull, J., White, A.: Pricing interest rate derivative securities. Rev. Financ. Stud. **3**(4), 573–592 (1990)
20. Li, H., Ye, X., Yu, F.: Unifying Gaussian dynamic term structure models from a Heath–Jarrow–Morton perspective. Eur. J. Oper. Res. **286**(3), 1153–1167 (2020)
21. Mainardi, F.: Why the Mittag-Leffler function can be considered the queen function of the fractional calculus? Entropy **22**, 1359 (2020)
22. Mercurio, F., Moraleda, J.: An analytically tractable interest rate model with humped volatility. Eur. J. Oper. Res. **120**(1), 109–121 (2000)
23. Moreno, M., Platania, F.: A cyclical square-root model for the term structure of interest rates. Eur. J. Oper. Res. **241**, 109–121 (2015)
24. Njike Leunga, C., Hainaut, D.: Interbank credit risk modelling with self-exciting jump processes. Int. J. Theor. Appl. Finance **23**(6), 2050039 (2020)
25. Schmidt, W.M.: Interest rate term structure modelling. Eur. J. Oper. Res. **214**(1), 1–14 (2011)

Chapter 9
Affine Volterra Processes and Rough Models

The previous chapters studied processes that depend on the convolution of a function and their past sample path. For instance, the fractional Brownian motion of Chap. 6 is proportional to $\int_0^t (t-u)^{H-\frac{1}{2}} \, dW_u$, where W_u is a Brownian motion. In a similar manner, the interest rate model of Chap. 8 in the Brownian case depends upon $\int_0^t g(t-u) \, dW_u$, where g is a decreasing kernel function. This chapter further studies the properties of processes of this type which are solutions of stochastic convolution equations with affine coefficients. What are called Volterra processes have gathered a broad family of dynamics: Ornstein–Uhlenbeck, Heston, and even the Rough Heston model of El Euch et al. [12]. In this last model, which has become a standard for options pricing, the volatility behaves like the mean reverting fractional motion studied in Chap. 6. The theoretical results presented in this Chapter are inspired by Abi Jaber et al. [2]. We conclude this chapter with an econometric analysis of the rough Heston model. For this purpose, we adapt the particle filter of Chap. 3 to fit the rough Heston model to the S&P 500 time series.

9.1 Introduction

We consider a class of d-dimensional affine Volterra processes $(X_t)_{t \geq 0}$ of the form

$$X_t = X_0 + \int_0^t K(t-s)b(X_s)ds + \int_0^t K(t-s)\sigma(X_s)dW_s, \qquad (9.1)$$

where $(W_t)_{t \geq 0}$ is a d-dimensional Brownian process defined on a probability space Ω, endowed with the filtration $\mathcal{F}$ and the probability measure $\mathbb{P}$. The kernel function

$K \in L^2_{\text{loc}}(\mathbb{R}_+, \mathbb{R}^{d \times d})$ is a locally square integrable matrix function, i.e.,

$$\int_E |K_{i,j}(u)|^2 du < \infty \ , \ i, j \in \{1, ...d\}$$

for any compact subset E of $\mathbb{R}^+$. If $\boldsymbol{x} = (x_1, ..., x_d)^\top$, the function $b(\boldsymbol{x})$ is the scalar product of $\boldsymbol{x}$ and of a matrix $B \in \mathbb{R}^{d \times d}$, plus a constant b_0:

$$b(\boldsymbol{x}) = b_0 + B\boldsymbol{x} \ .$$

The function $\sigma(\boldsymbol{x})$ is such that the $d \times d$-matrix $a(\boldsymbol{x})$ is

$$a(\boldsymbol{x}) := \sigma(\boldsymbol{x})\sigma(\boldsymbol{x})^\top = A_0 + \sum_{j=1}^d A_j x_j,$$

where $\left(A_j\right)_{j=0,...,d}$ belong to the space of symmetric positive definite matrices of dimension d, denoted $\mathbb{S}^d$. This modeling framework encompasses extensions of well-known financial models such as the Ornstein–Uhlenbeck or the square-root processes and the Heston model we already met in Chap. 3. We detail them hereunder.

(a) **The Volterra Ornstein–Uhlenbeck process**. In this case, the matrices $\left(A_j\right)_{j=1,...,d}$ are null and X_t is defined by the following sum of convolutions:

$$X_t = X_0 + \int_0^t K(t - s)(b_0 + BX_s)\,ds + \int_0^t K(t - s)\sigma\,dW_s \ , \qquad (9.2)$$

where $B \in \mathbb{R}^{d \times d}$ and σ is a $d \times d$ constant matrix such that $\sigma\sigma^\top = A_0$ is a constant positive matrix.

(b) **The Volterra square-root process**. The state space of this process is $\mathbb{R}^{d+}$. The matrix A_0 is null, whereas A_i is null except for the (i, i) element which is equal to σ_i^2 with $\sigma_i \geq 0$. The matrix K is diagonal with scalar L^2-kernels $(K_i)_{i=1,...,d}$. In this setup,

$$X_{i,t} = X_{i,0} + \int_0^t K_i(t - s)\left(b_{0,i} + B_{i,.}X_s\right)ds + \int_0^t K_i(t - s)\sigma_i\sqrt{X_{i,s}}\,dW_s \ , \qquad (9.3)$$

for $i = 1, ..., d$.

(c) **The Volterra Heston model**. Let us, respectively, denote by $(S_t)_{t \geq 0}$ and $(V_t)_{t \geq 0}$ the price of a stock and its instantaneous variance. The process X_t that we consider here is bivariate and equal to $X_t = (\log S_t, V_t)$. If $\rho \in [0, 1]$ is the correlation between the price and its variance, the dynamic of the stock price is given by

$$\frac{dS_t}{S_t} = \mu dt + \sqrt{V_t}\left(\sqrt{1 - \rho^2} dW_s^1 + \rho dW_s^2\right), \tag{9.4}$$

whereas the variance is such that

$$V_t = V_0 + \int_0^t g(t-s)\kappa \left(\gamma - V_s\right) ds + \int_0^t g(t-s)\sigma\sqrt{V_s} dW_s^2. \tag{9.5}$$

Here $g(\cdot)$ is a univariate kernel. When this kernel is an exponential decreasing function $g(t) = e^{-\kappa t}$, we retrieve the Heston model of Chap. 3. Using Itô's lemma, the log-price is such that

$$\ln S_t = \ln S_0 + \int_0^t \mu - \frac{V_s}{2} ds + \int_0^t \sqrt{V_s}\left(\sqrt{1 - \rho^2} dW_s^1 + \rho dW_s^2\right).$$

This model can be reframed as a bivariate affine Volterra process, $X_t = (\ln S_t, V_t)$. In this case, the 2×2 matrix of standard deviations is given by

$$\sigma(X_t) = \begin{pmatrix} \sqrt{V_t}\sqrt{1 - \rho^2} & \sqrt{V_t}\rho \\ 0 & \sigma\sqrt{V_t} \end{pmatrix}.$$

We may check that $A_0 = A_1 = 0$ and that

$$K = \begin{pmatrix} 1 & 0 \\ 0 & g \end{pmatrix} \quad , \quad b_0 = \begin{pmatrix} \mu \\ \kappa\gamma \end{pmatrix}$$

$$B = \begin{pmatrix} 0 & -\frac{1}{2} \\ 0 & -\kappa \end{pmatrix} \quad , \quad A_2 = \begin{pmatrix} 1 & \rho\sigma \\ \rho\sigma & \sigma^2 \end{pmatrix}$$

allow us to rewrite Eqs. (9.4) and (9.5) as a bivariate Volterra affine process.

9.2   Convolution and Resolvent

For a measurable function K on $\mathbb{R}^+$ and a measure L on $\mathbb{R}^+$ of locally bounded variation, the convolution $(K * L)(t)$ is defined as

$$(K * L)(t) = \int_0^t K(t-s)L(ds).$$

For two functions K and F from $\mathbb{R}^+$ to $\mathbb{R}^d$, we denote by $(K * F)(t)$ their convolution:

$$(K * F)(t) = \int_0^t K(t - s)F(s)ds .$$

If we denote by $(M_t)_{t \geq 0}$ a d-dimensional continuous local martingale, the convolution of $K \in L^2_{\text{loc}}(\mathbb{R}^+, \mathbb{R}^d)$ is defined as

$$(K * dM)_t = \int_0^t K(t - s)dM_s .$$

This is a well-defined Itô integral such that for any $t \geq 0$,

$$\int_0^t |K(t - s)|^2 d \langle M \rangle_s < \infty ,$$

where $\langle M \rangle_t$ is the angular bracket of M_t. This notation can be extended to the matrix case. Using it, we can rewrite the Volterra equation (9.1) in a concise manner

$$X_t = X_0 + (K * (b(X)dt + \sigma(X)dW))_t .$$

If $Z_t = \int_0^t b(X_s)ds + \int_0^t \sigma(X_s)dW_s$, X_t is equal to $X_0 + (K * Z)_t$.

Proposition 9.1 *Let $K \in L^2_{\text{loc}}(\mathbb{R}^+, \mathbb{R}^{d \times d})$ and L be an $\mathbb{R}^{d \times d}$-valued measure on $\mathbb{R}^+$ of locally bounded variation. Let $(M_t)_{t \geq 0}$ be a d-dimensional continuous local martingale with $\langle M \rangle_t = \int_0^t a_s ds$ for some locally bounded adapted process, $(a_t)_{t \geq 0}$. Then*

$$(L * (K * dM))_t = ((L * K) * dM)_t ,$$

for every $t \geq 0$. In particular, taking $F \in L^1_{\text{loc}}(\mathbb{R}^+)$ and $L(dt) = F(t)dt$, we have

$$(F * (K * dM))_t = ((F * K) * dM)_t .$$

Proof This result is proven with the stochastic Fubini theorem. In the case $d = 1$, we have indeed that

$$(L * (K * dM))_t = \int_0^t \int_0^{t-s} K(t - s - u)dM_u \, L(ds)$$

$$= \int_0^t \int_0^t 1_{\{u \leq t-s\}} K(t - s - u)dM_u \, L(ds) .$$

This integral is finite almost surely,

$$\int_0^t \left(\int_0^t 1_{\{u \le t-s\}} K(t-s-u) d\langle M \rangle_u \right)^2 L(\mathrm{d}s)$$

$$\le \max_{0 \le s \le t} |a_s|^{1/2} \|K\|_{L^2(0,t)} L([0,t]).$$

We can then apply Fubini's theorem and infer that

$$(L * (K * \mathrm{d}M))_t = \int_0^t \int_0^t 1_{\{u \le t-s\}} K(t-s-u) L(\mathrm{d}s) \, \mathrm{d}M_u$$

$$= ((L * K) * \mathrm{d}M)_t.$$

$\square$

We now introduce the concept of the resolvent of a kernel. This resolvent plays an important role in the reformulation of affine Volterra processes.

Definition 9.2 Let us consider a kernel function $K \in L^1_{\mathrm{loc}}(\mathbb{R}^+, \mathbb{R}^{d \times d})$. The *resolvent (of the second kind)* of K is the kernel $R \in L^1_{\mathrm{loc}}(\mathbb{R}^+, \mathbb{R}^{d \times d})$ such that

$$K * R = R * K = K - R.$$

It may be proven that the resolvent always exists and inherits properties from K such as integrability and continuity. We refer the interested reader to Gripenberg et al. [14] for further details on resolvents. Using this resolvent allows us to infer the variation of constants formula detailed in the next proposition.

Proposition 9.3 *Let $(X_t)_{t \ge 0}$ be a d-dimensional continuous process, $F : \mathbb{R}^+ \to \mathbb{R}^d$ a continuous function, $B \in \mathbb{R}^{d \times d}$, and $Z_t = \int_0^t b(X_s)\mathrm{d}s + \int_0^t \sigma(X_s)\mathrm{d}W_s$. Then*

$$X = F + (K B) * X + K * \mathrm{d}Z$$

$$\Updownarrow \qquad\qquad (9.6)$$

$$X = F - R_B * F + E_B * \mathrm{d}Z,$$

where R_B is the resolvent of $-K B$ (i.e., $-R_B * K B = -K B - R_B$) and $E_B = K - R_B * K$.

Proof Assume that $X = F + (K B) * X + K * \mathrm{d}Z$. Convolving this with R_B and using Proposition 9.1 give us

$$X - R_B * X = (F - R_B * F) + (K B - R_B * K B) * X + (K - R_B * K) * \mathrm{d}Z$$

$$= (F - R_B * F) - R_B * X + E_B * \mathrm{d}Z.$$

Conversely, we assume that $X = F - R_B * F + E_B * dZ$. By definition of R_B, we have that

$$(KB) * E_B = (KB) * (K - R_B * K) = -R_B * K .$$

Then the convolution of KB and X is

$$KB * X = KB * F - KB * R_B * F - (R_B * K) * dZ .$$

Since the difference $X - (KB) * X$ is equal to

$$X - (KB) * X = F + \underbrace{(-KB - R_B + KB * R_B)}_{=0} * F$$

$$+ \underbrace{(E_B + R_B * K)}_{=K} * dZ ,$$

we infer the result.   □

We apply this last proposition to the Volterra processes introduced in Sect. 9.1 in order to emphasize its utility.

(a) **The Volterra Ornstein–Uhlenbeck process**. We consider here a 1-dimensional process. If $Z_t = \int_0^t b_0 ds + \int_0^t \sigma(X_s) dW_s$, the 1-dimensional Volterra process defined by Eq. (9.2) is rewritten as

$$X_t = X_0 + (KB * X)_t + (K * Z)_t .$$

From relation (9.6), we infer that

$$X_t = \left(1 - \int_0^t R_B(s) ds\right) X_0 + b_0 \int_0^t E_B(s) ds + \int_0^t E_B(t-s) \sigma dW_s . \quad (9.7)$$

A particular case is the *Vasicek model*. In this univariate Ornstein–Uhlenbeck model, the process X_t is such that

$$dX_t = \kappa (\theta - X_t) dt + \sigma dW_t,$$

and therefore X_t is given by

$$X_t = X_0 + \int_0^t \kappa \theta ds - \int_0^t \kappa X_s ds + \int_0^t \sigma dW_t . \quad (9.8)$$

We identify that $b_0 = \kappa\theta$, $K(t) = 1$, $B = -\kappa$, $a = \sigma^2$. The resolvent of $-KB = \kappa$ is $R_B(t) = \kappa e^{-\kappa t}$ and

$$E_B(t) = 1 - \int_0^t \kappa e^{-\kappa(t-s)} ds$$

$$= e^{-\kappa t}.$$

Since $\int_0^t R_B(s)ds = 1 - e^{-\kappa t}$ and $\int_0^t E_B(s)ds = \frac{1}{\kappa}\left(1 - e^{-\kappa t}\right)$, from Eq. (9.7), we infer that

$$X_t = e^{-\kappa t} X_0 + \theta\left(1 - e^{-\kappa t}\right) + \int_0^t e^{-\kappa(t-s)}\sigma dW_s.$$

(b) **The Volterra square-root process**. We consider a 1-dimensional Volterra process as defined by Eq. (9.3). From relation (9.6), we obtain that

$$X_t = \left(1 - \int_0^t R_B(s)ds\right) X_0 + b_0 \int_0^t E_B(s)ds \qquad (9.9)$$

$$+ \int_0^t E_B(t-s)\sigma\sqrt{X_s}dW_s.$$

A particular case is the Cox–Ingersoll–Ross model. The process X_t is in this case

$$dX_t = \kappa\left(\theta - X_t\right)dt + \sigma\sqrt{X_t}dW_t$$

or after integration

$$X_t = X_0 + \int_0^t \kappa\theta ds - \int_0^t \kappa X_s ds + \int_0^t \sigma\sqrt{X_s}dW_s.$$

We identify that $b_0 = \kappa\theta$, $K(t) = 1$, $B = -\kappa$, $a(x) = \sigma^2 x$. The resolvent R_B is the same as the one obtained for the Vasicek model, and from Eq. (9.9), we deduce that

$$X_t = e^{-\kappa t} X_0 + \theta\left(1 - e^{-\kappa t}\right) + \int_0^t e^{-\kappa(t-s)}\sigma\sqrt{X_s}dW_s.$$

(c) **The Volterra Heston process**. We consider the Volterra Heston process as defined by Eq. (9.3). From relation (9.6), we obtain that

$$X_t = \left(I_2 - \int_0^t R_B(s)ds\right) X_0 + \int_0^t E_B(s)ds\, b_0 + \int_0^t E_B(t-s)\sigma\left(X_s\right)dW_s,$$

$$(9.10)$$

where I_2 is the 2×2 identity matrix. A particular case is the Heston model. In this case, K is the identity matrix, whereas $-KB$ and its resolvent R_B are such that

$$-R_B * KB = -KB - R_B.$$

Writing $R_B(t) = \begin{pmatrix} r_1(t) & r_2(t) \\ r_3(t) & r_4(t) \end{pmatrix}$, it solves the following equation:

$$\begin{pmatrix} 0 & \frac{1}{2}\int_0^t r_1(t-s)\mathrm{d}s + \int_0^t \kappa r_2(t-s)\mathrm{d}s \\ 0 & \frac{1}{2}\int_0^t r_3(t-s)\mathrm{d}s + \int_0^t \kappa r_4(t-s)\mathrm{d}s \end{pmatrix} = \begin{pmatrix} 0 & \frac{1}{2} \\ 0 & \kappa \end{pmatrix} - \begin{pmatrix} r_1(t) & r_2(t) \\ r_3(t) & r_4(t) \end{pmatrix}.$$

From this last relation, we immediately infer that $r_1(t) = r_3(t) = 0$, whereas $r_2(t)$ and $r_4(t)$ are solutions of

$$\int_0^t \kappa r_2(s)\mathrm{d}s = \frac{1}{2} - r_2(t),$$

$$\int_0^t \kappa r_4(s)\mathrm{d}s = \kappa - r_4(t).$$

Solving the differential equations obtained by differentiating these last expressions leads to $r_4(t) = \kappa e^{-\kappa t}$ and $r_2(t) = \frac{1}{2}e^{-\kappa t}$. A direct calculation of $E_B = K - R_B * K$ gives us

$$E_B = \begin{pmatrix} 1 & \frac{1}{2\kappa}\left(e^{-\kappa t} - 1\right) \\ 0 & e^{-\kappa t} \end{pmatrix}.$$

The integrals of R_B and E_B are, respectively, given by

$$\int_0^t R_B(s)\mathrm{d}s = \begin{pmatrix} 0 & \frac{1}{2\kappa}\left(1 - e^{-\kappa t}\right) \\ 0 & 1 - e^{-\kappa t} \end{pmatrix}$$

and

$$\int_0^t E_B(s)\mathrm{d}s = \begin{pmatrix} t & \int_0^t \frac{1}{2\kappa}\left(e^{-\kappa s} - 1\right)\mathrm{d}s \\ 0 & \frac{1}{\kappa}\left(1 - e^{-\kappa t}\right) \end{pmatrix}.$$

Injecting these relations into Eq. (9.10) allows us to rewrite the joint dynamic of $\ln S_t$ and V_t as follows:

$$
\begin{pmatrix} \ln S_t \\ V_t \end{pmatrix} = \begin{pmatrix} 1 & -\frac{1}{2\kappa}\left(1 - e^{-\kappa t}\right) \\ 0 & e^{-\kappa t} \end{pmatrix} \begin{pmatrix} \ln S_0 \\ V_0 \end{pmatrix}
$$

$$
+ \begin{pmatrix} t & \int_0^t \frac{1}{2\kappa}\left(e^{-\kappa s} - 1\right) ds \\ 0 & \frac{1}{\kappa}\left(1 - e^{-\kappa t}\right) \end{pmatrix} \begin{pmatrix} \mu \\ \kappa\gamma \end{pmatrix}
$$

$$
+ \int_0^t \begin{pmatrix} 1 & \frac{1}{2\kappa}\left(e^{-\kappa s} - 1\right) \\ 0 & e^{-\kappa s} \end{pmatrix} \begin{pmatrix} \sqrt{V_s}\sqrt{1 - \rho^2} & \sqrt{V_s}\rho \\ 0 & \sigma\sqrt{V_s} \end{pmatrix} dW_s \ .
$$

Rearranging the terms of the first equation leads to

$$
\ln S_t = \ln S_0 + \int_0^t \sqrt{V_s}\left(\sqrt{1 - \rho^2}dW_s^1 + \rho dW_s^2\right)
$$

$$
+ \mu t - \frac{1}{2}\int_0^t \underbrace{e^{-ks}V_0 + \gamma\left(1 - e^{-\kappa s}\right) + \int_0^s e^{-\kappa u}\sigma\sqrt{V_u}dW_u^2}_{V_s} ds \ .
$$

To conclude this section, we present in Table 9.1 some interesting kernels and their resolvents. Let us recall that the Mittag-Leffler function used in this table, denoted by $E_{\alpha,\beta}(x)$, where $\alpha, \beta \in \mathbb{R}^+$, is defined by the converging sum

$$
E_{\alpha,\beta}(x) = \sum_{n=0}^{\infty} \frac{(x)^n}{\Gamma(n\alpha + \beta)} \ .
$$

We met this function in Chap. 8, where it is used in the definition of the memory kernel of an interest rate model. The second kernel of Table 9.1 is called fractional because $\int_0^t \frac{s^{\alpha-1}}{\Gamma(\alpha)}dW_s$ shares properties of the fractional Brownian motion, as already mentioned in the introduction of this chapter. Later, we will study in detail the combination of this kernel with the Volterra Heston framework, also called the "rough" Heston model.

Table 9.1 Examples of kernel and resolvent functions

	$K(t)$	$R(t)$
Constant	c	ce^{-ct}
Fractional	$c\frac{t^{\alpha-1}}{\Gamma(\alpha)}$	$ct^{\alpha-1}E_{\alpha,\alpha}(-ct^\alpha)$
Exponential	$ce^{-\lambda t}$	$ce^{-\lambda t}e^{-ct}$

9.3 Moments and the Moment Generating Function

In this section, we consider an affine Volterra process as defined by Eq. (9.1) and study its moments with Proposition 9.3.

Proposition 9.4 *Let $(X_t)_{t \geq 0}$ be an affine Volterra process of dimension d. Then for all $t \leq T$, if I_d is the $d \times d$ identity matrix, then*

$$\mathbb{E}\left(X_T | \mathcal{F}_t\right) = \left(I_d - \int_0^T R_B(s)ds\right) X_0 + \left(\int_0^T E_B(s)ds\right) b_0$$

$$+ \int_0^t E_B(T-s)\,\sigma(X_s)\,dW_s \,,$$

where R_B is the resolvent of $-KB$ (i.e., $-R_B * KB = -KB - R_B$) and $E_B = K - R_B * K$.

Proof Let $Z_t = \int_0^t b_0 ds + \int_0^t \sigma(X_s)dW_s$. The Volterra process is rewritten as

$$X_t = X_0 + (KB * X)_t + (K * Z)_t \,.$$

From Proposition 9.3, we have that

$$X_t = (I - R_B * I)_t \, X_0 + (E_B * dZ)_t \,. \tag{9.11}$$

If we develop $(E_B * dZ)_t$, we obtain that

$$(E_B * dZ)_t = \int_0^t E_B(t-s)b_0 ds + \int_0^t E_B(t-s)\sigma(X_s)dW_s \,.$$

Since $M_t = \int_0^t E_B(t-s)\sigma(X_s)dW_s$ is a martingale (see Abi Jaber et al. [2, Lemma 3.1]), considering the conditional expectation of (9.11) leads to the result.   □

The next proposition provides the joint Laplace transform of an affine Volterra process and of a convolution of it. We use the following notation for $\omega \in \mathbb{C}^d$:

$$A(\omega) = \left(\omega^\top A_1 \omega, ..., \omega^\top A_d \omega\right)^\top \,,$$

that is, a d column vector. We also recall that $a(x) := \sigma(x)\sigma(x)^\top$.

Proposition 9.5 *Let $\omega \in \mathbb{C}^d$, $f \in L^1(\mathbb{R}^+, \mathbb{C}^d)$, and $T < \infty$. Let us assume that $\psi \in L^1([0, T], \mathbb{C}^d)$ solves the Riccati–Volterra equation*

$$\psi^\top = \omega^\top K + \left(f^\top + \psi^\top B + \frac{1}{2}A(\psi)^\top\right) * K \,. \tag{9.12}$$

We define the process $\{Y_t, 0 \le t \le T\}$ by

$$Y_t = Y_0 + \int_0^t \psi(T-s)^\top \sigma(X_s)dW_s \tag{9.13}$$

$$-\frac{1}{2}\int_0^t \psi(T-s)^\top a(X_s)\psi(T-s)ds,$$

$$Y_0 = \omega^\top X_0 + \int_0^T f(s)^\top X_0 ds \tag{9.14}$$

$$+\int_0^T \left(\psi(s)^\top b(X_0) + \frac{1}{2}\psi(s)^\top a(X_0)\psi(s) \right) ds,$$

which satisfies for all $0 \le t \le T$

$$Y_t = \mathbb{E}\left(\omega^\top X_T + (f^\top * X)_T | \mathcal{F}_t \right) \tag{9.15}$$

$$+\frac{1}{2}\int_t^T \psi(T-s)^\top a\left(\mathbb{E}(X_s|\mathcal{F}_t) \right) \psi(T-s)ds.$$

The process $\exp(Y_t)$ for $t \in [0, T]$ is a local martingale, and if it is a true martingale, one has the exponential affine transform formula

$$\mathbb{E}\left(\exp\left(\omega^\top X_T + (f^\top * X)_T \right) | \mathcal{F}_t \right) = \exp(Y_t).$$

The proof of this result requires the following proposition, which provides an equivalent formulation of the condition defining ψ.

Proposition 9.6 *The Riccati–Volterra equation (9.12) is equivalent to*

$$\psi^\top = \omega^\top E_B + \left(f^\top + \frac{1}{2}A(\psi)^\top \right) * E_B. \tag{9.16}$$

Proof If Eq. (9.16) holds, then from the identity $E_B * BK = -R_B * K$, we infer that

$$\psi^\top - \psi^\top * BK = \omega^\top (E_B + R_B * K) + \left(f^\top + \frac{1}{2}A(\psi)^\top \right) * (E_B + R_B * K),$$

which is equal to Eq. (9.12). We just sketch the proof of the converse. Let us assume that Eq. (9.12) holds. If $\tilde{R}_B$ is the resolvent of $-BK$, we obtain

$$\psi^\top - \psi^\top * \tilde{R}_B = \omega^\top \left(K - K * \tilde{R}_B \right) + \left(f^\top + \frac{1}{2}A(\psi)^\top \right) * \left(K - K * \tilde{R}_B \right)$$

$$-\psi^\top * \tilde{R}_B.$$

To infer Eq. (9.16), we need to show that $K * \tilde{R}_B = R_B * K$, which is the case. We refer the reader to Abi Jaber et al. [2, Lemma 4.4] for a proof of this statement.   □

We can now prove Proposition 9.5.

Proof The right-hand side of Eq. (9.15) is denoted by $\tilde{Y}_t$. We first prove that $\tilde{Y}_0 = Y_0$. Using the identity $v^\top a(\boldsymbol{x})v = v^\top A_0 v + A(v)^\top \boldsymbol{x}$ and the definition (9.14) yields

$$\tilde{Y}_t - Y_0 = \omega^\top \mathbb{E}\,(X_T - X_0) + \left(f^\top * \mathbb{E}\,(X - X_0)\right)(T) \tag{9.17}$$

$$+ \left(\frac{1}{2}A(\psi)^\top * \mathbb{E}\,(X - X_0)\right)(T) - \left(\psi^\top * (b_0 + BX_0)\right)(T),$$

where $\mathbb{E}\,(X_t - X_0) := \mathbb{E}\,(X_t - X_0|\mathcal{F}_0)$. By definition of X_t, this expectation is equal to $(K * (b_0 + B\mathbb{E}(X)))(t)$. Therefore, from Eq. (9.12), we obtain

$$\frac{1}{2}A(\psi)^\top * \mathbb{E}\,(X - X_0)$$

$$= \frac{1}{2}A(\psi)^\top * K * (b_0 + B\mathbb{E}(X))$$

$$= \frac{1}{2}\left(\psi^\top - \omega^\top K - (f^\top + \psi^\top B) * K\right) * (b_0 + B\mathbb{E}(X))$$

$$= \psi^\top * (b_0 + B\mathbb{E}\,(X)) - \omega^\top \mathbb{E}\,(X - X_0)$$

$$- (f^\top + \psi^\top B) * \mathbb{E}\,(X - X_0)\,.$$

Substituting this into Eq. (9.17) yields $\tilde{Y}_0 - Y_0 = 0$.

We prove that $\tilde{Y}_t = Y_t$ for $t > 0$. In the remainder of the proof, we denote by C a quantity not depending on t and changing from line to line. Using again the relation $v^\top a(\boldsymbol{x})v = v^\top A_0 v + A(v)^\top \boldsymbol{x}$, we obtain that

$$\tilde{Y}_t = C + \omega^\top \mathbb{E}\,(X_T|\mathcal{F}_t) + \int_0^T \left(f^\top + \frac{1}{2}A(\psi)^\top\right)(T-s)\mathbb{E}\,(X_s|\mathcal{F}_t)\,\mathrm{d}s \tag{9.18}$$

$$- \frac{1}{2}\int_0^t \psi(T-s)^\top a(X_s)\psi(T-s)\mathrm{d}s\,.$$

Using Fubini's theorem and a change of variable yields

$$\int_0^T \left(f^\top + \frac{1}{2}A(\psi)^\top\right)(T-s)\mathbb{E}\,(X_s|\mathcal{F}_t)\,\mathrm{d}s$$

$$= C + \int_0^T \left(f^\top + \frac{1}{2}A(\psi)^\top\right)(T-s)\int_0^t \mathbf{1}_{\{u<s\}}E_B(s-u)\sigma(X_u)\mathrm{d}W_u\,\mathrm{d}s$$

$$= C + \int_0^t \left(\int_u^t \left(f^\top + \frac{1}{2} A(\psi)^\top \right) (T - s) E_B(s - u) \, ds \right) \sigma(X_u) dW_u$$

$$= C + \int_0^t \left(\left(\left(f^\top + \frac{1}{2} A(\psi)^\top \right) * E_B \right) (T - u) \right) \sigma(X_u) dW_u \, .$$

From Proposition 9.4, $\mathbb{E}(X_T | \mathcal{F}_t) = C + \int_0^t E_B(T - u) \sigma(X_u) dW_u$ and then

$$\tilde{Y}_t = C + \int_0^T \left(\omega^\top E_B + \left(f + \frac{1}{2} A(\psi) \right)^\top * E_B \right) (T - u) \, \sigma(X_u) dW_u$$

$$- \frac{1}{2} \int_0^t \psi(T - s)^\top a(X_s) \psi(T - s) ds \, .$$

Evaluating this equation at $t = 0$ provides that $C = \tilde{Y}_0 = Y_0$. From Proposition 9.6 and Eq. (9.13), we deduce that $\tilde{Y} = Y$. The final statements are straightforward. By definition, Eq. (9.13), $Y + \frac{1}{2} \langle Y, Y \rangle$ is a local martingale so that $\exp(Y)$ is a local martingale too and $Y_T = \omega^\top X_T + (f^\top * X)_T$ from Eq. (9.15). $\square$

If $f = 0$ and $t = 0$, the moment generating function of X_t can be rewritten as

$$\mathbb{E}\left(\exp\left(\omega^\top X_T \right) | \mathcal{F}_0 \right) = \exp\left(\phi(T) + \varphi(T) X_0 \right) , \tag{9.19}$$

where

$$\begin{cases} \phi(T) &= \int_0^T \psi(s)^\top b_0 + \frac{1}{2} \psi(s)^\top A_0 \psi(s) ds \, , \\ \varphi(T)^\top &= \omega^\top + \int_0^T \psi(s)^\top B + \frac{1}{2} A(\psi(s))^\top \, ds \, . \end{cases} \tag{9.20}$$

An equivalent differential formulation is

$$\begin{cases} \frac{d\phi(t)}{dt} = \psi(t)^\top b_0 + \frac{1}{2} \psi(t)^\top A_0 \psi(t) & \phi(0) = 0 \, , \\ \frac{d\varphi(t)^\top}{dt} = \psi(t)^\top B + \frac{1}{2} A(\psi(t))^\top & \varphi(0) = \omega^\top . \end{cases} \tag{9.21}$$

If K admits an inverse by convolution, we can express φ as a function of ψ. Let us denote by L this inverse, also called the *resolvent of the first kind*, such that $K * L = I_d$, where I_d is the identity matrix. If we convolve L with the Riccati–Volterra Equation (9.12) and compare with Eq. (9.21), we directly infer that

$$\psi^\top * L = \omega^\top I_d + \left(\psi^\top B + \frac{1}{2} A(\psi)^\top \right) * I_d$$

$$= \varphi^\top .$$

This relation will be particularly useful when calculating the moment generating function of state variables in the rough Heston model of Sect. 9.4.

For a Volterra Ornstein–Uhlenbeck process, the matrices $(A_j)_{j=1,\ldots,d}$ are null, $B \in \mathbb{R}^{d \times d}$ and σ is a $d \times d$-matrix such that $\sigma \sigma^\top = a$. We infer from Proposition 9.6 that ψ satisfies the Riccati–Volterra equation:

$$\psi = \omega^\top E_B + f^\top * E_B .$$

For instance, in the one-dimensional Vasicek model, $b_0 = \kappa\theta$, $K(t) = 1$, $B = -\kappa$, $a = \sigma^2$, $b(x) = \kappa(\theta - x)$. The resolvent of $-KB = \kappa$ is $R_B(t) = \kappa e^{-\kappa t}$ and $E_B(t) = e^{-\kappa t}$. Therefore

$$\psi(t) = \omega e^{-\kappa t} + f * e^{-\kappa t} .$$

If $\omega = 0$ and $f = -1$, then $f * e^{-\kappa t} = -\int_0^t e^{-\kappa s} ds = \frac{1}{\kappa}\left(e^{-\kappa t} - 1\right)$. As $\psi(t) = -\frac{1}{\kappa}\left(1 - e^{-\kappa t}\right)$, expressions of Y_t and Y_0 are given by

$$Y_t = Y_0 - \frac{1}{2\kappa^2} \int_0^t \left(1 - e^{-\kappa(T-s)}\right)^2 \sigma^2 ds - \frac{1}{\kappa} \int_0^t \left(1 - e^{-\kappa(T-s)}\right) \sigma \, dW_s ,$$

$$Y_0 = -\int_0^T X_0 e^{-\kappa s} + \left(1 - e^{-\kappa s}\right)\theta - \frac{1}{2\kappa^2}\left(1 - e^{-\kappa s}\right)^2 \sigma^2 ds ,$$

and $\mathbb{E}\left(\exp\left(-\int_0^T X_s ds\right) | \mathcal{F}_t\right) = \exp(Y_t)$. In a similar manner, we infer that the Riccati–Volterra equation for a 1-dimensional Volterra square-root process is

$$\psi = \omega E_B + \left(f + \frac{\sigma^2}{2}\psi^2\right) * E_B .$$

In particular, for the Cox–Ingersoll–Ross model (9.8), $E_B(t) = e^{-\kappa t}$ and ψ is the solution of

$$\psi(t) = \omega e^{-\kappa t} + \int_0^t f(s-t)e^{-\kappa s} + \frac{\sigma^2}{2}\psi(s-t)^2 e^{-\kappa s} ds .$$

The case of the Volterra Heston model is detailed in the next section.

9.4 The Volterra and Rough Heston Model

Let us consider the Volterra Heston process as defined by Eq. (9.3). As $A_1 = 0$ and

$$\begin{pmatrix} \psi_1 \\ \psi_2 \end{pmatrix}^\top \begin{pmatrix} 1 & \rho\sigma \\ \rho\sigma & \sigma^2 \end{pmatrix} \begin{pmatrix} \psi_1 \\ \psi_2 \end{pmatrix} = \psi_1^2 + 2\rho\sigma\psi_2\psi_1 + \sigma^2\psi_2^2 ,$$

the pair (ψ_1, ψ_2) satisfies the following equality:

$$\begin{pmatrix} \psi_1 \\ \psi_2 \end{pmatrix}^\top = \begin{pmatrix} \omega_1 \\ \omega_2 g \end{pmatrix}^\top + \left(\begin{pmatrix} f_1 \\ f_2 \end{pmatrix}^\top + \begin{pmatrix} \psi_1 \\ \psi_2 \end{pmatrix}^\top \begin{pmatrix} 0 & -\frac{1}{2} \\ 0 & -\kappa \end{pmatrix} \right.$$

$$\left. + \frac{1}{2} \begin{pmatrix} 0 \\ \psi_1^2 + 2\rho\sigma\psi_2\psi_1 + \sigma^2\psi_2^2 \end{pmatrix}^\top \right) * \begin{pmatrix} 1 & 0 \\ 0 & g \end{pmatrix}.$$

From this, we obtain the equivalent formulation

$$\psi_1 = \omega_1 + 1 * f_1 , \tag{9.22}$$

$$\psi_2 = \omega_2 g + g * \left(f_2 - \kappa\psi_2 + \frac{1}{2}\left(\psi_1^2 - \psi_1 \right) + \rho\sigma\psi_2\psi_1 + \frac{1}{2}\sigma^2\psi_2^2 \right).$$

Let us again consider the classical Heston model. In this case, $g(t) = 1$ and we assume that $\omega_2 = f_1 = f_2 = 0$. We immediately infer that $\psi_1 = \omega_1$ and

$$\psi_2(t) = \int_0^t \left(\frac{1}{2}\left(\omega_1^2 - \omega_1 \right) + (\rho\sigma\omega_1 - \kappa)\,\psi_2(s) + \frac{1}{2}\sigma^2\psi_2^2(s) \right) ds .$$

Notice that

$$\frac{d\psi_2(t)}{dt} = \frac{1}{2}\left(\omega_1^2 - \omega_1 \right) + (\rho\sigma\omega_1 - \kappa)\,\psi_2(t) + \frac{1}{2}\sigma^2\psi_2^2(t)$$

$$= -\frac{\partial B(\omega_1, t, s)}{\partial t} ,$$

where $B(\omega_1, t, s)$ is the function involved in the mgf of $\ln \frac{S_t}{S_0}$ defined in Chap. 3, Eq. (3.14).

El Euch et al. [12] have developed the rough Heston model. In this model, the volatility is driven by an integral of the type $\int_0^t \frac{(t-s)^{H-\frac{1}{2}}}{\Gamma(H+\frac{1}{2})}dW_s$ with $H \in (0, \frac{1}{2})$. This integral shares many properties of the fractional Brownian motion with a Hurst index smaller than $\frac{1}{2}$. This model can be reframed as a Volterra Heston model with a fractional kernel

$$g(t) = \frac{t^{\alpha-1}}{\Gamma(\alpha)} ,$$

where $\alpha \in (\frac{1}{2}, 1]$. Notice that the operator

$$I_0^\alpha f(t) = \frac{1}{\Gamma(\alpha)} \int_0^t (t - u)^{\alpha-1} f(u)\, du$$

$$= (g * f)(t)$$

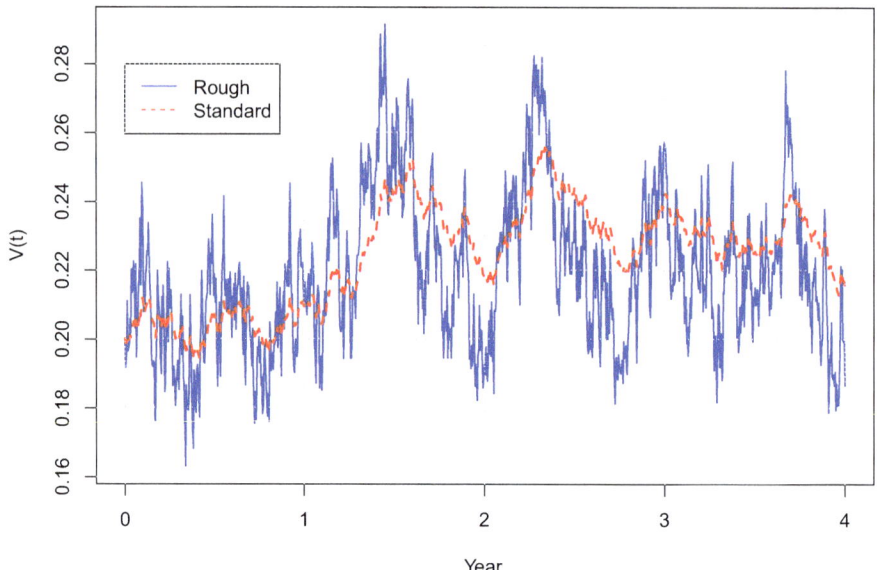

Fig. 9.1 Simulated sample paths of V_t in the rough and standard Heston models. Parameters: $\kappa = 0.8$, $\gamma = 0.20$, $\sigma = 0.05$, $\alpha = 0.58$, and $V_0 = \gamma$

is the left fractional Riemann–Liouville integral of order α. In this setting, the dynamic of the stock price under the risk neutral measure $\mathbb{Q}$ is given by

$$\frac{dS_t}{S_t} = r dt + \sqrt{V_t} \left(\sqrt{1 - \rho^2} dW_s^1 + \rho dW_s^2 \right), \tag{9.23}$$

whereas the variance is defined by the next expression

$$V_t = V_0 + \int_0^t \frac{(t - s)^{\alpha - 1}}{\Gamma(\alpha)} \kappa \left(\gamma - V_s \right) ds + \int_0^t \frac{(t - s)^{\alpha - 1}}{\Gamma(\alpha)} \sigma \sqrt{V_s} dW_s^2. \tag{9.24}$$

Figure 9.1 compares the sample path of V_t in the rough and classical Heston models, with the same occurrences for $\left(W_t^2 \right)_{t \geq 0}$. This graph emphasizes that the variance in the rough Heston model has a much more chaotic behavior than the variance in the classical setting. Gatheral et al. [13] provide empirical evidence that the volatility under the real measure has a rough dynamic.

Options and the pdf of S_t in the rough Heston model can be computed by a discrete Fourier transform (DFT) algorithm as detailed in Chap. 1, Sect. 1.6. Such a method requires the moment generating function of $\ln S_t$ to be determined. This is obtained by setting $\omega_2 = 0$ and $f_1 = f_2 = 0$ in the system of equations (9.22). Furthermore, it admits an elegant reformulation in terms of the left fractional

Riemann–Liouville derivative,

$$D^\alpha f(t) = \frac{dI^{1-\alpha} f(t)}{dt}$$

$$= \frac{1}{\Gamma(1-\alpha)} \frac{d}{dt} \int_0^t \frac{f(u)}{(t-u)^\alpha} \, du \,,$$

that is, the inverse operator of the fractional integral, i.e., $D^\alpha I^\alpha f(t) = f(t)$. Recalling Eq. (9.21) and noticing that $\psi_1 = \omega_1$, the joint moment generating function of the log-return and of the variance is equal to

$$\mathbb{E}\left(e^{\omega_1 \ln S_T} | \mathcal{F}_0\right) = \exp\left(\phi(T) + \varphi_1(T) \ln S_0 + \varphi_2(T) V_0\right), \qquad (9.25)$$

where

$$\begin{cases} \frac{d\phi(t)}{dt} = \omega_1 r + \kappa \gamma \psi_2(t) & \phi(0) = 0, \\ \frac{d\varphi_1(t)}{dt} = 0 & \varphi_1(0) = \omega_1, \\ \frac{d\varphi_2(t)}{dt} = \frac{1}{2}\left(\omega_1^2 - \omega_1\right) + (\rho\sigma\omega_1 - \kappa)\,\psi_2(t) + \frac{1}{2}\sigma^2\psi_2(t)^2 & \varphi_2(0) = 0. \end{cases}$$

$$(9.26)$$

Given that $L = \frac{t^{-\alpha}}{\Gamma(1-\alpha)}$ is such that $g * L = 1$, and $f * L = L * f = \int_0^t \frac{(t-u)^{(1-\alpha)-1}}{\Gamma(1-\alpha)} f(u) du = I^{1-\alpha} f(t)$, we have that $\varphi_2(t) = I^{1-\alpha}\psi_2(t)$ and

$$\frac{d\varphi_2(t)}{dt} = \frac{dI^{1-\alpha}\psi_2(t)}{dt} = D^\alpha \psi_2(t)\,.$$

From the system (9.26), we infer that $\varphi_1(t) = \omega_1$ and that $\psi_2(t)$ is such that

$$D^\alpha \psi_2(t) = \frac{1}{2}\left(\omega_1^2 - \omega_1\right) + (\rho\sigma\omega_1 - \kappa)\,\psi_2(t) + \frac{1}{2}\sigma^2\psi_2(t)^2\,, \qquad (9.27)$$

with the initial condition $I^{1-\alpha}\psi_2(0) = 0$ or $\psi_2(0) = 0$.

To conclude this section, we develop a numerical method for calculating the functions $\psi_2(t)$, $\varphi_2(t)$, and $\phi(t)$ involved in the mgf (9.25). Since $D_0^\alpha I_0^\alpha f(t) = f(t)$, the fractional integral of Eq. (9.27) provides the following expression for ψ_2:

$$\psi_2(t) = \int_0^t \frac{(t-s)^{\alpha-1}}{\Gamma(\alpha)} \left(\frac{1}{2}\left(\omega_1^2 - \omega_1\right) + (\rho\sigma\omega_1 - \kappa)\,\psi_2(s) + \frac{1}{2}\sigma^2\psi_2(s)^2\right) ds\,,$$

that we discretize. We choose a time step, Δ, and define $t_k = k\,\Delta$ for $k = 0, ..., n_T$, where $n_T = \lfloor \frac{T}{\Delta} \rfloor$. To lighten the notation, we denote by $\psi_2(k)$ the approached value of ψ_2 at time t_k, computed using the recursion

$$\psi_2(k) = \sum_{j=0}^{k-1} \frac{(t_k - t_j)^{\alpha-1}}{\Gamma(\alpha)} \left(\frac{1}{2} \left(\omega_1^2 - \omega_1 \right) + (\rho\sigma\omega_1 - \kappa)\,\psi_2(j) + \frac{1}{2}\sigma^2\psi_2(j)^2 \right) \Delta,$$

for $k = 1, ..., n_T$ and an initial value $\psi_2(0) = 0$. As $\varphi_2(t) = I^{1-\alpha}\psi_2(t)$ and $D^{\alpha}\psi_2(t) = \frac{dI^{1-\alpha}\psi_2(t)}{dt}$, we approach $\varphi(t)$ by the sum

$$\varphi_2(k) = \sum_{j=0}^{k-1} \left(\frac{1}{2} \left(\omega_1^2 - \omega_1 \right) + (\rho\sigma\omega_1 - \kappa)\,\psi_2(j) + \frac{1}{2}\sigma^2\psi_2(j)^2 \right) \Delta.$$

On the other hand, the function $\phi(t_k)$ is calculated as

$$\phi(k) = \sum_{j=0}^{k-1} (\omega_1 r + \kappa\gamma\psi_2(j))\,\Delta.$$

El Euch et al. [12] propose instead an implicit method that converges faster than our explicit scheme but that requires a careful initialization.

9.5   Filtering

In this section, we adapt the technique introduced in Chap. 3 in order to filter the volatility from a time series, in a rough Heston framework. We consider a sample of discrete observations of log-returns $X_t = \ln \frac{S_{t+\Delta}}{S_t}$. The sample is denoted by $x = \{x_1, x_2, ..., x_T\}$. The interval of time between two observations is Δ and the times of sampling are $\{t_1, ..., t_T\}$. We adopt the notations $V_j = V_{t_j}$ and $X_j = X_{t_j}$, whereas v_j and x_j are the realizations of V_j. The dynamics of log-returns is approached by discretizing equation (9.4):

$$X_{j+1} = \left(\mu - \frac{V_j}{2} \right) \Delta + \sqrt{1 - \rho^2}\sqrt{V_j}\,\Delta W^1_{j+1} + \rho\sqrt{V_j}\,\Delta W^2_{j+1},$$

where $\Delta W_j^2 \sim N\left(0, \sqrt{\Delta}\right)$ and $\Delta W_j^1 \sim N\left(0, \sqrt{\Delta}\right)$. The expression (9.5) of V_t is approached in the same manner by

$$
\begin{aligned}
V_{j+1} = V_0 &+ \sum_{k=0}^{j} \frac{\left(t_{j+1} - t_k\right)^{\alpha-1}}{\Gamma(\alpha)} \kappa \left(\gamma - V_k\right) \Delta \\
&+ \sum_{k=0}^{j} \frac{\left(t_{j+1} - t_k\right)^{\alpha-1}}{\Gamma(\alpha)} \sigma \sqrt{V_k}\, \Delta W_{k+1}^2 .
\end{aligned}
\tag{9.28}
$$

The set of parameters is denoted by $\boldsymbol{\theta} = \{\alpha, \mu, \kappa, \gamma, \sigma, \rho\}$ and it is assumed to be known. The particle filter estimates the most likely sample path of the variance process.

We denote by Δw_j^2 the realizations of Δw_j^2. The information $V_k = v_k$ and $\Delta W_k^2 = \Delta w_k^2$ for $k = 1, ..., j$ is stored in a "particle", denoted $\boldsymbol{u}_j$. In contrast to the classical Heston model, we have to keep track of the whole sample path of the variance and of the related Brownian increments.

The visible information up to time t_j is stored in a vector $\boldsymbol{x}_{1:j} = \{x_1, \ldots, x_j\}$ and the density of the "measurement" $f(x_j | \boldsymbol{u}_j)$, conditionally to the information in the particle, is a normal probability density function. If we denote by $\phi(\cdot)$ the pdf of an $N(0, 1)$, we have that

$$
f(x_j | \boldsymbol{u}_j) = \frac{1}{\sqrt{1-\rho^2}\sqrt{v_j \Delta}} \phi\left(\frac{x_j - \left(\mu - \frac{1}{2}v_{j-1}\right)\Delta - \rho\sqrt{v_{j-1}}\,\Delta w_j^2}{\sqrt{1-\rho^2}\sqrt{v_{j-1}\Delta}}\right).
$$

On the other hand, the transition density $f(\boldsymbol{u}_{j+1} | \boldsymbol{u}_j)$ is simulated by Eq. (9.28). The pdf of $\boldsymbol{u}_0$ is $f(\boldsymbol{u}_0)$. The posterior distribution of $\boldsymbol{u}_j$ conditionally to the available information at t_j is then rewritten as follows:

$$
f(\boldsymbol{u}_j | \boldsymbol{x}_{1:j}) = \frac{f(x_j | \boldsymbol{u}_j)}{\int f(x_j | \boldsymbol{u}_j) f(\boldsymbol{u}_j | \boldsymbol{x}_{1:j-1}) \mathrm{d}v_j} f(\boldsymbol{u}_j | \boldsymbol{x}_{1:j-1}),
\tag{9.29}
$$

where

$$
f(\boldsymbol{u}_j | \boldsymbol{x}_{1:j-1}) = \int f(\boldsymbol{u}_j | \boldsymbol{u}_{j-1}) f(\boldsymbol{u}_{j-1} | \boldsymbol{x}_{1:j-1}) \mathrm{d}\boldsymbol{u}_{j-1}.
\tag{9.30}
$$

The particle filter is presented in Algorithm 9.1.

Algorithm 9.1 Particle filtering of the rough volatility process

---

Initial step:

    draw N values of $v_0^{(i)}$ for $i = 1, \dots, N$, from an initial distribution $f(v_0)$.

Main procedure:

    For $j = 1$ to maximum epoch, T

    Prediction step

        Draw a sample $\Delta w_j^{2\,(i)}$ from a $N(0, \sqrt{\Delta})$, $i = 1, \dots N$.

        Update $v_j^{(i)}$ using the relation

$$v_j^{(i)} = v_0^{(i)} + \sum_{k=0}^{j-1} \frac{(t_j - t_k)^{\alpha-1}}{\Gamma(\alpha)} \kappa \left(\gamma - v_k^{(i)} \right) \Delta + \sum_{k=0}^{j-1} \frac{(t_j - t_k)^{\alpha-1}}{\Gamma(\alpha)} \sigma \sqrt{v_k^{(i)}} \, \Delta w_{k+1}^{2\,(i)}$$

    Correction step:

        The "particle" $u_j^{(i)} = \left(v_k^{(i)}, \Delta w_k^{2\,(i)} \right)_{k=1:j}$ has a probability of occurrence equal

to

$$p_j^{(i)} = \frac{f(x_j \mid u_j^{(i)})}{\sum_{i=1:N} f(x_j \mid u_j^{(i)})} \,,$$

    where

$$f(x_j \mid u_j^{(i)}) = \frac{1}{\sqrt{1-\rho^2}\sqrt{v_{j-1}^{(i)}\Delta}}$$

$$\times \phi \left(\frac{x_j - \left(\mu - \frac{1}{2}v_{j-1}^{(i)} \right) \Delta - \rho \sqrt{v_{j-1}^{(i)}} \, \Delta w_j^{2(i)}}{\sqrt{1-\rho^2}\sqrt{v_{j-1}^{(i)}\Delta}} \right)$$

    Resampling step:

        Resample with replacement N particles according to
        the importance weights $p_j^{(i)}$.

        The new importance weights are set to $p_j^{(i)} = \frac{1}{N}$.

    End loop on epochs

The filtered volatility for the period j is computed as

$$\widehat{V}_j = \mathbb{E}\left(V_j \mid x_{1:j} \right) = \sum_{i=1}^{N} v_j^{(i)} p_j^{(i)} \,,$$

whereas the log-likelihood of the whole sample is approached before resampling by the sum:

$$\log f(x \mid \theta) = \sum_{j=1}^{T} \log \left(\sum_{i=1}^{N} p_j^{(i)} f(x_j \mid u_j^{(i)}) \right).$$

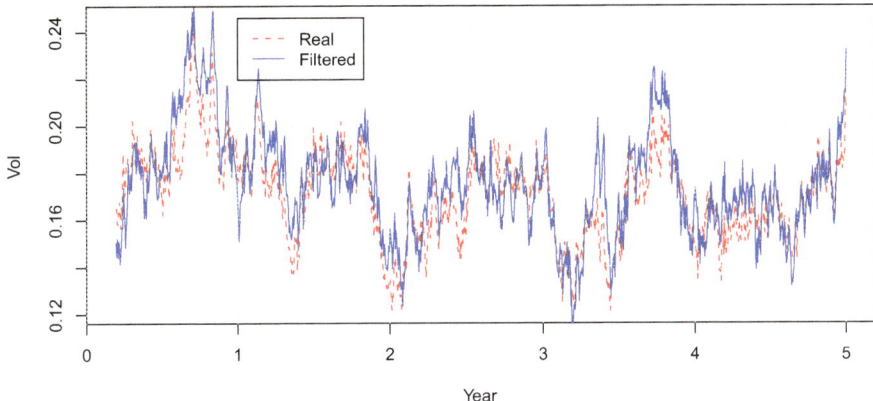

Fig. 9.2 Simulated sample path of $\sqrt{V_t}$ and its filtered estimate $\sqrt{\widetilde{V}_t}$. Parameters are $\kappa = 1.8$, $\gamma = 0.15^2$, $\sigma = 0.10$, $\alpha = 0.85$, and $V_0 = \gamma$

To test the reliability of the filter, we simulate a daily sample path over 5 years of a stock price and filter the volatility with 500 particles. Figure 9.2 compares the real and estimated volatilities.

Next, we fit the rough Heston model to the time series of S&P500 daily log-returns. We use the Metropolis–Hastings algorithm of Chap. 2, in a similar manner to what is done in Chap. 3 to estimate the classical Heston model. The rough version of this model is non-stationary, and updating the variance at each iteration with Eq. (9.28) is computationally intensive. For this reason, we consider a dataset with only 750 daily observations that range from 6/2/2017 to 29/1/2020. We also limit the number of MCMC iterations to 500 with 60 particles. As a benchmark, we fit a classical Heston model with the same procedure. Table 9.2 reports the estimates with a burn-in period of 400 iterations.

These estimates are next used to filter the volatility with 200 particles. The results of this procedure are illustrated in Fig. 9.3, which compares filtered sample paths of V_t in the rough and non-rough Heston model to the 50-days trailing volatility. The burn-in period is 50 days. We do not observe wide deviations between rough and non-rough filtered volatilities. Nevertheless, the log-likelihoods of the rough

Table 9.2 Parameter estimates of the rough Heston model fitted to daily S&P 500 data, 6/2/2017 to 29/1/2020

Estimates			
Rough		Non-rough	
$\widehat{\mu}$	0.1247	$\widehat{\mu}$	0.1626
$\widehat{\kappa}$	0.8329	$\widehat{\kappa}$	0.8835
$\widehat{\gamma}$	0.0460	$\widehat{\gamma}$	0.0273
$\widehat{\sigma}$	0.1989	$\widehat{\sigma}$	0.2031
$\widehat{\rho}$	−0.2874	$\widehat{\rho}$	−0.4800
$\widehat{\alpha}$	0.7677		

Filtered volatilities S&P 500

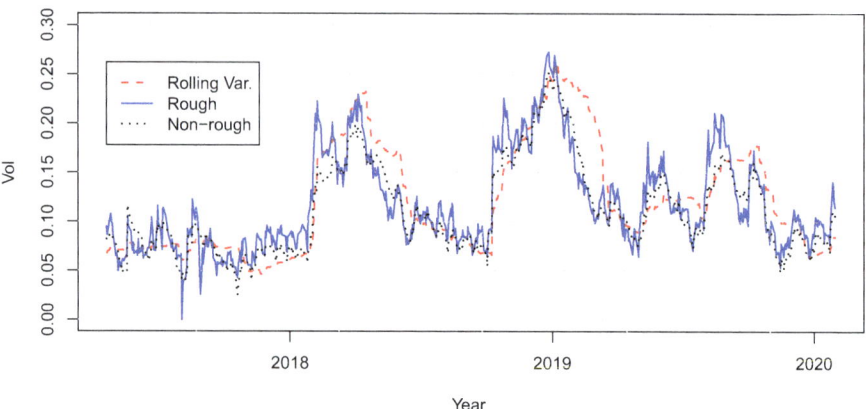

Fig. 9.3 Filtered sample paths of volatilities in the rough and non-rough Heston model, compared to the 50-day trailing volatility

and non-rough models are, respectively, equal to 2497 and 2264. This significant difference of likelihood confirms that the rough model provides a better fit from a statistical point of view.

To conclude this section, we compute the pdf of $\ln S_t$ for $t = 4$ years with a discrete Fourier transform and the numerical scheme of Sect. 9.4. The parameters are those estimated for the S&P 500 in Table 9.2 and V_0 is set to γ. The left and right plots of Fig. 9.4 compare the rough pdf with $\alpha = 0.77$ (estimated for the S&P 500), $\alpha = 0.51$, and $\alpha = 1$ (non-rough case). These graphs emphasize that the log-return density in the rough model does not widely differ from the one obtained with a non-rough dynamic, at least with S&P 500 parameter estimates. The rough Heston model better explains the sample path of S&P 500 volatility, but the impact on roughness on, for instance, the value at risk of the S&P 500 is limited (compared to the non-rough Heston model).

9.6 Further Reading

Affine Volterra processes extend the exponential affine framework for classical diffusions as detailed by Duffie et al. [8]. First systematic treatments of stochastic Volterra integral equations can be found in Berger and Mizel [5, 6], which in Protter [17] are extended to a semimartingale setting. In Pardoux and Protter [16], anticipative coefficients are introduced, and Volterra equations in an infinite-dimensional context are analyzed in Zhang [18]. Rough diffusions belong to the family of Volterra processes and accurately reproduce the behavior of historical volatility time series. There is a flourishing literature on rough models. In addition

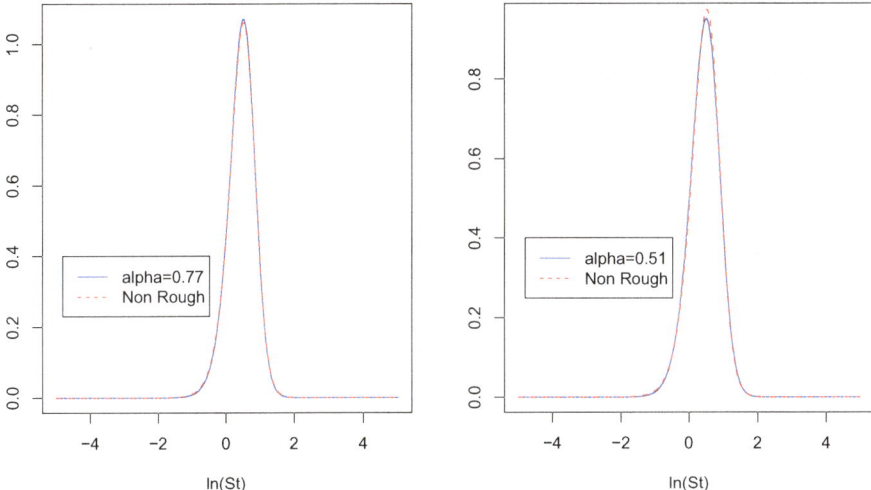

Fig. 9.4 Comparison of rough and non-rough pdf's of $\ln S_t$ for $t = 4$ years and parameters of Table 9.2

to contributions cited in previous sections, we quote Bayer et al. [4], who discuss the pricing of contingent claims on the underlying and integrated volatility. El Euch et al. [10] propose a microstructural approach based on Hawkes processes and show that it behaves in the long run as a rough Heston model. El Euch and Rosenbaum [11] study the hedging strategies in this framework. Horvath et al. [15] discuss the pricing and hedging of volatility options in a rough volatility setting. Abi Jaber [1] proposes a Markov multi-factor version of the rough Heston model using a similar approach to those developed in Chaps. 5 and 8. Bäuerle and Desmettre [3] consider a variant of the rough Heston model and solve a portfolio optimization problem. Dupret and Hainaut [9] study portfolio insurance under rough volatility and Volterra processes.

References

1. Abi Jaber, E.: Lifting the Heston model. Quant. Finance **19**(12), 1995–2013 (2019)
2. Abi Jaber, E., Larsson, M., Pulido, S.: Affine volterra processes. Ann. Appl. Probab. **29**(5), 3155–3200 (2019)
3. Bäuerle, N., Desmettre, S.: Portfolio optimization in fractional and rough heston models. SIAM J. Financ. Math. **11**(1), 437–469 (2020)
4. Bayer, C., Friz, P., Gatheral, J.: Pricing under rough volatility. Quant. Finance **16**(6), 887–904 (2016)
5. Berger, M.A., Mizel, V.J.: Volterra equations with Itô integrals I. J. Integral Equ. **2**(3), 187–245 (1980)
6. Berger, M.A., Mizel, V.J.: Volterra equations with Itô integrals II. J. Integral Equ. **2**(4), 319–337 (1980)

7. Cox, D.R.: Some statistical methods connected with series of events. J. R. Stat. Soc. B **17**, 129–164 (1955)
8. Duffie, D., Filipovic, D., Schachermayer, W.: Affine processes and applications in finance. Ann. Appl. Probabil. **13**(3), 984–1053 (2003)
9. Dupret, J.L., Hainaut, D.: Portfolio insurance under rough volatility and Volterra processes. Int. J. Theor. Appl. Financ. **24**(6), 2150036 (2021)
10. El Euch, O., Fukasawa, M., Rosenbaum, M.: The microstructural foundations of leverage effect and rough volatility. Finance Stoch. **22**(2), 241–281 (2018)
11. El Euch, O., Rosenbaum, M.: Perfect hedging in rough Heston models. Ann. Appl. Probab. **28**, 3813–3856 (2018)
12. El Euch, O., Rosenbaum, M.: The characteristic function of rough Heston models. Math. Finance **29**, 3–38 (2019)
13. Gatheral, J., Jaisson, T., Rosenbaum, M.: Volatility is rough. Quant. Finance **18**(6), 933–949 (2014)
14. Gripenberg, G., Londen, S.O., Staffans, O.: Volterra Integral and Functional Equations. Encyclopedia of mathematics and its applications. Cambridge University Press, Cambridge (1990)
15. Horvath, B., Jacquier, A., Tankov, P.: Volatility options in rough volatility models. SIAM J. Finance Math. **11**(2), 437–469 (2018)
16. Pardoux, E., Protter, P.: Stochastic Volterra equations with anticipating coefficients. The Ann. Probab. **18**(4), 1635–1655 (1990)
17. Protter, P.: Volterra equations driven by semimartingales. Ann. Probab. **13**(2), 519–530 (1985)
18. Zhang, X.: Stochastic volterra equations in banach spaces and stochastic partial differential equation. J. Funct. Anal. **258**(4), 1361–1425 (2010)

Chapter 10
Sub-diffusion for Illiquid Markets

The previous chapters have initiated our journey into the world of processes that are not martingales. We introduce in this chapter a new category of processes perfectly adapted for modeling illiquidity. In emerging or in small cap markets, the number of participants is often low and thus transactions are sparse. The time series of stock prices in such conditions display characteristic periods in which they stay motionless. This phenomenon is also visible at high frequency: the sample paths of stock prices look like stepwise random processes rather than continuous ones. A similar behavior is observed in physical systems exhibiting sub-diffusion. The constant periods of financial processes correspond to the trapping events in which a heavy particle gets immobilized, see, e.g., Eliazar and Klafter [7] or Metzler and Klafter [14]. In statistical physics, this type of dynamic is modeled by a sub-diffusive Brownian motion. This process is obtained by observing a standard Brownian motion on a different scale of time. In this chapter, after introducing the detailed features of this stochastic clock, we show that the density of a sub-diffusion is described in terms of a fractional Fokker–Planck (FFP) equation. We evaluate European options in this setting. The chapter concludes by estimating the model using the particle MCMC algorithm presented in Chap. 3.

10.1 The Stochastic Clock of Sub-diffusions

A sub-diffusion is built by observing a standard Brownian motion on a random time scale. The process defining the speed of time, denoted by $(S_t)_{t \geq 0}$, is the inverse of a subordinator. A *subordinator* is a stochastic process, denoted $(U_t)_{t \geq 0}$, with positive, non-decreasing sample paths and taking values in $\mathbb{R}^+$. We consider here α-stable Lévy subordinators, which are Lévy processes with independent and homogeneously distributed increments. Subordinators can be used directly as stochastic clocks, but since the increments are independent, the type of memory is

D. Hainaut, *Continuous Time Processes for Finance*,
Bocconi & Springer Series 12, https://doi.org/10.1007/978-3-031-06361-9_10

limited (conditionally to U_t, the duration of sojourn in a state is still exponential). For this reason, we instead consider a stochastic clock based on an inverse α-stable subordinator. In order to understand how we build this inverse, we first review the main features of Lévy subordinators. The process U_t is a $\frac{1}{\alpha}$-self-similar process, meaning that for all $0 < a < 1$

$$U_{at} \overset{d}{=} (at)^{\frac{1}{\alpha}} U_1 \, .$$

From the Lévy–Khintchine formula, the Laplace transform of α-stable Lévy subordinators has the following structure:

$$\mathbb{E}\left(e^{-\omega U_t}\right) = e^{-t\omega^\alpha} \, . \tag{10.1}$$

Notice that the process U_t is a pure jump process with an infinite activity. Its inverse is a process denoted by $(S_t)_{t\geq 0}$ defined by the following relation:

$$S_t = \inf\{\tau > 0 \, : \, U_\tau \geq t\} \, .$$

It corresponds to the hitting time of U_t to the barrier t. This inverse is no longer a Lévy process. However, $(S_t)_{t\geq 0}$ is positive and non-decreasing and is then also a subordinator. In the next sections, we use S_t as time change for a Brownian motion, denoted $\left(W_t^P\right)_{t\geq 0}$. By construction, the discontinuities of U_t imply that the process S_t may be constant over relatively short periods of time. Therefore, any process subordinated by S_t exhibits motionless periods.

The probability space is Ω, whereas $\mathbb{P}$ is the probability measure. The natural filtration of $(S_t)_{t\geq 0}$ is denoted by $(\mathcal{G}_t)_{t\geq 0}$. The probability density functions of U_t and S_t are denoted by $p_U(t, u) = \frac{d}{du}\mathbb{P}(u \leq U_t \leq u + du)$ and $g(t, \tau) = \frac{d}{d\tau}\mathbb{P}(\tau \leq S_t \leq \tau + d\tau)$. On the other hand, the self-similarity leads to the relation:

$$\mathbb{P}(U_\tau \leq t) = \mathbb{P}(\tau^{\frac{1}{\alpha}} U_1 \leq t)$$
$$= \mathbb{P}(U_1 \leq t\tau^{-\frac{1}{\alpha}}) \, .$$

If we differentiate this last expression with respect to t, we obtain the following scaling property:

$$p_U(\tau, t) = \tau^{-\frac{1}{\alpha}} p_U(1, t\tau^{-\frac{1}{\alpha}}). \tag{10.2}$$

This distribution is required for pricing derivatives at their inception. However for their evaluation at a later time $s > 0$, we have to establish the features of $S_t - S_s$ conditionally to the available information up to s. To the best of our knowledge, the literature only focuses on properties of S_t seen from time 0, when $U_0 = S_0 = 0$. We propose then to extend these features when the process S_t is considered from a later time. However, S_t is by nature not Markov, and its filtration up to time $s < t$

is not sufficient to draw any conclusions about $S_t - S_s$. We have then to adjoin additional information about the α-stable process that is inverted. In the following, we show that most of properties available at time 0 admit an extension at later time s if the available information is the value of S_s and U_{S_s}. We will see later that in this case the price of a European option admits an analytical expression. If U_{S_s} is not observed, an average price can be calculated by simulating U_t and S_t. Nevertheless, in comparison to an evaluation fully based on a Monte Carlo method, we do not have to perform simulations in simulations, which are computationally intensive.

Let us recall that $\mathbb{P}(S_t \leq \tau) = \mathbb{P}(U_\tau \geq t)$. Therefore, for $0 \leq s \leq u \leq t$, the cdf of $S_t - S_s$ is related to the pdf of U_t through the relation

$$\mathbb{P}(S_t - S_s \leq \tau \mid S_s = v , \, U_{S_s} = u) \tag{10.3}$$

$$= \mathbb{P}(S_t \leq \tau + v \mid S_s = v , \, U_{S_s} = u)$$

$$= \mathbb{P}(U_{\tau + v} \geq t \mid S_s = v , \, U_{S_s} = u)$$

$$= \mathbb{P}(U_\tau \geq t - u) .$$

If $u > t$, then the clock is stopped between t and s and

$$\mathbb{P}(S_t - S_s \leq \tau \mid S_s = v , \, U_{S_s} = u) = 1$$

for all $\tau \geq 0$. As $\mathbb{P}(U_\tau \geq t - u)$ does not depend on v, this leaves us wondering about the utility of the condition $S_s = v$. To understand this, we rewrite the conditional probability (10.3) as

$$\mathbb{P}(S_t - v \leq \tau \mid U_v = u) = \mathbb{P}(U_\tau \geq t - u) .$$

This last equation emphasizes that $\mathbb{P}(U_\tau \geq t - u)$ is the cdf of $S_t - v$, conditionally to $U_v = u$ for all $v \in \mathbb{R}^+$. This also explains why v does not explicitly appear in many results detailed in the remainder of this section. The conditional pdf of $S_t - S_s$, denoted by $g(t, \tau \mid s, v , u)$ for $t \geq s$, is the derivative of this last expression with respect to τ:

$$g(t, \tau \mid s, v , u) = \frac{\partial}{\partial \tau} \mathbb{P}(S_t - S_s \leq \tau \mid S_s = v , \, U_{S_s} = u) \tag{10.4}$$

$$= -\frac{\partial}{\partial \tau} \mathbb{P}(U_\tau \leq t - u)$$

$$= -\frac{\partial}{\partial \tau} \int_0^{t-u} p_U(\tau, x) \mathrm{d}x .$$

The Laplace transform of $g(\cdot \mid \cdot)$ with respect to time t is therefore equal to

$$\tilde{g}(\omega, \tau \mid s, v , u) = -\frac{\partial}{\partial \tau} \mathcal{L}\left(\int_0^{t-u} p_U(\tau, x) \mathrm{d}x \right)(\omega) , \tag{10.5}$$

where $\mathcal{L}\left(\int_0^{t-u} p_U(\tau, x)\mathrm{d}x\right)(\omega)$ is the Laplace transform of the integral of $p_U(\cdot)$. The process $(U_t)_{t\geq 0}$ is being strictly positive, and its density is null on $\mathbb{R}^-$. Integrating by parts leads to the following expression for the Laplace transform of $p_U(\cdot)$:

$$\mathcal{L}\left(\int_0^{t-u} p_U(\tau, x)\mathrm{d}x\right)(\omega) = \int_u^\infty \mathrm{e}^{-\omega t}\int_0^{t-u} p_U(\tau, x)\mathrm{d}x\,\mathrm{d}t$$

$$= \left[-\frac{\mathrm{e}^{-\omega t}}{\omega}\int_0^{t-u} p_U(\tau, x)\mathrm{d}x\right]_{t=u}^{t=\infty} + \frac{1}{\omega}\int_u^\infty \mathrm{e}^{-\omega t} p_U(\tau, t-u)\,\mathrm{d}t$$

$$= \frac{1}{\omega}\mathrm{e}^{-\omega u}\int_0^\infty \mathrm{e}^{-\omega z} p_U(\tau, z)\,\mathrm{d}z$$

$$= \frac{1}{\omega}\mathrm{e}^{-\omega u}\,\widetilde{p_U(\tau, \omega)}\,,$$

where $\widetilde{p_U(\tau, \omega)} = \mathrm{e}^{-\tau\omega^\alpha}$. From Eq. (10.5), we infer that the Laplace transform of $g(\cdot)$ is equal to

$$\tilde{g}(\omega, \tau\,|s, v, u) = \omega^{\alpha-1}\mathrm{e}^{-\omega u - \tau\omega^\alpha}\,. \tag{10.6}$$

On the other hand, from the self-similarity of U_t, we know that $\mathbb{P}(U_\tau \leq t - u) = \mathbb{P}(U_1 \leq (t-u)\tau^{-\frac{1}{\alpha}})$. If we differentiate this last expression with respect to τ and use Eq. (10.2), we obtain that

$$\frac{\partial}{\partial \tau}\mathbb{P}\left(U_1 \geq (t-u)(\tau)^{-\frac{1}{\alpha}}\right) = \frac{(t-u)}{\alpha\tau}p_U(\tau, t-u) \tag{10.7}$$

because U_t is α-stable. From Eqs. (10.4) and (10.7), we infer an important relation linking the conditional pdf of increments to the pdf of U_t:

$$g(t, \tau\,|s, v, u) = \frac{(t-u)}{\alpha\tau}p_U(\tau, t-u)\,. \tag{10.8}$$

In the next proposition, we establish a kind of conditional self-similarity for the density $g(\cdot|\cdot)$:

Proposition 10.1 *For $0 \leq s \leq u \leq t$, the pdf of $S_t - S_s\,|\,S_s = v$, $U_{S_s} = u$, is self-similar in the following sense:*

$$g(t, \tau\,|s, v, u) = \frac{1}{(t-u)^\alpha}g\left(u+1, \frac{\tau}{(t-u)^\alpha}\,|s, v, u\right)\,. \tag{10.9}$$

Proof Applying Eq. (10.8) to $g\left(u+1, \frac{\tau}{(t-u)^\alpha} \mid s, v, u\right)$ leads to the relation

$$g\left(u+1, \frac{\tau}{(t-u)^\alpha} \mid s, v, u\right) = \frac{(t-u)^\alpha}{\alpha\tau} p_U\left(\frac{\tau}{(t-u)^\alpha}, 1\right). \qquad (10.10)$$

Using the self-similarity of $p_U(\tau, t) = \tau^{-\frac{1}{\alpha}} p_U(1, t\tau^{-\frac{1}{\alpha}})$ twice allows us to infer that

$$p_U\left(\frac{\tau}{(t-u)^\alpha}, 1\right) = (t-u)\tau^{-\frac{1}{\alpha}} p_U\left(1, (t-u)\tau^{-\frac{1}{\alpha}}\right) \qquad (10.11)$$

$$= (t-u) p_U(\tau, (t-u)).$$

Combining Equations (10.10) and (10.11) provides the equality:

$$g\left(u+1, \frac{\tau}{(t-u)^\alpha} \mid s, v, u\right) = (t-u)^\alpha \frac{(t-u)}{\alpha\tau} p_U(\tau, (t-u))$$

$$= (t-u)^\alpha g(t, \tau \mid s, v, u).$$

$\square$

Notice that for $0 \le s \le t \le u$, a time-changed Brownian motion $\left(W^P_{S_t}\right)_{t\ge0}$ is motionless since the stochastic clock is stopped. Using this last result allows us to calculate the moments of $S_{u+1} - S_s$.

Proposition 10.2 *For $0 \le s \le u \le t$, the moments of $S_{u+1} - S_s$, conditionally to S_s and U_{S_s}, for $n \in \mathbb{N}$ are given by*

$$\mathbb{E}\left((S_{u+1} - S_s)^n \mid S_s = v, U_{S_s} = u\right) = \frac{\Gamma(n+1)}{\Gamma(n\alpha+1)}. \qquad (10.12)$$

Proof Using the self-similarity property of Proposition 10.1, the moment generating function of U_{t-u} is rewritten as

$$e^{-(t-u)\omega^\alpha} = \int_0^\infty e^{-\omega x} p_U(t-u, x)\, dx$$

$$= \int_0^\infty e^{-\omega x} (t-u)^{-\frac{1}{\alpha}} p_U(1, x(t-u)^{-\frac{1}{\alpha}})\, dx.$$

From Eqs. (10.8) and (10.2), we infer that

$$p_U(1, x(t-u)^{-\frac{1}{\alpha}}) = (t-u)^{\frac{1}{\alpha}} p_U(t-u, x)$$

$$= \frac{\alpha}{x}(t-u)^{\frac{1}{\alpha}+1} g(u+x, (t-u) \mid s, v, u).$$

The Laplace transform of U_{t-u} then becomes

$$e^{-(t-u)\,\omega^\alpha} = \alpha \int_0^\infty e^{-\omega x} \frac{(t-u)}{x} g(u+x, t-u\,|s,\,v\,,u)\mathrm{d}x.$$

From Eq. (10.9) of Proposition 10.1, we rewrite this last integral as

$$e^{-(t-u)\,\omega^\alpha} = \alpha \int_0^\infty e^{-\omega x} \frac{(t-u)}{x^{\alpha+1}} g(u+1, (t-u)x^{-\alpha}\,|s,\,v\,,u)\mathrm{d}x. \quad (10.13)$$

Multiplying the left-hand side by $(t-u)^{n-1}$ and integrating it with respect to t give us (where $\Gamma(z) = \int t^{z-1} e^{-t}\mathrm{d}t$)

$$\int_u^\infty (t-u)^{n-1} e^{-(t-u)\,\omega^\alpha}\mathrm{d}t = \int_0^\infty t^{n-1} e^{-t\,\omega^\alpha}\mathrm{d}t$$

$$= \frac{1}{\omega^{n\alpha}} \int_0^\infty (\omega^\alpha t)^{n-1} e^{-t\,\omega^\alpha}\mathrm{d}\left(\omega^\alpha t\right)$$

$$= \frac{\Gamma(n)}{\omega^{n\alpha}}. \quad (10.14)$$

We multiply the right-hand side of Eq. (10.13) by $(t-u)^{n-1}$ and integrate it, using again the self-similarity Eq. (10.9):

$$\alpha \int_u^\infty \int_0^\infty e^{-\omega x} \frac{(t-u)^n}{x^{\alpha+1}} g(u+1, (t-u)x^{-\alpha}\,|s,\,v\,,u)\mathrm{d}x\,\mathrm{d}t$$

$$= \alpha \int_0^\infty \int_0^\infty e^{-\omega x} \frac{\tau^n}{x^{\alpha+1}} g(u+1, \tau x^{-\alpha}\,|s,\,v\,,u)\mathrm{d}x\,\mathrm{d}\tau$$

$$= \alpha \int_0^\infty \int_0^\infty e^{-\omega x} \frac{x^{n\alpha}}{x^{\alpha+1}} \left(\tau x^{-\alpha}\right)^n g(u+1, \tau x^{-\alpha}\,|s,\,v\,,u)\,\mathrm{d}x\,\mathrm{d}\tau.$$

We change the order of integration and obtain

$$\alpha \int_u^\infty \int_0^\infty e^{-\omega x} \frac{(t-u)^n}{x^{\alpha+1}} g(u+1, (t-u)x^{-\alpha}\,|s,\,v\,,u)\mathrm{d}x\,\mathrm{d}t$$

$$= \alpha \int_0^\infty e^{-\omega x} \frac{x^{n\alpha-1}}{x^\alpha} \int_0^\infty \left(\tau x^{-\alpha}\right)^n g(u+1, \tau x^{-\alpha}\,|s,\,v\,,u)\,\mathrm{d}\tau\,\mathrm{d}x$$

$$= \alpha \int_0^\infty e^{-\omega x} x^{n\alpha-1} \int_0^\infty z^n g(u+1, z\,|s,\,v\,,u)\,\mathrm{d}z\,\mathrm{d}x$$

$$= \alpha \int_0^\infty e^{-\omega x} x^{n\alpha-1}\,\mathrm{d}x\,\mathbb{E}\left((S_{u+1} - S_v)^n\,|\,S_s = v\,,\,U_{S_s} = u\right)$$

$$= \alpha \frac{\Gamma(n\alpha)}{\omega^{n\alpha}}\,\mathbb{E}\left((S_{u+1} - S_v)^n\,|\,S_s = v\,,\,U_{S_s} = u\right).$$

Combining this result with Eq. (10.13), we obtain that

$$\mathbb{E}\left((S_{u+1} - S_s)^n \mid S_s = v, U_{S_s} = u\right) = \frac{\Gamma(n)}{\Gamma(n\alpha)\alpha} = \frac{\Gamma(n+1)}{\Gamma(n\alpha + 1)}.$$

☐

From the moments of $S_{u+1} - S_s$, we infer the Laplace transform of the stochastic clock increments.

Corollary 10.3 *For $0 \leq s \leq u \leq t$, the Laplace transform of $S_t - S_s$ conditionally to S_s and U_{S_s} is*

$$\mathbb{E}\left(e^{-\omega(S_t - S_s)} \mid S_s = v, U_{S_s} = u\right) = E_\alpha\left(-\omega(t-u)^\alpha\right), \qquad (10.15)$$

where $E_\alpha(X) = \sum_{n=0}^\infty \frac{X^n}{\Gamma(n\alpha+1)}$ is the Mittag–Leffler function. The moments of order $n \in \mathbb{N}$ are given by

$$\mathbb{E}\left((S_t - S_s)^n \mid S_s = v, U_{S_s} = u\right) = \frac{\Gamma(n+1)}{\Gamma(n\alpha + 1)}(t-u)^{n\alpha}. \qquad (10.16)$$

If A is an $l \times l$ matrix such that $\lim_{t \to \infty} A^t$ exists, then

$$\mathbb{E}\left(e^{A(S_t - S_s)} \mid S_s = v, U_{S_s} = u\right) = \mathbf{E}_\alpha\left(A(t-u)^\alpha\right), \qquad (10.17)$$

where $\mathbf{E}_\alpha(X) = \sum_{n=0}^\infty \frac{X^n}{\Gamma(n\alpha+1)}$ is the matrix Mittag–Leffler function.

Proof Using again the self-similarity property of Proposition 10.1 and performing a change of variable, we rewrite the Laplace transform as

$$\mathbb{E}\left(e^{-\omega(S_t - S_s)} \mid S_s = v, U_{S_s} = u\right) = \int_0^\infty e^{-\omega\tau} g(t, \tau \mid s, v, u)d\tau$$

$$= \int_0^\infty e^{-\omega\tau} \frac{1}{(t-u)^\alpha} g\left(u+1, \frac{\tau}{(t-u)^\alpha} \mid s, v, u\right)d\tau$$

$$= \int_0^\infty e^{-\omega\gamma(t-u)^\alpha} g(u+1, \gamma \mid s, v, u)\,d\gamma$$

$$= \mathbb{E}\left(e^{-\omega(t-u)^\alpha(S_{u+1} - S_s)} \mid S_s = v, U_{S_s} = u\right).$$

The Taylor series of the Laplace transform around $\omega = 0$, combined with Proposition 10.2, allows us to rewrite this transform as the following infinite sum:

$$\mathbb{E}\left(e^{-\omega(t-u)^\alpha(S_{u+1}-S_s)} \mid S_s = v, \, U_{S_s} = u\right)$$

$$= \sum_{n=0}^{\infty} \frac{\mathbb{E}\left((S_{u+1} - S_s)^n \mid S_s = v, \, U_{S_s} = u\right)}{\Gamma(n+1)} \left(-\omega(t-u)^\alpha\right)^n$$

$$= \sum_{n=0}^{\infty} \frac{\left(-\omega(t-u)^\alpha\right)^n}{\Gamma(n\alpha + 1)},$$

which is precisely the Mittag–Leffler function at the point $-\omega(t-u)^\alpha$. Moments are obtained by differentiating the Laplace transform. Equation (10.15) is proved in the same way as the first statement.   □

Notice that for $0 \le s \le t \le u$, $\mathbb{E}\left(e^{-\omega(S_t - S_s)} \mid S_s = v, \, U_{S_s} = u\right) = 1$. The probability density function of S_t, conditionally to S_s and U_{S_s}, can be represented as an infinite sum:

Proposition 10.4 *The pdf of $S_t - S_s \mid S_s = v$, $U_{S_s} = u$ for $0 \le s \le u \le t$ is of the form:*

$$g(t, \tau \mid s, v, u) = \frac{\partial}{\partial \tau} \mathbb{P}(S_t - S_s \le \tau \mid S_s = v, \, U_{S_s} = u) \tag{10.18}$$

$$= \frac{1}{\alpha \pi} \sum_{k=1}^{\infty} (-1)^{k+1} \frac{\Gamma(\alpha k + 1) \sin(\pi k \alpha)}{k!} \frac{\tau^{k-1}}{(t-u)^{\alpha k}}.$$

Proof From Eqs. (10.8) and (10.2), we infer that the conditional density of the stochastic clock is

$$g(t, \tau \mid s, v, u) = \frac{(t-u)}{\alpha \tau} p_U(\tau, t-u) \tag{10.19}$$

$$= \frac{(t-u)}{\alpha} \tau^{-1-\frac{1}{\alpha}} p_U\left(1, (t-u)\tau^{-\frac{1}{\alpha}}\right).$$

From Uchaikin and Zolotarev [16], the density $p_U(1, x)$ of any α-stable process admits the following representation:

$$p_U(1, x) = \frac{1}{\pi} \sum_{k=1}^{\infty} (-1)^{k+1} \frac{\Gamma(\alpha k + 1)}{k!} \frac{1}{x^{\alpha k + 1}} \sin(\pi k \alpha).$$

□

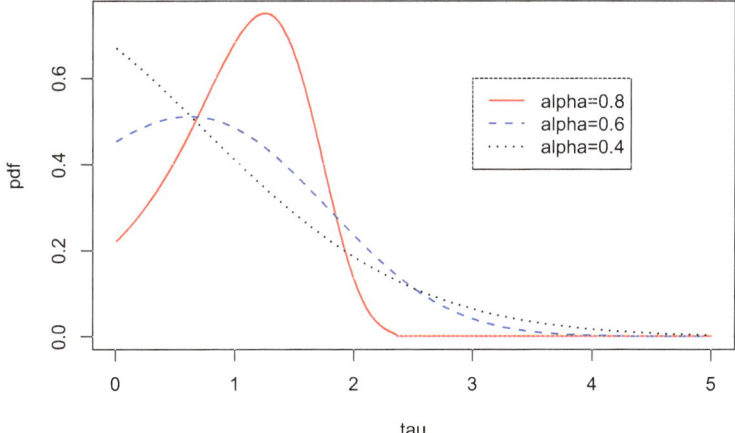

Fig. 10.1 Pdf of $S_t \mid S_s = v$, $U_{S_s} = u$, with $u = 3$, $t = 4$, and $\alpha \in \{0.4, 0.6, 0.8\}$

In practice, we truncate the series (10.18) in order to approximate the conditional pdf of S_t. Figure 10.1 compares the conditional pdf's of $S_t - S_s$ for different values of α, obtained by truncating this sum up to the first 100 terms. We can also calculate the conditional expectation of the infinitesimal variation of S_t, which takes a very simple form.

Proposition 10.5 *For $0 \leq s \leq u \leq t$, $v \geq 0$, the expectation of $\frac{dS_t}{dt}$ is equal to*

$$\mathbb{E}\left(\frac{dS_t}{dt} \mid S_s = v, \ U_{S_s} = u\right) = \frac{\alpha\,(t-u)^{\alpha-1}}{\Gamma(\alpha+1)}. \tag{10.20}$$

Proof From Corollary 10.3, we deduce that

$$\mathbb{E}\left(\frac{S_{t+\Delta} - S_t}{\Delta} \mid S_s = v, \ U_{S_s} = u\right)$$

$$= \mathbb{E}\left(\frac{S_{t+\Delta} - S_s}{\Delta} - \frac{S_t - S_s}{\Delta} \mid S_s = v, \ U_{S_s} = u\right)$$

$$= \frac{(t + \Delta - u)^{\alpha} - (t - u)^{\alpha}}{\Delta\,\Gamma(\alpha+1)}.$$

Taking the limit leads to Eq. (10.20).   □

Magdziarz [13] proposes an efficient algorithm for simulating sample paths of S_t. We denote by $S_\delta(t)$ the piecewise constant approximation of S_t, where $\delta \in \mathbb{R}^+$ is the step size:

$$S_\delta(t) = (\min\{n \in \mathbb{N} \mid U_{\delta n} > t\} - 1)\,\delta. \tag{10.21}$$

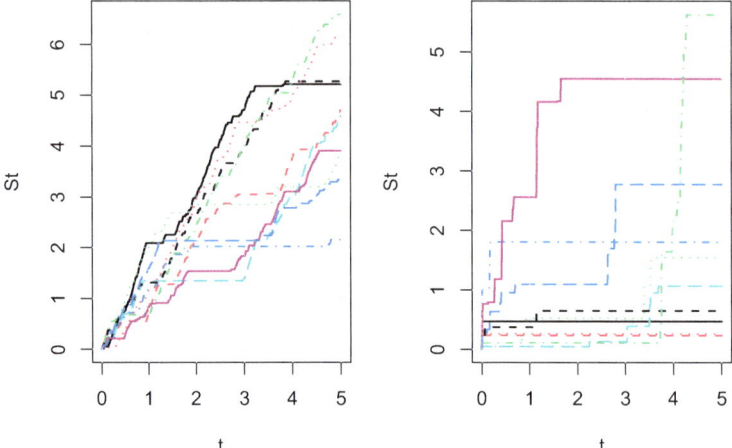

Fig. 10.2 Ten simulated sample paths of an inverse α-stable process: left plot $\alpha = 0.5$ and right plot $\alpha = 0.8$

The subordinator is simulated using Euler's method and summing up its increments:

$$
\begin{cases}
U_0 = 0 \\
U_{\delta n} = U_{\delta(n-1)} + \delta^{\frac{1}{\alpha}} \xi_n \quad n \geq 1
\end{cases},
$$

where ξ_n are i.i.d. random variables. The ξ_n are defined by

$$
\xi_n = \frac{\sin\left(\alpha\left(V + \frac{\pi}{2}\right)\right)}{(\cos(V))^{\frac{1}{\alpha}}} \left(\frac{\cos\left(V - \alpha\left(V + \frac{\pi}{2}\right)\right)}{B}\right)^{\frac{1-\alpha}{\alpha}},
$$

where V is a uniform random variable on $(-\frac{\pi}{2} ; \frac{\pi}{2})$ and B is an exponential random variable with mean equal to one.

Figure 10.2 presents simulated sample paths of inverse α-stable processes. As expected, we observe periods over which S_t is constant. These periods are due to sharp increases of U_t.

10.2 The Non-fractional Market

We first consider a non-time-changed financial market in which a risk-free bond and a stock ruled by a Brownian motion are traded. The risk-free bond, denoted by B_t, earns a constant interest rate $r \in \mathbb{R}$ and satisfies the differential equation

$$
\frac{\mathrm{d}B_t}{B_t} = r\mathrm{d}t \quad B_0 = 1 , \, t \geq 0 .
$$

A geometric diffusion, denoted by $(A_t)_{t\geq0}$, represents the price of the stock and is a solution of the SDE

$$\frac{\mathrm{d}A_t}{A_t} = \mu\,\mathrm{d}t + \sigma\,\mathrm{d}W_t^P, \tag{10.22}$$

where $\mu, \sigma \in \mathbb{R}^+$. The price process is defined on $\left(\Omega, (\mathcal{F}_t)_{t\geq0}, \mathbb{P}\right)$, where $\mathcal{F}_t$ is the natural filtration of the Brownian motion. From Itô's lemma, the stock price at time t is related to its value at time $s \leq t$ by the relation

$$A_t = A_s \exp\left(\left(\mu - \frac{\sigma^2}{2}\right)(t-s) + \sigma\left(W_t^P - W_s^P\right)\right). \tag{10.23}$$

As A_t has a log-normal distribution, its conditional expectation is equal to $\mathbb{E}\left(A_t \mid \mathcal{F}_s\right) = A_s e^{\mu(t-s)}$, whereas its conditional variance is

$$\mathbb{V}\left(A_t \mid \mathcal{F}_s\right) = A_s^2 \left(e^{(\sigma^2+2\mu)(t-s)} - e^{2\mu(t-s)}\right).$$

The pdf of A_t, conditionally to $\mathcal{F}_s$, exclusively depends upon A_s. Thus, we denote by $p(t, x|s, y)$ its probability density function, which is defined as follows:

$$p(t, x|s, y)\mathrm{d}x = \mathbb{P}\left(A_t \in [x, x+\mathrm{d}x] \mid A_s = y\right), \forall x \in \mathbb{R}^+,$$

for $s \leq t$ and $x \in \mathbb{R}^+$. By construction, $p(t, x|s, y) = p(t-s, x|0, y)$. This pdf is a solution of a forward differential equation, as stated in the next proposition.

Proposition 10.6 *The pdf of A_t is a solution of the following Fokker–Planck equation, also called a forward Kolmogorov equation:*

$$\frac{\partial p(t, x|s, y)}{\partial t} = -\frac{\partial}{\partial x}\left(\mu\,x\,p(t, x|s, y)\right) + \frac{1}{2}\frac{\partial^2}{\partial x^2}\left[\sigma^2 x^2\,p(t, x|s, y)\right] \tag{10.24}$$

with the initial condition: $p(s, x|s, y) = \delta_{\{x-y\}}$, where δ_z is the Dirac measure located at z.

Proof From the Kramers–Moyal forward expansion, we know that the pdf's $p(t + \Delta, x|s, y)$ at time $t + \Delta$ and $p(t, x|s, y)$ at time t are linked to the moments of A_t by the relation

$$p(t + \Delta, x|s, y) - p(t, x|s, y) = \tag{10.25}$$

$$\sum_{n=1}^{\infty} \frac{(-1)^n}{n!} \frac{\partial^n}{\partial x^n}\left[M_n(t, x, \Delta)p(t, x|s, y)\right],$$

where $M_n(t, x, \Delta)$ is the moment of order n of $\Delta A_t = A_{t+\Delta} - A_t$:

$$
M_n(t, x, \Delta) = \mathbb{E}\left[(A_{t+\Delta} - A_t)^n \mid A_t = x\right]
$$
$$
= \int_{-\infty}^{+\infty} (z - x)^n \, p(t + \Delta, z \mid t, x) \, \mathrm{d}z .
$$

On the other hand, the centered moments of A_t may be expanded as follows:

$$
\mathbb{E}\left[A_{t+\Delta} - A_t \mid A_t = x\right] = \mu x \Delta + O\left(\Delta_t^2\right) ,
$$
$$
\mathbb{E}\left[(A_{t+\Delta} - A_t)^2 \mid A_t = x\right] = \sigma^2 x^2 \Delta + O\left(\Delta_t^2\right) ,
$$
$$
\mathbb{E}\left[(A_{t+\Delta} - A_t)^n \mid A_t = x\right] = O\left(\Delta_t^{n/2}\right) \; \forall n \geq 3 .
$$

Injecting these expansions into Eq. (10.25) gives us

$$
\frac{p(t + \Delta, x \mid s, y) - p(t, x \mid s, y)}{\Delta} = -\frac{\partial}{\partial x}\left[\mu \, x \, p(t, x \mid s, y)\right] \qquad (10.26)
$$
$$
+ \frac{1}{2}\frac{\partial^2}{\partial x^2}\left[\sigma^2 x^2 \, p(t, x \mid s, y)\right] + O(\Delta) .
$$

The limit of this last equation when $\Delta \to 0$ leads to the desired result. $\square$

The market model introduced in this section is complete, and the absence of arbitrage entails the existence of an equivalent probability measure, $\mathbb{Q}$, called the "risk neutral" measure. Under this measure, discounted asset prices are martingales, and the stock earns on average the risk-free rate. This measure is defined by the following Radon–Nikodym derivative:

$$
Z_t = \left.\frac{\mathrm{d}\mathbb{Q}}{\mathrm{d}\mathbb{P}}\right|_t = \exp\left(-\frac{1}{2}\theta^2 t - \theta \, W_t^P\right) , \qquad (10.27)
$$

where $\theta = \frac{\mu - r}{\sigma}$ is the cost of market risk. Under $\mathbb{P}$, the process $\mathrm{d}W_t = \mathrm{d}W_t^P + \theta \mathrm{d}t$ is a Brownian motion, and the stock price dynamic becomes

$$
\frac{\mathrm{d}A_t}{A_t} = r \, \mathrm{d}t + \sigma \, \mathrm{d}W_t .
$$

A European call option gives the buyer the right, but not the obligation, to buy from the seller the underlying stock, A_T, at a specified strike price K within a specified maturity time T. The value of the call option at maturity is equal to $\max(A_T - K, 0)$. A put option is the opposite of a call option, and it gives the holder the right to sell shares. In order to avoid arbitrage opportunities, their values before expiry are equal to the expected discounted payoff under the risk neutral measure. Their prices at time t are functions of the underlying asset and of the remaining time before expiry, denoted $\tau = T - t$:

$$C_{BS}(A_t, \tau) = \mathbb{E}^{\mathbb{Q}} \left(e^{-r\tau} \max(A_T - K, 0) \,|\, \mathcal{F}_{T-\tau} \right),$$

$$P_{BS}(A_t, \tau) = \mathbb{E}^{\mathbb{Q}} \left(e^{-r\tau} \max(K - A_T, 0) \,|\, \mathcal{F}_{T-\tau} \right).$$

Black and Scholes [3] have found closed form expressions of these prices. If $\Phi(\cdot)$ is the pdf of an $N(0, 1)$, they are respectively given by

$$C_{BS}(A_t, \tau) = A_t \Phi\left(d_1(A_t, \tau)\right) - K e^{-r\tau} \Phi(d_2(A_t, \tau)), \tag{10.28}$$

$$P_{BS}(A_t, \tau) = K e^{-r\tau} \Phi(-d_2(A_t, \tau)) - A_t \Phi\left(-d_1(A_t, \tau)\right), \tag{10.29}$$

where $d_1(A_t, \tau)$ and $d_2(A_t, \tau)$ are equal to

$$d_1(A_t, \tau) = \frac{\ln\left(\frac{A_t}{K}\right) + \left(r + \frac{\sigma^2}{2}\right)\tau}{\sigma\sqrt{\tau}},$$

$$d_2(A_t, \tau) = d_1(A_t, \tau) - \sigma\sqrt{\tau}.$$

A short position in a European call or put can be hedged by purchasing a quantity Δ of the underlying asset. We briefly recall the principle behind this delta hedging strategy. For this purpose, we consider a short position in $C_{BS}(A_t, \tau)$. We hedge this position with Δ shares between t and $t+\delta$, partly funded by a borrowing. We denote by $\delta A_t = A_{t+\delta} - A_t$ the variation of prices. The profit and loss (P&L) account over the period $[t, t + \delta]$ is in this case:

$$P\&L = -\left(C_{BS}(A_{t+\delta}, \tau - \delta) - C_{BS}(A_t, \tau)\right)$$
$$+ r\, C_{BS}(A_t, \tau)\,\delta + \Delta\,(\delta A_t - r A_t \delta).$$

We expand this P&L in powers of δA_t and δ with a Taylor series. We limit this expansion to the lowest non-trivial orders for δ and δA_t, i.e., order one.

$$C_{BS}(A_{t+\delta}, \tau - \delta) \approx C_{BS}(A_t, \tau) + \frac{\partial C_{BS}(A_t, \tau)}{\partial A_t}\delta A_t - \frac{\partial C_{BS}(A_t, \tau)}{\partial \tau}\delta.$$

Therefore, the P&L is as follows:

$$P\&L \approx \frac{\partial C_{BS}(A_t, \tau)}{\partial \tau}\delta - \frac{\partial C_{BS}(A_t, \tau)}{\partial A_t}\delta A_t$$
$$+ r\, C_{BS}(A_t, \tau)\,\delta + \Delta\,(\delta A_t - r A_t \delta).$$

Choosing $\Delta = \frac{\partial C_{BS}(A_t, \tau)}{\partial A_t}$ cancels the first-order term in δA_t and removes at first order the randomness in the P&L. In the Black & Scholes model, this Δ is equal to $\Phi(d_1(A_t, \tau))$ for a call and to $-\Phi(-d_1(A_t, \tau)) = \Phi(d_1(A_t, \tau)) - 1$ for a put. We will use a similar reasoning to establish the delta hedging strategy in the fractional market.

To conclude this section, we present the characteristic function of the asset log-return, defined as $X_t = \ln\left(\frac{A_t}{A_0}\right)$. This function is also the Fourier transform of the pdf of X_t. By construction, $X_t \sim N\left(\left(\mu - \frac{\sigma^2}{2}\right)t\,;\,\sigma^2 t\right)$, the characteristic function of the log-return over $[s, t]$, i.e., $X_t - X_s$, is then

$$\mathcal{F}_{X_t - X_s}(\omega) = \mathbb{E}\left(e^{i\,\omega\,(X_t - X_s)} \mid \mathcal{F}_s\right) \tag{10.30}$$

$$= \exp\left(\left(i\,\omega\left(\mu - \frac{\sigma^2}{2}\right) - \omega^2 \frac{\sigma^2}{2}\right)(t - s)\right).$$

We will use this feature in the next section to construct the characteristic function of the log-return in the fractional market.

10.3   The Fractional Market

The inverse alpha-stable process S_t is used as a stochastic time scale for observing the asset prices. Let us recall that the filtrations of S_t and A_t are, respectively, $(\mathcal{G}_t)_{t \geq 0}$ and $(\mathcal{F}_t)_{t \geq 0}$. We denote by $\mathcal{H}_t$ the augmented filtration $\mathcal{G}_t \vee \mathcal{F}_{S_t}$. This is the smallest filtration at the intersection of $\mathcal{G}_t$ and $\mathcal{F}_{S_t}$. The time-changed stock price is governed by the following equation:

$$A_{S_t} = A_{S_s} \exp\left(\left(\mu - \frac{\sigma^2}{2}\right)(S_t - S_s) + \sigma\left(W_{S_t}^P - W_{S_s}^P\right)\right), \tag{10.31}$$

whereas the bond price is B_{S_t}. Figure 10.3 presents simulated sample paths A_{S_t} for $\alpha = 0.8$ and $\alpha = 0.9$. As expected, we observe periods over which the stock price A_{S_t} is constant.

The next proposition presents its first two centered moments.

Proposition 10.7 *For $0 \leq s \leq u \leq t$, $v \geq 0$, the expectation of A_{S_t} conditionally to S_s, U_{S_s}, and $\mathcal{F}_{S_s}$ is given by*

$$\mathbb{E}\left(A_{S_t} \mid \{S_s = v\,,\, U_{S_s} = u\} \vee \mathcal{F}_{S_s}\right) = A_v\, E_\alpha\left(\mu\,(t - u)^\alpha\right), \tag{10.32}$$

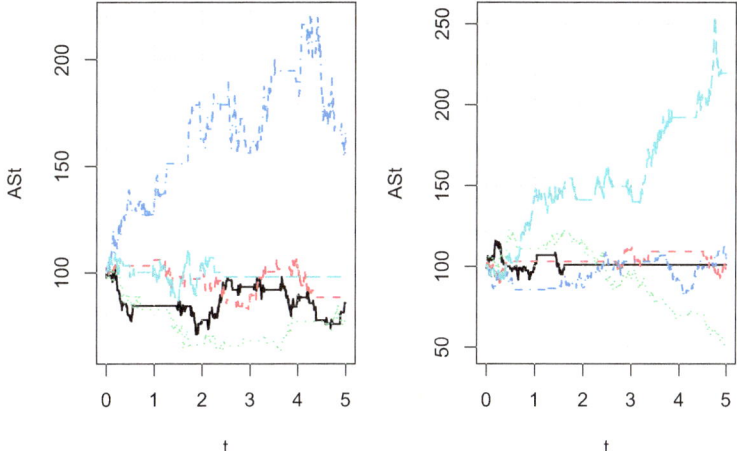

Fig. 10.3 Ten simulated sample paths of A_{S_t}: left plot $\alpha = 0.8$ and right plot $\alpha = 0.9$

whereas the conditional variance is equal to

$$\mathbb{V}\left(A_{S_t} \mid \{S_s = v, \, U_{S_s} = u\} \vee \mathcal{F}_{S_s}\right) =$$
$$A_v^2 \left(E_\alpha \left(\left(\sigma^2 + 2\mu\right)(t - u)^\alpha\right) - \left(E_\alpha \left(\mu \, (t - u)^\alpha\right)\right)^2\right). \qquad (10.33)$$

Proof Conditionally to S_t, $\dfrac{A_{S_t}}{A_{S_s}}$ has a log-normal distribution, and the conditional
expectation with respect to $\mathcal{G}_t \vee \mathcal{F}_{S_s}$ is equal to $\mathbb{E}\left(A_{S_t} \mid \mathcal{G}_t \vee \mathcal{F}_{S_s}\right) = A_{S_s} e^{\mu(S_t - S_s)}$,
whereas its conditional variance is

$$\mathbb{V}\left(A_{S_t} \mid \mathcal{G}_t \vee \mathcal{F}_{S_s}\right) = A_{S_s}^2 \left(e^{(\sigma^2 + 2\mu)(S_t - S_s)} - e^{2\mu(S_t - S_s)}\right). \qquad (10.34)$$

Using nested expectations and Corollary 10.3, for $0 \leq s \leq u \leq t$, the Laplace
transform of $S_t - S_s$ conditionally to S_s and U_{S_s} is then

$$\mathbb{E}\left(A_{S_t} \mid \{S_s, \, U_{S_s}\} \vee \mathcal{F}_{S_s}\right)$$
$$= \mathbb{E}\left(\mathbb{E}\left(A_{S_t} \mid \mathcal{G}_t \vee \{U_{S_s}\} \vee \mathcal{F}_{S_s}\right) \mid \{S_s, \, U_{S_s}\} \vee \mathcal{F}_{S_s}\right)$$
$$= \mathbb{E}\left(A_{S_s} e^{\mu(S_t - S_s)} \mid \{S_s, \, U_{S_s}\} \vee \mathcal{F}_{S_s}\right)$$
$$= A_{S_s} E_\alpha \left(\mu \left(t - U_{S_s}\right)^\alpha\right).$$

The conditional variance is rewritten as the sum of the expected variance and the variance of the expectation:

$$
\mathbb{V}\left(A_{S_t} \mid \{S_s,\, U_{S_s}\} \vee \mathcal{F}_{S_s}\right)
$$
$$
= \mathbb{E}\left(\mathbb{V}\left(A_{S_t} \mid \mathcal{G}_t \vee \{U_{S_s}\} \vee \mathcal{F}_{S_s}\right) \mid \{S_s,\, U_{S_s}\} \vee \mathcal{F}_{S_s}\right)
$$
$$
+ \mathbb{V}\left(\mathbb{E}\left(A_{S_t} \mid \mathcal{G}_t \vee \{U_{S_s}\} \vee \mathcal{F}_{S_s}\right) \mid \{S_s,\, U_{S_s}\} \vee \mathcal{F}_{S_s}\right).
$$

Equation (10.34) allows us to develop the first term in this equation as follows:

$$
\mathbb{E}\left(\mathbb{V}\left(A_{S_t} \mid \mathcal{G}_t \vee \{U_{S_s}\} \vee \mathcal{F}_{S_s}\right) \mid \{S_s,\, U_{S_s}\} \vee \mathcal{F}_{S_s}\right)
$$
$$
= \mathbb{E}\left(A_{S_s}^2 \left(e^{(\sigma^2 + 2\mu)(S_t - S_s)} - e^{2\mu(S_t - S_s)}\right) \mid \{S_s,\, U_{S_s}\} \vee \mathcal{F}_{S_s}\right)
$$
$$
= A_{S_s}^2 \left(E_\alpha\left((\sigma^2 + 2\mu)(t - U_{S_s})^\alpha\right) - E_\alpha\left(2\mu(t - U_{S_s})^\alpha\right)\right).
$$

By definition of the variance and Eq. (10.34), the second term is

$$
\mathbb{V}\left(\mathbb{E}\left(A_{S_t} \mid \mathcal{G}_t \vee \{U_{S_s}\} \vee \mathcal{F}_{S_s}\right) \mid \{S_s,\, U_{S_s}\} \vee \mathcal{F}_{S_s}\right)
$$
$$
= \mathbb{E}\left(A_{S_s}^2 e^{2\mu(S_t - S_s)} \mid \{S_s,\, U_{S_s}\} \vee \mathcal{F}_{S_s}\right)
$$
$$
- \left(\mathbb{E}\left(A_{S_s} e^{\mu(S_t - S_s)} \mid \{S_s,\, U_{S_s}\} \vee \mathcal{F}_{S_s}\right)\right)^2
$$
$$
= A_{S_s}^2 \left(E_\alpha\left(2\mu(t - U_{S_s})^\alpha\right) - \left(E_\alpha\left(\mu(t - U_{S_s})^\alpha\right)\right)^2\right).
$$

Combining these last two results leads to the stated equation.   □

The log-return of the risky asset is $X_{S_t} = \ln\left(\frac{A_{S_t}}{A_0}\right)$. From Eqs. (10.30) and (10.15), we deduce that the Fourier transform of this return between times s and t is given by

$$
\mathcal{F}_{X_{S_t} - X_{S_s}}(\omega) = \mathbb{E}\left(e^{i\omega\,(X_{S_t} - X_{S_s})} \mid \{S_s = v,\, U_{S_s} = u\} \vee \mathcal{F}_{S_s}\right) \qquad (10.35)
$$
$$
= E_\alpha\left(\left(i\omega\left(\mu - \frac{\sigma^2}{2}\right) - \omega^2\frac{\sigma^2}{2}\right)(t - u)^\alpha\right).
$$

The pdf of $X_{S_t} - X_{S_s}$ is the inverse Fourier transform of this Mittag–Leffler function. Following a similar procedure to what is done in Chap. 1, this pdf is numerically computable with the Discrete Fourier Transform (DFT) algorithm.

10.4   A Fractional Fokker–Planck Equation

The previous sections have revealed the important role played by the Mittag–Leffler function. This function is closely related to the concept of the *fractional* or *Caputo derivative*. Caputo's derivative of order $\alpha \in]0, 1[$ for a function $h(t, x) : \mathbb{R}^+ \times \mathbb{R} \to \mathbb{R}, C^1$ with respect to t is defined by

$$\frac{\partial^\alpha}{\partial t^\alpha} h(t, x) = \frac{1}{\Gamma(1 - \alpha)} \left(\frac{\partial}{\partial t} \int_0^t (t - s)^{-\alpha} h(s, x) ds - \frac{h(0, x)}{t^\alpha} \right). \qquad (10.36)$$

An alternative formulation is the following:

$$\frac{\partial^\alpha}{\partial t^\alpha} h(t, x) = \frac{1}{\Gamma(1 - \alpha)} \int_0^t (t - s)^{-\alpha} \frac{\partial}{\partial s} h(s, x) ds . \qquad (10.37)$$

When $\alpha = 1$, this derivative corresponds to the derivative with respect to time. Let $\tilde{h}(\omega, x)$ be the usual Laplace transform of a function $h(t, x)$ with respect to time t. A direct calculation shows that the Laplace transform of Caputo's derivative $\frac{\partial^\alpha}{\partial t^\alpha} h(t, x)$ is equal to

$$\frac{\widetilde{\partial^\alpha h}}{\partial t^\alpha}(\omega, x) = \omega^\alpha \tilde{h}(\omega, x) - \omega^{\alpha-1} h(0, x) ,$$

which reduces to the familiar form when $\alpha = 1$. Notice that Caputo's fractional derivative of a power function is given by

$$\frac{\partial^\alpha}{\partial t^\alpha} t^p = \begin{cases} \frac{\Gamma(p+1)}{\Gamma(p-\alpha+1)} t^{p-\alpha} & p \geq 1 , \ p \in \mathbb{R} \\ 0 & p \leq 0 , \ p \in \mathbb{N} \end{cases} .$$

On the other hand, the solution of the fractional differential equation:

$$\frac{\partial^\alpha}{\partial t^\alpha} y(t) = \lambda y(t) \qquad 0 < \alpha < 1$$

with the initial condition $y(0) = b_0$ is precisely the Mittag–Leffler function $y(t) = E_\alpha(\lambda t^\alpha)$.

Fractional derivatives are essentially an analytic continuation of the concept of the differential operator (or the antiderivative) into a unified differintegral operator, in the same way that the gamma function is an analytic continuation of the factorial function, so grasping for intuitive explanations is not easy. It is used in the physics of materials and replaces the gradient, for example for porous materials. These materials present by nature strong irregularities, and when we want to explain the evolution of heat in them, the derivative with respect to time ($\alpha = 1$) in the heat equation is replaced by a fractional derivative with $\alpha < 1$. Brociek et al. [4] illustrate this for porous aluminum. Compared to non-porous aluminum, its conductivity is lower, and the heat diffuses less quickly than in the porous metal.

We show in this section that the pdf of the time-changed risky asset price is a solution of a fractional Fokker–Planck equation, similar to (10.24), in which the derivative with respect to time is replaced by a Caputo derivative.

For $0 \le s \le u \le t$, the pdf of A_{S_t}, conditionally to $\{S_s = v,\ U_{S_s} = u,\ A_{S_s} = y\}$, is denoted by $p_\alpha(t, x|s, v, u, y)$. This function is such that

$$p_\alpha(t, x|s, v, u, y)\mathrm{d}x =$$

$$\mathbb{P}\left(A_{S_t} \in [x, x + \mathrm{d}x] \mid S_s = v,\ U_{S_s} = u,\ A_s = y\right).$$

This pdf is a solution of a forward differential equation, as stated in the next proposition.

Proposition 10.8 *For $0 \le s \le u \le t$, the conditional pdf $p_\alpha(t, x|s, v, u, y)$ of A_t is a solution of the following Fokker–Planck equation, also called a forward Kolmogorov equation:*

$$\frac{\partial^\alpha p_\alpha(t, x|\cdot)}{\partial t^\alpha} = -\frac{\partial}{\partial x}\left(\mu\, x\, p_\alpha(t, x|\cdot)\right) + \frac{1}{2}\frac{\partial^2}{\partial x^2}\left[\sigma^2 x^2\, p_\alpha(t, x|\cdot)\right],\ (10.38)$$

with the initial condition: $p(u, x|s, v, u, y) = \delta_{\{x-y\}}$, where δ_z is the Dirac measure located at z. For all times t such that $0 \le s \le t < u$, we have $p_\alpha(t, x|s, v, u, y) = \delta_{\{x-y\}}$.

Proof To lighten the notation, we momentarily adopt the following definitions: $p_\alpha(t, x) := p_\alpha(t, x|s, v, u, y)$ and $g(t, \tau) = g(t, \tau|s, v, u)$. As $g(t, \tau)$ is the pdf of $S_t - S_s \mid S_s,\ U_{S_s}$, we infer that

$$p_\alpha(t, x) = \int_0^\infty p(v + \tau, x \mid v, y)g(t, \tau)\mathrm{d}\tau,$$

for $0 \le s \le u \le t$. If $t < u$, $p_\alpha(t, x) = \delta_{\{x-y\}}$. Notice that

$$p(v + \tau, x|v, y) = p(\tau, x \mid 0, y).$$

The Laplace transform of $p_\alpha(t, x)$ with respect to time t is thus given by

$$\tilde{p}_\alpha(\omega, x) = \int_0^\infty \int_0^\infty p(\tau, x \mid 0, y)e^{-\omega t} g(t, \tau)\mathrm{d}\tau\mathrm{d}t$$

$$= \int_0^\infty p(\tau, x \mid 0, y)\,\tilde{g}(\omega, \tau)\,\mathrm{d}\tau,$$

where $\tilde{g}(\omega, \tau) = \int_0^\infty e^{-\omega t} g(t, \tau) dt$ is the Laplace transform of $g(t, \tau)$ with respect to time. Since this transform is given by Eq. (10.6), $\tilde{p}_\alpha(\cdot)$ is equal to

$$\tilde{p}_\alpha(\omega, x) = \omega^{\alpha-1} e^{-\omega u} \int_0^\infty p(\tau, x \mid 0, y) e^{-\tau \omega^\alpha} d\tau$$

$$= \omega^{\alpha-1} e^{-\omega u} \tilde{p}(\omega^\alpha, x),$$

where $\tilde{p}(\omega, x) := \int_0^\infty e^{-\omega \tau} p(\tau, x \mid 0, y) d\tau$ is the Laplace transform of $p(\tau, x \mid 0, y)$ with respect to time. From the FPE (10.24) and using the property of the Laplace transform, we have that

$$\omega \tilde{p}(\omega, x \mid 0, y) - p(0, x \mid 0, y) =$$

$$-\frac{\partial}{\partial x} (\mu x \tilde{p}(\omega, x \mid 0, y)) + \frac{1}{2} \frac{\partial^2}{\partial x^2} \left[\sigma^2 x^2 \tilde{p}(\omega, x \mid 0, y) \right].$$

As $\tilde{p}_\alpha(\omega, x) = \omega^{\alpha-1} e^{-\omega u} \tilde{p}(\omega^\alpha, x)$, replacing ω by ω^α leads to

$$\omega^\alpha \tilde{p}(\omega^\alpha, x) - p(0, x \mid 0, y) = -\frac{\partial}{\partial x} (\mu x \tilde{p}(\omega^\alpha, x)) + \frac{1}{2} \frac{\partial^2}{\partial x^2} \left[\sigma^2 x^2 \tilde{p}(\omega^\alpha, x) \right].$$

If we multiply this last equation by $\omega^{\alpha-1} e^{-\omega u}$, we obtain that

$$\omega^\alpha \left(\omega^{\alpha-1} e^{-\omega u} \tilde{p}(\omega^\alpha, x) \right) - \omega^{\alpha-1} e^{-\omega u} p(0, x \mid 0, y) = \qquad (10.39)$$

$$-\frac{\partial}{\partial x} \left(\mu x \omega^{\alpha-1} e^{-\omega u} \tilde{p}(\omega^\alpha, x) \right) + \frac{1}{2} \frac{\partial^2}{\partial x^2} \left[\sigma^2 x^2 \omega^{\alpha-1} e^{-\omega u} \tilde{p}(\omega^\alpha, x) \right].$$

We rewrite this last equation as follows:

$$\omega^\alpha \tilde{p}_\alpha(\omega, x) - \omega^{\alpha-1} e^{-\omega u} p(0, x \mid 0, y) =$$

$$-\frac{\partial}{\partial x} (\mu x \tilde{p}_\alpha(\omega, x)) + \frac{1}{2} \frac{\partial^2}{\partial x^2} \left[\sigma^2 x^2 \tilde{p}_\alpha(\omega, x) \right].$$

The left-hand term is the Laplace transform of the Caputo derivative of $p_\alpha(t, x)$. Indeed, this transform is equal to

$$\int_0^\infty e^{-\omega t} \frac{\partial^\alpha}{\partial t^\alpha} p_\alpha(t, x) dt = \frac{1}{\Gamma(1-\alpha)} \int_u^\infty e^{-\omega t} \int_0^t (t-z)^{-\alpha} \frac{\partial}{\partial z} p_\alpha(z, x) dz \, dt$$

$$= \frac{1}{\Gamma(1-\alpha)} \int_u^\infty \frac{\partial}{\partial z} p_\alpha(z, x) \int_z^\infty e^{-\omega t} (t-z)^{-\alpha} dt \, dz$$

$$= \frac{1}{\Gamma(1-\alpha)} \int_u^\infty \frac{\partial}{\partial z} p_\alpha(z, x) e^{-\omega z} \int_0^\infty e^{-\omega t'} t'^{-\alpha} dt' \, dz$$

$$= \int_u^\infty \frac{\partial}{\partial z} p_\alpha(z, x) e^{-\omega z} \omega^{\alpha-1} dz.$$

Given that $p_\alpha(z, x) = 0$ for $z < u$, integrating by parts leads to

$$\int_0^\infty e^{-\omega t} \frac{\partial^\alpha}{\partial t^\alpha} p_\alpha(t, x)\, dt = \left[p_\alpha(z, x)\omega^{\alpha-1} e^{-\omega z} \right]_{z=u}^\infty + \omega^\alpha \int_u^\infty p_\alpha(z, x) e^{-\omega z}\, dz$$

$$= \omega^\alpha \tilde{p}_\alpha(\omega, x) - p_\alpha(u, x)\omega^{\alpha-1} e^{-\omega u},$$

which is the left-hand side of Eq. (10.39) since $p_\alpha(u, x) = p(0, x \mid 0, y)$. $\square$

10.5   Option Pricing in the Fractional Setting

The fractional market is incomplete by construction. Intuitively, the incompleteness arises from the risk induced by the stochastic clock, which cannot be hedged by any financial instrument. The risk neutral measure $\mathbb{Q}$ is therefore not unique, and in theory, there exist multiple equivalent measures to $\mathbb{P}$ under which discounted asset prices are martingales. Nevertheless, the time-changed version of the Radon–Nikodym derivative (10.27) is a natural candidate that preserves the analytical features of the model.

We consider an $\mathcal{H}_t$-adapted and square integrable process, denoted $(\theta_t)_{t\geq 0}$. We define the time-changed Radon–Nikodym derivative $(Z_{S_t})_{t\geq 0}$ as follows:

$$Z_{S_t} = \exp\left(-\frac{1}{2} \int_0^{S_t} \theta_s^2 ds - \int_0^{S_t} \theta_s dW_s^P \right), \tag{10.40}$$

which admits an equivalent infinitesimal representation $dZ_{S_t} = -Z_{S_t}\theta_{S_t} dW_{S_t}^P$. We can check that $\mathbb{E}(dZ_t|\mathcal{H}_t \vee \mathcal{G}_T) = 0$, and therefore, $\mathbb{E}(Z_T|\mathcal{H}_t \vee \mathcal{G}_T) = Z_t + \int_t^T \mathbb{E}(dZ_t|\mathcal{H}_t \vee \mathcal{G}_T) = Z_t$. Using nested expectations, the process Z_t is then a martingale with $Z_0 = 1$. Since Z_t is $\mathcal{H}_t$-adapted, it is a Radon–Nikodym derivative $Z_{S_t} = \frac{d\mathbb{P}^b}{d\mathbb{P}}\big|_t$ defining a new measure $\mathbb{P}^b$. The next proposition states the dynamics of Brownian motion and jump process under $\mathbb{P}^b$.

Proposition 10.9 *Under the equivalent measure $\mathbb{P}^b$ defined by the Radon–Nikodym derivative (10.40), $dW_{S_t} = dW_{S_t}^P + \theta_{S_t} dS_t$ is a time-changed Brownian motion.*

Proof The moment generating function (mgf) of W_t under the equivalent measure is

$$\mathbb{E}^{\mathbb{P}^b}\left(e^{\omega W_{S_t}} |\mathcal{H}_0 \right) = \mathbb{E}^{\mathbb{P}^b}\left(e^{\omega W_{S_t}^P + \omega \int_0^{S_t} \theta_s ds} |\mathcal{H}_0 \right)$$

$$= \mathbb{E}^{\mathbb{P}^b}\left(\mathbb{E}^{\mathbb{P}^b}\left(e^{\omega W_{S_t}^P + \omega \int_0^{S_t} \theta_s ds} |\mathcal{H}_0 \vee \mathcal{G}_t \right) |\mathcal{H}_0 \right).$$

We have that

$$\mathbb{E}^{\mathbb{P}^b}\left(e^{\omega W_{S_t}^P + \omega \int_0^{S_t} \theta_s \, ds} | \mathcal{H}_0 \vee \mathcal{G}_t\right)$$

$$= e^{\frac{1}{2}\omega^2 S_t} \mathbb{E}\left(e^{-\frac{1}{2}\int_0^{S_t}(\theta_s - \omega)^2 ds - \int_0^{S_t}(\theta_s - \omega) dW_s^P} | \mathcal{H}_0 \vee \mathcal{G}_t\right)$$

$$= e^{\frac{1}{2}\omega^2 S_t} .$$

To pass from the second to last to the last line, we use the property that the Doléans-Dade exponential of a martingale is a martingale. We recognize the mgf of a Brownian motion, time-changed by S_t. Therefore, we conclude that $W_{S_t} = W_{S_t}^P + \int_0^{S_t} \theta_s \, ds$ is a time-changed Brownian motion under $\mathbb{P}^b$.   □

From the previous proposition, we infer that the dynamics of the risky asset under $\mathbb{P}^b$ is given by

$$\frac{dA_{S_t}}{A_{S_t}} = \left(\mu - \theta_{S_t}\sigma\right) dS_t + \sigma \, dW_{S_t} .$$

The clock being independent from the filtration $(\mathcal{F}_t)_{t \geq 0}$, the expectation of the infinitesimal variation of A_t is equal to

$$\mathbb{E}^{\mathbb{P}^b}\left(\frac{dA_{S_t}}{A_{S_t}} | \mathcal{G}_t\right) = \left(\mu - \theta_{S_t}\sigma\right) dS_t .$$

Therefore, we obtain the condition on $\left(\theta_{S_t}\right)_{t \geq 0}$ that leads to a risk neutral measure

$$\theta_{S_t} = \frac{\mu - r}{\sigma} . \tag{10.41}$$

In the fractional Black and Scholes (FBS) market, a European call option of maturity T, written on the time-changed asset, is the discounted expected payoff under $\mathbb{Q}$. From Eqs. (10.28) and (10.29), we infer that conditionally to a given clock increment $\tau = S_t - S_s$, the call and put option prices are

$$C_{FBS}(A_{S_t}, \tau) = \mathbb{E}^{\mathbb{Q}}\left(e^{-r\tau}\left(A_{S_T} - K\right)_+ | \{S_T - S_t = \tau\} \vee \mathcal{F}_{S_t}\right)$$

$$= A_{S_t} \Phi\left(d_1(A_{S_t}, \tau)\right) - K e^{-r\tau} \Phi(d_2(A_{S_t}, \tau)) ,$$

and

$$P_{FBS}(A_{S_t}, \tau) = \mathbb{E}^{\mathbb{Q}}\left(e^{-r\tau}\left(K - A_{S_T}\right)_+ | \{S_T - S_t = \tau\} \vee \mathcal{F}_{S_t}\right)$$

$$= K e^{-r\tau} \Phi(-d_2(A_t, \tau)) - A_t \Phi\left(-d_1(A_t, \tau)\right) .$$

Let us recall that the pdf of $S_T - S_t \mid S_t = v$, $U_{S_t} = u$ for $0 \le t \le u \le T$ is denoted by $g(T, \tau \mid t, v, u)$. Option prices at time t, conditionally to some information about the stochastic clock and its inverse, are therefore

$$C_{FBS}(A_{S_t}, v, u) = \mathbb{E}^{\mathbb{Q}} \left(e^{-r(S_T - S_t)} \left(A_{S_T} - K \right)_+ \mid \{S_t = v, \, U_{S_t} = u\} \vee \mathcal{F}_{S_t} \right)$$

$$= \int_0^\infty g(T, \tau \mid t, v, u) \, C_{FBS}(A_{S_t}, \tau) \, d\tau, \qquad (10.42)$$

and

$$P_{FBS}(A_{S_t}, v, u) = \mathbb{E}^{\mathbb{Q}} \left(e^{-r(S_T - S_t)} \left(K - A_{S_T} \right)_+ \mid \{S_t = v, \, U_{S_t} = u\} \vee \mathcal{F}_{S_t} \right)$$

$$= \int_0^\infty g(T, \tau \mid t, v, u) \, P_{FBS}(A_{S_t}, \tau) \, d\tau. \qquad (10.43)$$

Proposition 10.4 provides a series expansion of the density $g(\cdot)$. We can therefore evaluate the fractional call and put prices by computing numerically the integrals in the above equations. An alternative consists in calculating the price by Monte Carlo simulations. These solutions are explored by Magdziarz [13] in a Brownian setting.

Figure 10.4 compares call and put prices for different values of α. The case $\alpha = 1$ corresponds to the Black and Scholes market. The strike price is $K = 95$, whereas the initial stock value is $A_0 = 100$. The volatility and risk-free rates are, respectively, set to 1 and 20%. At short term, we observe that both types of options

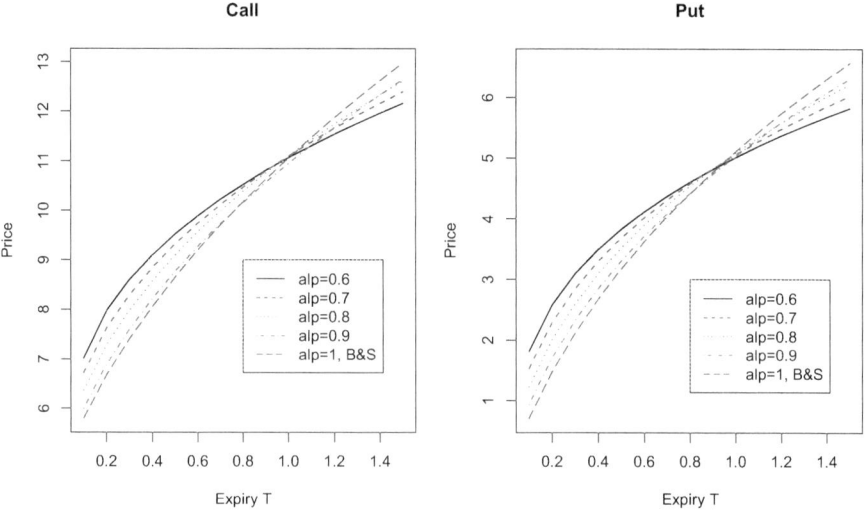

Fig. 10.4 Call and put prices : $t = 0$, $A_0 = 100$, $K = 95$, $r = 0.01$, $\sigma = 0.20$. The expiry date and α range from 0.1 to 1.5 years and from 0.6 to 1

are more expensive in the fractional market than in the pure Brownian setting. This trend reverts for longer maturities.

We end this section with a discussion about the hedging of options in the fractional market. We again consider a short position in $C_{FBS}(A_{S_t}, S_t, U_{S_t})$. We wish to hedge this position with Δ_{FBS} shares between t and $t + \delta$, partly funded by a borrowing. We define $\delta A_{S_t} = A_{S_{t+\delta}} - A_{S_t}$, the variation of stock prices, $\delta S_t = S_{t+\delta} - S_t$, the increment of the clock, and $\delta U_{S_t} = U_{S_{t+\delta}} - U_{S_t}$. The profit and loss (P&L) account over the period $[t, t + \delta]$ is then equal to

$$P\&L = - \big(C_{FBS}(A_{S_{t+\Delta}}, S_{t+\Delta}, U_{S_{t+\Delta}}) - C_{FBS}(A_{S_t}, S_t, U_{S_t})\big)$$
$$+ r\, C_{FBS}(A_{S_t}, S_t, U_t)\, \delta S_t + \Delta_{FBS}\left(\delta A_{S_t} - r\, A_t\, \delta S_t\right).$$

We again expand this P&L in powers of δA_t and δ with a first-order Taylor expansion of the call price

$$C_{FBS}(A_{S_{t+\Delta}}, S_{t+\Delta}, U_{S_{t+\Delta}}) = C_{FBS}(A_{S_t}, S_t, U_{S_t}) + \frac{\partial C_{FBS}(\cdot)}{\partial A_{S_t}} \delta A_{S_t}$$
$$+ \frac{\partial C_{FBS}(\cdot)}{\partial S_t} \delta S_t + \frac{\partial C_{FBS}(\cdot)}{\partial U_{S_t}} \delta U_{S_t}.$$

In view of Eqs. (10.42) and (10.18), the call price does not explicitly depend upon S_t (the main driving factor is U_{S_t}) and $\frac{\partial C_{FBS}(\cdot)}{\partial S_t} = 0$. The first-order approximation of the P&L is then equal to

$$P\&L \approx - \left(\frac{\partial C_{FBS}(\cdot)}{\partial S_t} \delta S_t + \frac{\partial C_{FBS}(\cdot)}{\partial A_{S_t}} \delta A_{S_t} + \frac{\partial C_{FBS}(\cdot)}{\partial U_{S_t}} \delta U_{S_t}\right)$$
$$+ r\, C_{FBS}(A_{S_t}, S_t, U_t)\, \delta S_t + \Delta_{FBS}\left(\delta A_{S_t} - r\, A_t\, \delta S_t\right).$$

Choosing $\Delta_{FBS} = \frac{\partial C_{FBS}(A_{S_t}, S_t, U_{S_t})}{\partial A_{S_t}}$ cancels the first-order term in δA_{S_t} and removes at first order the dependence between the P&L and the underlying asset. In contrast to the Black and Scholes case, the first-order approximation of the P&L is still exposed to joint random fluctuations of the stochastic clock and of the associated subordinator. If $S_t = v$ and $U_{S_t} = u$, the delta is given by the following:

$$\Delta_{FBS} = \int_0^\infty g(T, \tau \mid t, v, u)\, \frac{\partial C_{FBS}(A_{S_t}, \tau)}{\partial A_{S_t}} d\tau,$$

where $\frac{\partial C_{FBS}(A_{S_t}, \tau)}{\partial A_{S_t}} = \Phi(d_1(A_{S_t}, \tau))$ is the delta in the Black and Scholes market. For a put option, the fractional delta is still computable using the previous equation, in which $\frac{\partial C_{FBS}(A_{S_t}, \tau)}{\partial A_{S_t}} = \Phi(d_1(A_{S_t}, \tau)) - 1$.

Figure 10.5 shows the Δ_{FBS} for call and put options. Let us recall that it should be interpreted as the amount of stocks to hold ($\Delta_{FBS} > 0$) or to short sale

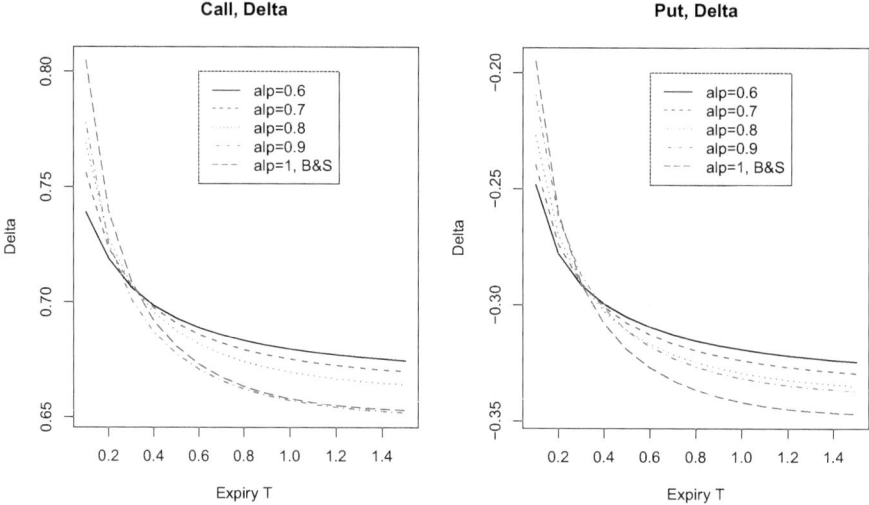

Fig. 10.5 Δ_{FBS} for call and put options: $t = 0$, $A_0 = 100$, $K = 95$, $r = 0.01$, $\sigma = 0.20$. The expiry date and α range from 0.1 to 1.5 years and from 0.6 to 1

($\Delta_{FBS} < 0$) for hedging an option. For call options close to expiry, the delta is lower in the fractional case than in a Brownian framework. For options away to expiry, we observe that the hedge is required to hold more stocks than in a Black and Scholes market. On the other hand, hedging short- and long-term put options respectively requires larger and smaller short sales than in the Brownian setting.

10.6   A Particle Filter

In the fractional model, the asset price is ruled by a stochastic clock that is not observable. This section presents a particle filtering method based on simulations that is used to determine the most likely evolution of hidden processes. A similar approach is introduced in Chap. 3 for filtering the stochastic volatility in a Heston model. We consider a sample of discrete observations of log-returns $X_t = \ln \frac{A_{S_{t+\Delta}}}{A_{S_t}}$. The sample is a vector denoted by $\boldsymbol{x} = \{x_1, x_2, ..., x_T\}$. The lag of time between two consecutive observations is Δ, and the times of sampling are $\{t_1, ..., t_T\}$. We adopt the notations $S_j = S_{t_j}$, $\Delta S_j = S_j - S_{j-1}$, and $X_j = X_{t_j}$, whereas s_j and x_j are the realizations of S_j, X_j. The dynamics of log-returns is approached by discretizing Equation (10.22):

$$X_j = \left(\mu - \frac{\sigma^2}{2} \right) \Delta S_j + \sigma \, \Delta W_j^P ,$$

where $\Delta W_j^P \sim N\left(0, \sqrt{\Delta S_j}\right)$. The dynamics of S_j is tuned by α and can be simulated with Eq. (10.21). The vector of parameters $\boldsymbol{\theta} = \{\mu, \sigma, \alpha\}$ is assumed to be known. The purpose of the particle filter is to estimate the most likely sample path of the stochastic clock, based on observed log-returns.

We denote by Δs_j the realizations of ΔS_j. The information $S_j = s_j$, $U_{S_j} = u_j$, and $\Delta s_j = \Delta S_j$ is stored in a "particle" denoted by $\boldsymbol{p}_j = \left(u_j, s_j, \Delta s_j\right)$. The trivariate process $\left(\boldsymbol{p}_j\right)_{j=1,\dots,T}$ is a discrete Markov chain. The visible information up to time t_j is contained in a vector $\boldsymbol{x}_{1:j} = \{x_1, \dots, x_j\}$, and the density of "measurement" $f(x_j \mid \boldsymbol{p}_j)$, conditionally to the information in the particle, is the Gaussian pdf. The pdf of an $N(0, 1)$ is denoted by $\phi(\cdot)$. We have that

$$f(x_j \mid \boldsymbol{p}_j) = \frac{1}{\sigma\sqrt{\Delta s_j}} \phi\left(\frac{x_j - \left(\mu - \frac{\sigma^2}{2}\right)\Delta s_j}{\sigma\sqrt{\Delta s_j}}\right).$$

On the other hand, we can simulate the transition density $f(\boldsymbol{p}_{j+1} \mid \boldsymbol{p}_j)$ with Eq. (10.21). The pdf of $\boldsymbol{p}_0$ is $f(\boldsymbol{p}_0)$, and the posterior distribution of $\boldsymbol{p}_j$ is denoted by $f(\boldsymbol{p}_j \mid \boldsymbol{x}_{1:j})$. The posterior distribution is equal to the ratio:

$$f(\boldsymbol{p}_j \mid \boldsymbol{x}_{1:j}) = \frac{f(\boldsymbol{x}_{1:j}, \boldsymbol{p}_j)}{f(\boldsymbol{x}_{1:j})}, \tag{10.44}$$

and according to Bayes' rule, the denominator satisfies the equality:

$$f(\boldsymbol{x}_{1:j}) = f(\boldsymbol{x}_{1:j-1}, x_j) = f(x_j \mid \boldsymbol{x}_{1:j-1})f(\boldsymbol{x}_{1:j-1}).$$

Since $f(x_j \mid \boldsymbol{x}_{1:j-1}, \boldsymbol{p}_j) = f(x_j \mid \boldsymbol{p}_j)$, the numerator of Eq. (10.44) is also equal to

$$f(\boldsymbol{x}_{1:j}, \boldsymbol{p}_j) = f(x_j \mid \boldsymbol{x}_{1:j-1}, \boldsymbol{p}_j)f(\boldsymbol{x}_{1:j-1}, \boldsymbol{p}_j)$$
$$= f(x_j \mid \boldsymbol{p}_j)f(\boldsymbol{p}_j \mid \boldsymbol{x}_{1:j-1})f(\boldsymbol{x}_{1:j-1}).$$

The posterior distribution of $\boldsymbol{p}_j$ conditionally to the available information at t_j is then rewritten as follows:

$$f(\boldsymbol{p}_j \mid \boldsymbol{x}_{1:j}) = \frac{f(x_j \mid \boldsymbol{p}_j)}{\int f(x_j \mid \boldsymbol{p}_j)f(\boldsymbol{p}_j \mid \boldsymbol{x}_{1:j-1})\mathrm{d}v_j} f(\boldsymbol{p}_j \mid \boldsymbol{x}_{1:j-1}), \tag{10.48}$$

where

$$f(\boldsymbol{p}_j \mid \boldsymbol{x}_{1:j-1}) = \int f(\boldsymbol{p}_j \mid \boldsymbol{p}_{j-1})f(\boldsymbol{p}_{j-1} \mid \boldsymbol{x}_{1:j-1})\mathrm{d}\boldsymbol{p}_{j-1}. \tag{10.49}$$

Algorithm 10.1 Particle filtering of the stochastic clock

Main procedure :
> **For** $j = 1$ to maximum epoch, T
> **Prediction step**
>> Simulate N times, $u_j^{(i)}, s_j^{(i)}$ and $\Delta s_j^{(i)} = s_j^{(i)} - s_{j-1}^{(i)}, i = 1, ...N$.
>>
>> Draw a sample $\Delta w_j^{(i)}$ from a $N\left(0, \sqrt{\Delta s_j^{(i)}}\right), i = 1, ...N$.
>
> **Correction step**:
>> The "particle" $\boldsymbol{p}_j^{(i)} = \left(u_j^{(i)}, s_j^{(i)}, \Delta s_j^{(i)}\right)$ has a probability of occurrence equal to
>>
>> $$\beta_j^{(i)} = \frac{f(x_j \mid \boldsymbol{p}_j^{(i)})}{\sum_{i=1:N} f(x_j \mid \boldsymbol{p}_j^{(i)})}, \tag{10.45}$$
>>
>> where
>>
>> $$f(x_j \mid \boldsymbol{p}_j^{(i)}) = \frac{1}{\sigma\sqrt{\Delta s_j^{(i)}}}\phi\left(\frac{x_j - \left(\mu - \frac{\sigma^2}{2}\right)\Delta s_j^{(i)}}{\sigma\sqrt{\Delta s_j^{(i)}}}\right). \tag{10.46}$$
>>
>> Calculate the effective sample size, $N_{\text{eff}}(j)$:
>>
>> $$N_{\text{eff}}(j) = 1 \left/ \sum_{i=1}^{N} \left(\beta_j^{(i)}\right)^2 \right. . \tag{10.47}$$
>
> **Resampling step**: if $N_{\text{eff}}(j) < N_{eff,up}$
>> Resample with replacement $n_r < N$ particles according to
>> the importance weights $\beta_j^{(i)}$.
>> Recalculate importance weights with Eqs. (10.45) and (10.46).
> **End loop** on epochs

The particle filter is presented in Algorithm 10.1. The calculation of $f(\boldsymbol{p}_j \mid \boldsymbol{x}_{1:j})$ is performed in three steps. The first one is a prediction step in which we simulate $f(\boldsymbol{p}_j \mid \boldsymbol{x}_{1:j-1})$. In the correction step, we evaluate the probabilities $f(\boldsymbol{p}_j \mid \boldsymbol{x}_{1:j})$ using Eq. (10.48). The integral in the prediction step is replaced by a Monte Carlo simulation, of N "particles," denoted $\boldsymbol{p}_j^{(i)}$ for $i = 1, \ldots, N$. The importance weights $\left(\beta_j^{(i)}\right)_{i=1,\ldots N}$ in Eq. (10.45) provide a discrete approximation of the probability density function of $\boldsymbol{p}_j^{(i)}$. If $\Delta s_j^{(i)} = 0$, the pdf $f(x_j \mid \boldsymbol{p}_j^{(i)})$ is a Dirac measure located at x_j. In practice, we replace this measure by a normal pdf with a small variance.

We next calculate an estimate of the effective sample size, $N_{\text{eff}}(j)$, with Eq. (10.47), which is defined as the equivalent number of independent samples generated directly from the target distribution. In the third step, we perform a partial resampling in order to keep track of the most likely sample paths of the stochastic clock and recalculate importance sampling weights.

The effective sample size may be equal to N when the observed process is frozen over a long period of time. In this case, the resampling exclusively selects particles with $\Delta s_j^{(i)} = 0$. This may lead to an over-representation of some particles after a few iterations. To avoid this, we use partial resampling with $n_r = \frac{1}{4}N$ and skip the resampling step if, for example, N_{eff} is, say, above 95% of N.

Finally, the filtered time scale for the period j is computed as the sum of particles weighted by their probability of occurrence:

$$\widehat{s}_j = \mathbb{E}\left(s_j \mid \boldsymbol{x}_{1:j}\right) = \sum_{i=1:N} s_j^{(i)} \beta_j^{(i)}.$$

In theory, the log-likelihood of the whole sample is approached before resampling by the sum:

$$\log f(\boldsymbol{x}|\boldsymbol{\theta}) = \sum_{j=1}^{T} \log f\left(x_j \mid x_{j-1}\right) \qquad (10.50)$$

$$= \sum_{j=1}^{T} \log \left(\int f(x_j \mid \boldsymbol{p}_j) f(\boldsymbol{p}_j | x_{j-1}) \mathrm{d}\boldsymbol{p}_j\right)$$

$$= \sum_{j=1}^{T} \log \left(\sum_{i=1}^{N} \beta_j^{(i)} f(x_j \mid \boldsymbol{p}_j^{(i)})\right).$$

In practice, this estimate is numerically unreliable because when $\Delta s_j^{(i)} = 0$, the density $f(x_j \mid \boldsymbol{p}_j^{(i)})$ is a Dirac measure approached by a normal pdf with a small variance. We cannot then consider maximizing this log-likelihood to estimate parameters when they are unknown. The next section presents an alternative calibration method.

To illustrate this section, we have simulated the sample path of A_t at high frequency. Under the assumption that $\mu = 5\%$, $\sigma = 15\%$, and $\alpha = 0.8$, we consider 1000 observations of A_{S_t} over one day of trading (duration set to $1/250$ year). The particle filter is run with 200, 500, 1000, and 1500 particles. Figure 10.6 compares the sample path of S_t to its filtered estimate for each configuration of the filter. This confirms the ability of the filter to estimate the sample path of the stochastic clock.

We may think to estimate U_{S_t} by $\widehat{u}_j = \sum_{i=1:N} u_j^{(i)} \beta_j^{(i)}$. However, this approach is not efficient. To explain why, let us recall that the process is frozen at time t_j if $t_j < U_{S_{t_j}} = u_j$. Therefore, in the resampling step, the filter mainly selects particles with $t_j < u_j^{(i)}$ but does not provide a hint about the exact value of u_j. Consequence: the filtered values of S_t are close to the sample path of the clock but may be paired to values $u_j^{(i)}$ different from the real u_j. Happily, the sample path of U_{S_t} may be

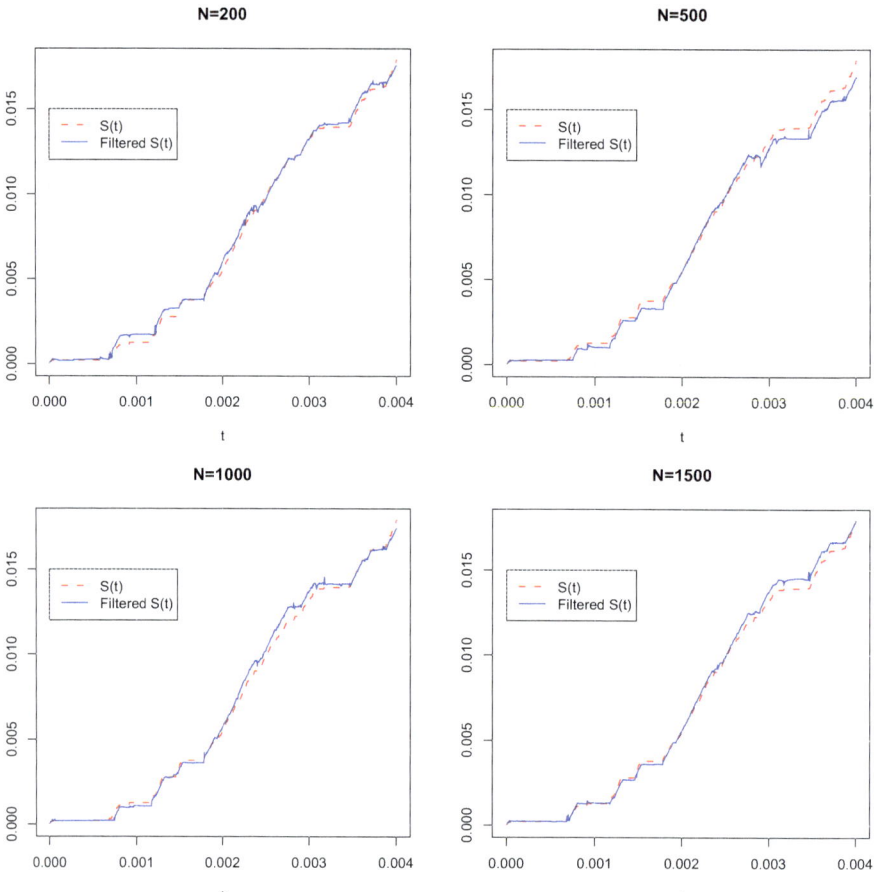

Fig. 10.6 Comparison of s_j and estimated $\hat{s}_j$. The particle filter is run with 200, 500, 1000, and 1500 particles

easily approximated since U_{S_t} is by definition an upper bound of t, close to t when the process is not frozen. Hence, we may assume that

$$\begin{cases} \widehat{u}_j = t_j & \text{if } x_j \neq x_{j+1} \\ \widehat{u}_j = t_k & \text{if } x_j = x_{j+1.} = \cdots = x_k \neq x_{k+1} \end{cases} . \tag{10.51}$$

The left plot of Fig. 10.7 compares the estimated sample path of U_{S_t}, obtained with the particle filter, to its real instances. As mentioned, this approach performs poorly. The right plot of the same figure confirms the efficiency of the approximation (10.51).

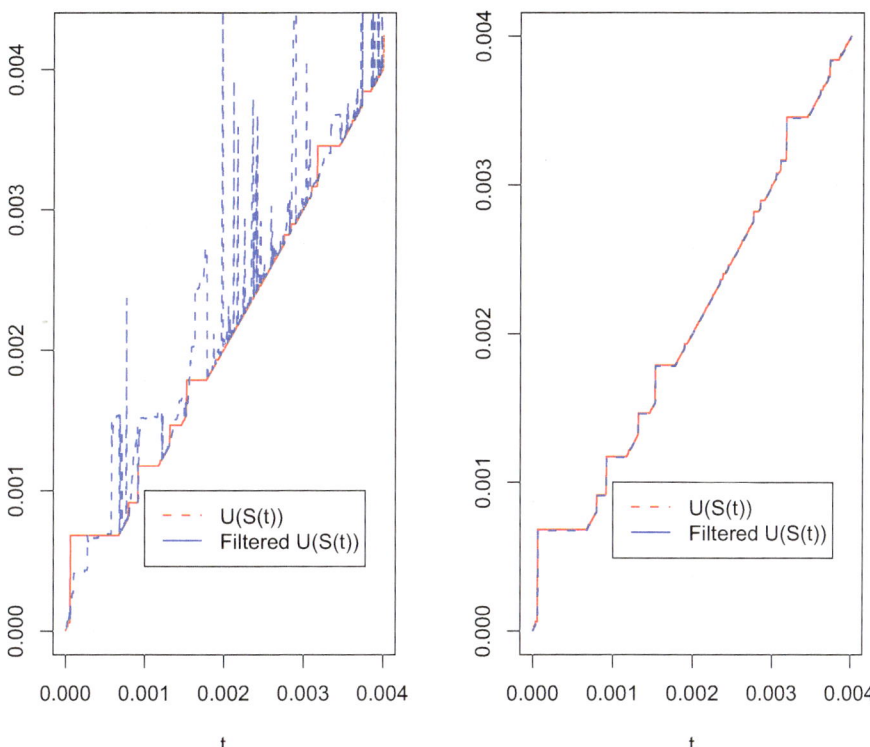

Fig. 10.7 Left plot: comparison of $u_j = U_{S_{t_j}}$ and estimated $\hat{u}_j$ obtained with the particle filter ($N =1500$). Right plot: comparison of u_j with estimates built with Eq. (10.51)

10.7   Estimation of Parameters

We consider again a sample of log-returns given by a vector $x = \{x_1, x_2, ..., x_T\}$. The lag of time between two consecutive observations is Δ, and the times of sampling are $\{t_1, ..., t_T\}$. We adopt the notations $S_j = S_{t_j}$, $\Delta S_j = S_j - S_{j-1}$, and $X_j = X_{t_j}$, whereas s_j and x_j are the realizations of S_j, X_j. The dynamics of log-returns is approached by discretizing Equation (10.22):

$$X_j = \left(\mu - \frac{\sigma^2}{2}\right)\Delta S_j + \sigma\,\Delta W_j^P \ , \ j = 1, ..., T \ .$$

Instead of filtering the sample path of S_j, we consider a quasi-likelihood approach. From Proposition 10.5, the expectation of X_j, conditioned by the information available at inception of the process, is denoted by μ_j^X and is approached by

$$
\mathbb{E}\left(X_j \mid \mathcal{H}_0\right) = \left(\mu - \frac{\sigma^2}{2}\right) \mathbb{E}\left(\Delta S_j \mid \mathcal{H}_0\right)
$$

$$
\approx \left(\mu - \frac{\sigma^2}{2}\right) \frac{\alpha t_j^{\alpha-1}}{\Gamma(\alpha+1)} \Delta
$$

$$
:= \mu_j^X ,
$$

whereas its variance, denoted by σ_j^X, is approximated by

$$
\mathbb{V}\left(X_j \mid \mathcal{H}_0\right) = \sigma^2 \mathbb{E}\left(\Delta S_j \mid \mathcal{H}_0\right)
$$

$$
\approx \sigma^2 \frac{\alpha t_j^{\alpha-1}}{\Gamma(\alpha+1)} \Delta
$$

$$
:= \left(\sigma_j^X\right)^2 .
$$

As $X_j \mid \mathcal{H}_0 \sim \mathcal{N}\left(\mu_j^X, \left(\sigma_j^X\right)^2\right)$, the quasi-log-likelihood of the instance x_j is equal to the logarithm of the normal pdf:

$$
\ln f(x_j \mid \mathcal{H}_0) = -\frac{\left(x_j - \mu_j^X\right)^2}{2\left(\sigma_j^X\right)^2} - \ln\left(\sigma_j^X \sqrt{2\pi}\right) .
$$

The estimates of μ, σ, and α are found by maximization of this quasi-log-likelihood:

$$
\{\widehat{\mu}, \widehat{\sigma}, \widehat{\alpha}\} = \arg\min_{\mu,\sigma,\alpha} \sum_{j=1}^{T} \left(\left(x_j - \left(\mu - \frac{\sigma^2}{2}\right) \frac{\alpha t_j^{\alpha-1}}{\Gamma(\alpha+1)} \Delta\right)^2 \left(2\sigma^2 \frac{\alpha t_j^{\alpha-1} \Delta}{\Gamma(\alpha+1)}\right)^{-1}\right.
$$

$$
\tag{10.52}
$$

$$
\left. + \ln\left(\sigma \sqrt{\frac{2\pi \alpha t_j^{\alpha-1} \Delta}{\Gamma(\alpha+1)}}\right)\right) .
$$

In order to illustrate this section, we simulate a sample path of A_{S_t} with parameters reported in Table 10.1 over a period of 10 years. We consider daily simulations, and as a year counts approximatively 250 days of trading, we set $\Delta = \frac{1}{250}$. Parameter estimates maximize the quasi-log-likelihood (10.52) and are

Table 10.1 Parameters and their estimates by quasi-likelihood maximization

μ	5%	$\hat{\mu}$	4.3986%
σ	10%	$\hat{\sigma}$	10.9642%
α	0.80	$\hat{\alpha}$	0.8405

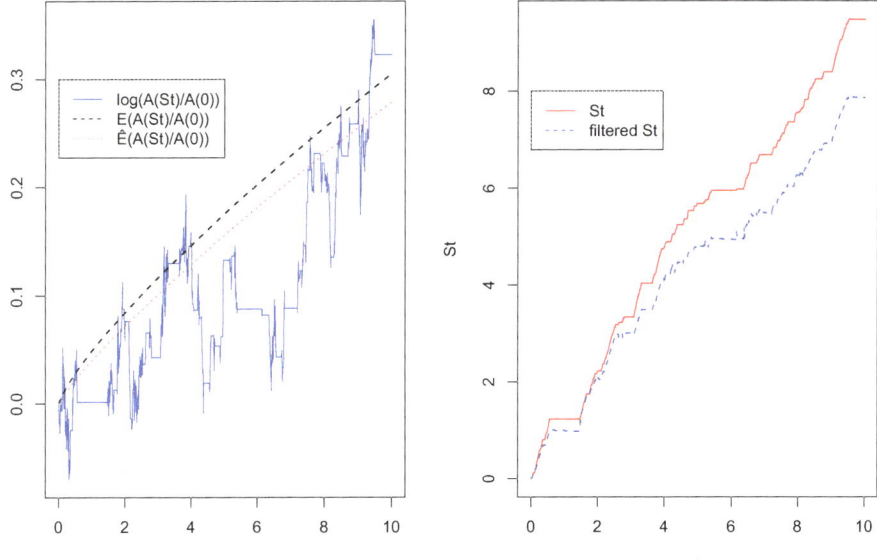

Fig. 10.8 Left plot: simulated sample path of $\log\left(\frac{A_{S_t}}{A_0}\right)$ and comparison of their exact and estimated expectations. Right plot: real and filtered sample paths of S_t

reported in the same table. They are in line with the real parameters. The left plot of Fig. 10.8 presents the simulated sample path of cumulated daily log-returns, i.e., $\log\frac{A_{S_t}}{A_0}$, and compares their expectations computed with estimated and exact parameters. In this case, we slightly underestimate the average growth rate of the stock. Next, we use the estimated parameters to filter the sample path of the stochastic clock with the particle filter (run with 2000 particles). The right plot of Fig. 10.8 compares the real and filtered S_t, which present clear similarities.

10.8 Further Reading

Subordination is an efficient technique to construct new processes. Furthermore, an empirical study carried by Ané and Geman [1] emphasizes that stock prices may be represented by a Brownian motion observed on a time scale that is proportional to trading volumes. Geman et al. [8] introduce a new class of Lévy processes that are written as time-changed Brownian motions. When the stochastic clock, also called the subordinator, is stable, the time-changed process is purely discontinuous with an infinite activity. Carr et al. [6] survey 25 different realizations of time-changed Lévy processes. Carr and Wu [5] subordinate Lévy processes by an integrated mean reverting process and propose a general framework to evaluate options. In a similar manner, Hainaut [9] uses as a clock the integrated intensity of a self-exciting process.

The fractional Fokker–Planck equation for sub-diffusions is studied in Barkai et al. [2]. Sub-diffusions are also popular in econophysics, and we refer the reader to Scalas [15] for a survey. Leonenko et al. [11, 12] study fractional Pearson diffusions and their correlation. Hainaut [10] proposes a time-changed version of a self-exciting process whose dynamics is explainable by a fractional Fokker–Planck equation. In the next chapter, we extend the Black and Scholes market with jumps and consider a broader class of stochastic clocks.

References

1. Ané, T., Geman, H.: Transaction clock, and normality of asset returns. J. Financ. **55**(5), 2259–2284 (2000)
2. Barkai, E., Metzler, R., Klafter, J.: From continuous time random walks to the fractional Fokker–Planck equation. Phys. Rev. E **61**, 132–138 (2000)
3. Black, F., Scholes, M.: The pricing of options and corporate liabilities. J. Polit. Econ. **81**, 637–659 (1973)
4. Brociek, R., Slota, D., Krol, M., Matula, G., Kwasny, W.: Modeling of heat distribution in porous aluminum using fractional differential equation. Fract. Fract. **1**(1), 1–9 (2017)
5. Carr, P., Wu, L.: Time-changed Lévy processes and option pricing. J. Financ. Econ. **71**, 113–141 (2004)
6. Carr, P., Geman, H., Madan, D., Yor, M.: Stochastic volatility for Lévy processes. Math. Financ. **13**, 345–382 (2003)
7. Eliazar, I., Klafter, J.: Spatial gliding, temporal trapping and anomalous transport. Physica D **187**, 30–50 (2004)
8. Geman, H., Madan, D., Yor, M.: Time changes for Lévy processes. Math. Financ. **11**, 79–96 (2001)
9. Hainaut, D.: Clustered Lévy processes and their financial applications. J. Comput. Appl. Math. **319**, 117–140 (2017)
10. Hainaut, D.: Fractional Hawkes processes. Phys. A Statist. Mech. Appl. **549**, 124330 (2020)
11. Leonenko, N., Meerschaert, M., Sikorskii, A.: Fractional Pearson diffusions. J. Math. Analy. Appl. **403**, 532–546 (2013)
12. Leonenko, N., Meerschaert, M., Sikorskii, A.: Correlation structure of fractional Pearson diffusions. Comput. Math. Appl. **66**, 737–745 (2013)
13. Magdziarz, M.: Stochastic representation of subdiffusion processes with time-dependent drift. Stoch. Process. Appl. **119**, 3238–3252 (2009)
14. Metzler R., Klafter J.: The restaurant at the end of the random walk: recent developments in the description of anomalous transport by fractional dynamics. J. Phys. A Math. General **37**(31), R161 (2004)
15. Scalas, E.: Five years of continuous-time random walks in econophysics. In: Namatame, A., Kaizouji, T., Aruka, Y. (eds.) The Complex Networks of Economic Interactions, pp. 3–16. Springer, New York (2006)
16. Uchaikin, V.V., Zolotarev, V.M.: Chance and Stability. Modern Probability and Statistics. VSP, Utrecht. Stable Distributions and Their Applications, With a foreword by V.Yu. Korolev and Zolotarev (1999). https://doi.org/10.1515/9783110935974

Chapter 11
A Fractional Dupire Equation
for Jump-Diffusions

In Chap. 10, we managed illiquidity in a Black and Scholes framework with an appropriate time change. Such an approach can be extended to jump-diffusions. Nevertheless, option pricing is a challenging task in this framework mainly because there is no analytical formula for options in the non-time-changed model. This chapter explores a new approach based on a fractional version of what is called Dupire's equation (Dupire [9]), which is a forward partial differential equation (PDE) for option prices. This PDE is built on the assumption that the stock value is ruled by a geometric Brownian diffusion with a volatility that is a function of time and price.

We extend Dupire's framework in two directions. Firstly, we consider a fractional jump-diffusion with stochastic volatility for the asset return. The sample path of such a process behaves like a Brownian motion with motionless periods but exhibits discontinuities caused by jumps. Secondly, we consider a wider family of fractional dynamics. In the previous chapter, fractional processes are built by replacing the time scale by a random clock that is the inverse of an α-stable Lévy process. Here, we find a very general form of a fractional Dupire equation valid for all invertible Lévy subordinators. As an illustration, we compare fractional dynamics based on inverted Poisson and α-stable subordinators. Finally, we propose a numerical scheme to evaluate options in these two cases.

11.1 Non-fractional Jump-Diffusion Model

This section reviews some well-known results about option valuation in a jump-diffusion framework. Later we will consider a time-changed version of this model. For the moment, we consider a financial market in which is traded a risk-free bond and a stock. The risk-free bond, denoted B_t, earns a constant interest rate r and satisfies the differential equation:

$$\frac{\mathrm{d}B_t}{B_t} = r\,\mathrm{d}t \quad B_0 = 1\,,\ t \geq 0\,.$$

The return of the stock is ruled by a Brownian motion $\left(W_t^P\right)_{t\geq0}$ and by a compound Poisson process, $(J_t)_{t\geq0}$. This compound Poisson process is defined as

$$J_t = \sum_{k=1}^{N_t^J} Y_k\,,$$

where N_t^J is a Poisson process with parameter λ_J. Jumps are independent and identically distributed random variables, $Y_k \sim Y$, on $[-1, \infty)$. We denote by $f_Y(\cdot)$ the probability density function (pdf) of jumps and the jump expectation is denoted by $\xi = \mathbb{E}(Y)$. We assume that the stock price, A_t, is ruled by the following geometric jump-diffusion:

$$\frac{\mathrm{d}A_t}{A_t} = \mu\,\mathrm{d}t + \sigma_t\,\mathrm{d}W_t^P + \mathrm{d}J_t\,, \tag{11.1}$$

where $\mu \in \mathbb{R}^+$ and $(\sigma_t)_{t\geq0}$ is a positive process. All processes are defined on a probability space Ω, endowed with their natural filtration $(\mathcal{F}_t)_{t\geq0}$ and a probability measure, $\mathbb{P}$. The volatility process is $\mathcal{F}_t$-adapted and square integrable, i.e., $\int_0^t \sigma_s^2(\omega)\mathrm{d}s < \infty$ for all $\omega \in \Omega$ and $t > 0$. By construction, the expected instantaneous return of the stock price is equal to $\mathbb{E}\,(\mathrm{d}A_t|\mathcal{F}_{t-}) = (\mu + \lambda_J\xi)\,A_t\mathrm{d}t$. Applying Itô's lemma to $\mathrm{d}\ln A_t$ leads after integration to the following expression for the stock price:

$$A_t = A_0 \exp\left(\int_0^t \mu - \frac{\sigma_s^2}{2}\mathrm{d}s + \int_0^t \sigma_s\,\mathrm{d}W_s^P\right)\prod_{k=1}^{N_t^J}(1 + Y_k)\,. \tag{11.2}$$

This is the dynamics of the risky asset under the real measure $\mathbb{P}$. Nevertheless, the pricing of financial derivatives is performed under the risk neutral measure so as to exclude arbitrages. Under this measure, risky assets earn on average the risk-free rate whatever their volatility. Equivalent probability measures are constructed as follows. Firstly, we define a function $\phi\,(\cdot, \kappa) = \ln\left(\kappa\frac{f_Y^b(\cdot)}{f_Y(\cdot)}\right)$, where $\kappa \in \mathbb{R}^+$ and $f_Y^b(\cdot)$ is a pdf on $[-1, \infty)$, eventually null on the same subdomain as $f_Y(\cdot)$. Secondly, we introduce an $\mathcal{F}_t$-adapted and square integrable process, denoted by $(\theta_t)_{t\geq0}$. Thirdly, we define a process $(Z_t)_{t\geq0}$ as follows:

$$Z_t = \exp\left(-\frac{1}{2}\int_0^t \theta_s^2\mathrm{d}s - \int_0^t \theta_s\mathrm{d}W_s^P + \sum_{k=1}^{N_t^J}\phi\,(Y_k, \kappa) - (\kappa - 1)\lambda_J t\right) \tag{11.3}$$

with $Z_0 = 1$. If we apply Itô's lemma to Z_t, we immediately infer its infinitesimal dynamics:

$$dZ_t = Z_t \left(-\theta_t dW_t^P - (\kappa - 1)\lambda_J dt + \left(\kappa \frac{f_Y^b(Y)}{f_Y(Y)} - 1 \right) dN_t^J \right).$$

Given that $\mathbb{E}\left(\frac{f_Y^b(Y)}{f_Y(Y)} \right) = 1$, $\mathbb{E}(dZ_s | \mathcal{F}_t) = 0$ and $\mathbb{E}(Z_s | \mathcal{F}_t) = Z_t + \int_t^s \mathbb{E}(dZ_s | \mathcal{F}_t)$ $= Z_t$. This proves that Z_t is a martingale. The process $(Z_t)_{t \geq 0}$ is a Radon–Nikodym derivative $Z_t = \frac{d\mathbb{P}^b}{d\mathbb{P}}\Big|_t$, which defines an equivalent probability measure, denoted $\mathbb{P}^b$. Under this measure, the process J_t is still a Poisson process, but its frequency of jumps is equal to $\kappa \lambda_J$, whereas the jump pdf becomes $f_Y^b(\cdot)$. Under $\mathbb{P}^b$, the process $dW_t = dW_t^P + \theta_t dt$ is a Brownian motion. We refer the reader to Shreve [13, Chapter 11, Section 6] for detailed explanations. The equivalent measure $\mathbb{P}^b$ is a risk neutral one, denoted $\mathbb{Q}$, if and only if discounted prices are martingales. If we denote the expected jump size by $\xi^b = \mathbb{E}(Y^b)$, the condition

$$r = \mu + \kappa \lambda_J \xi^b - \sigma_t \theta_t \tag{11.4}$$

must be fulfilled to ensure that the equivalent measure is a risk neutral one. As this condition is satisfied for an infinity of combinations of $(\theta_t)_{t \geq 0}$, κ and $f_Y^b(\cdot)$, the risk neutral measure is not unique and the market is said to be incomplete. In practice, the risk neutral measure is chosen in order to replicate at best quoted prices of options.

In order to lighten the notation, we assume without loss of generality that the distribution and frequency of jumps is identical under $\mathbb{P}$ and $\mathbb{Q}$ ($\kappa = 1$ and $f_Y^b(\cdot) = f_Y(\cdot)$). In this case, the non-arbitrage condition imposes that $\theta_s = \frac{\mu + \lambda_J \xi - r}{\sigma_s}$. Under the risk neutral measure, the drift of the risky asset is equal to the risk-free rate:

$$\frac{dA_t}{A_t} = (r - \lambda_J \xi) \, dt + \sigma_t \, dW_t + dJ_t. \tag{11.5}$$

We denote by $C(t, K)$ the value of a European call option paying the positive difference between the stock and strike prices at expiry (time t). According to the fundamental theorem of asset pricing, the value of this call option is the expected payoff under the risk neutral measure:

$$C(t, K) = \mathbb{E}^{\mathbb{Q}}\left(e^{-rt} (A_t - K)_+ \right).$$

Similarly, the European put option of maturity t and strike price K is equal to

$$D(t, K) = \mathbb{E}^{\mathbb{Q}}\left(e^{-rt} (K - A_t)_+ \right).$$

The call and put prices are solutions of a forward partial integro-differential equation (PIDE) presented in the next proposition.

Proposition 11.1 *The call price is the solution of the forward partial integro-differential equation (PIDE):*

$$\frac{\partial C(t, K)}{\partial t} + (r - \lambda_J \xi) K \frac{\partial C(t, K)}{\partial K} - \frac{\mathbb{E}^{\mathbb{Q}}\left(\sigma_t^2 \mid A_t = K\right) K^2}{2} \tag{11.6}$$

$$\times \frac{\partial^2 C(t, K)}{\partial K^2} - \lambda_J \left(\mathbb{E}^{\mathbb{Q}}\left((1 + Y)C\left(t, \frac{K}{1 + Y}\right)\right) - (1 + \xi) C(t, K)\right) = 0$$

with initial condition $C(0, K) = (S_0 - K)_+$. The put price is the solution of the same PIDE but with the initial condition $D(0, K) = (K - S_0)_+$.

Proof The Heaviside function, denoted $\upsilon(x)$, is defined as

$$\upsilon(x) = \begin{cases} 1 & x \geq 0 \\ 0 & x < 0, \end{cases}$$

and we denote the Dirac delta function by $\delta(\cdot)$. Using the Itô–Tanaka lemma for $e^{-rt}(A_t - K)_+$, we immediately obtain that

$$d\left(e^{-rt}(A_t - K)_+\right) = e^{-rt}\upsilon(A_t - K)\left((r - \lambda_J \xi) A_t dt + \sigma_t A_t dW_t\right) \tag{11.7}$$

$$-re^{-rt}(A_t - K)_+ dt + \frac{1}{2}e^{-rt}\delta(A_t - K)\sigma_t^2 A_t^2 dt$$

$$+e^{-rt}\left((A_t(1 + Y) - K)_+ - (A_t - K)_+\right) dN_t^J .$$

The next step consists in calculating the expectation of this infinitesimal variation. By definition of the call price, we have the useful relations:

$$\mathbb{E}^{\mathbb{Q}}\left(e^{-rt}\upsilon(A_t - K)\right) = -\frac{\partial C(t, K)}{\partial K}, \tag{11.8}$$

$$\mathbb{E}^{\mathbb{Q}}\left(e^{-rt}\delta(A_t - K)\right) = \frac{\partial^2 C(t, K)}{\partial K^2} .$$

On the other hand, the call price can be rewritten with the Heaviside function as follows:

$$C(t, K) = \mathbb{E}^{\mathbb{Q}}\left(e^{-rt}A_t \upsilon(A_t - K)\right) - K\, \mathbb{E}^{\mathbb{Q}}\left(e^{-rt}\upsilon(A_t - K)\right) .$$

From this last equation, the discounted expectation of the product A_t and the Heaviside function is therefore equal to

$$\mathbb{E}^{\mathbb{Q}}\left(e^{-rt}A_t \upsilon(A_t - K)\right) = C(t, K) - K\frac{\partial C(t, K)}{\partial K} . \tag{11.9}$$

Given that jumps are independent from N_t^J, we have that

$$\mathbb{E}^{\mathbb{Q}}\left(e^{-rt}\left((A_t(1+Y) - K)_+ - (A_t - K)_+\right) dN_t^J\right) \tag{11.10}$$

$$= \lambda_J \left(\mathbb{E}^{\mathbb{Q}}\left((1+Y)C\left(t, \frac{K}{1+Y}\right)\right) - C(t, K)\right) dt.$$

Combining Equations (11.7), (11.8), (11.9), and (11.10) leads to the following forward equation:

$$\frac{\partial C(t, K)}{\partial t} = -(r - \lambda_J \xi) K \frac{\partial C(t, K)}{\partial K} + \frac{K^2}{2} e^{-rt} \mathbb{E}^{\mathbb{Q}}\left(\delta(A_t - K)\sigma_t^2\right)$$

$$+ \lambda_J \left(\mathbb{E}^{\mathbb{Q}}\left((1+Y)C\left(t, \frac{K}{1+Y}\right)\right) - (1+\xi)C(t, K)\right). \tag{11.11}$$

The expected variance of the Brownian term, conditioned by the asset value, is equal to

$$\mathbb{E}^{\mathbb{Q}}\left(\sigma_t^2 \mid A_t = K\right) = \frac{e^{-rt}\mathbb{E}^{\mathbb{Q}}\left(\delta(A_t - K)\sigma_t^2\right)}{e^{-rt}\mathbb{E}^{\mathbb{Q}}\left(\delta(A_t - K)\right)}.$$

Since $e^{-rt}\mathbb{E}^{\mathbb{Q}}\left(\delta(A_t - K)\right) = \frac{\partial^2 C(t,K)}{\partial K^2}$, the PIDE (11.11) becomes Eq. (11.6). The same reasoning holds for a put option. □

The proof is similar to that of Bergomi [5] for a diffusion. In the absence of jumps, Eq. (11.6) is called *Dupire's equation* (see Dupire [9]). When the $\mathcal{F}_t$-adapted process σ_t is a function of time and of the asset price, $\sigma_t = \sigma(t, A_t)$, we say that the volatility is *local*. For example, in the constant elasticity volatility model (CEV), the volatility function is set to

$$\sigma(t, A_t)^2 = \sigma_0^2 A_t^{2\gamma - 2}, \tag{11.12}$$

where $\gamma \in \mathbb{R}$. The CEV process, introduced by Cox [8], became popular due to its ability to capture the implied volatility skew exhibited by options prices. A possible alternative local volatility with time dependence is

$$\sigma(t, A_t)^2 = \beta_0 + \beta_1 \left(\frac{A_t}{F_t}\right)^{\gamma} + \beta_1 \left(\frac{A_t}{F_t}\right)^{2\gamma}, \tag{11.13}$$

where $F_t = A_0 e^{rt}$ is the forward stock price and $\gamma \in \mathbb{R}^+$. This choice is motivated by the fact that, in practice, implied volatilities reach their minimum when the strike is close or equal to the forward price ("at the money" options). Bergomi [5] studies a similar local volatility function in a Brownian setting. In the remainder of this chapter, we explore the properties of a time-changed version of this model. The next section introduces the stochastic clocks that we use as time changes.

11.2 Subordinators

As recalled in Chap. 10, a *subordinator* is a stochastic process, denoted $(U_t)_{t\geq 0}$, with positive, non-decreasing sample paths and taking values in $\mathbb{R}^+$. Instead of focusing on α-stable process, we consider here general Lévy subordinators for which increments are independent and homogeneously distributed. From the Lévy–Khintchine formula, we know that the Laplace transform of a Lévy subordinator has the following form:

$$\mathbb{E}\left(e^{-\omega U_t}\right) = e^{-t f(\omega)}, \tag{11.14}$$

where $f(\omega) = b\omega + \int_0^\infty \left(1 - e^{-\omega z}\right) \bar{\nu}(\mathrm{d}z)$ and $b \in \mathbb{R}^+$. The function $\bar{\nu}(\cdot)$ is a non-negative measure on $(0, \infty)$, referred to as the *Lévy measure*, satisfying the integrability condition

$$\int_0^\infty (z \wedge 1)\, \bar{\nu}(\mathrm{d}z) < \infty.$$

The function $f(\omega)$ is also called a *Bernstein function*. The inverse of the subordinator is a process denoted by $(S_t)_{t\geq 0}$ that is defined as follows:

$$S_t = \inf\{\tau > 0 : U_\tau \geq t\}.$$

This is the time at which U_t hits the barrier t. This inverted Lévy subordinator is no longer a Lévy process but is positive and non-decreasing and has all the requisite properties to be used as a stochastic clock. By construction, the inverted process may be constant. Therefore, any process subordinated by S_t exhibits motionless periods. This point is illustrated later in this section. The natural filtration of $(S_t)_{t\geq 0}$ is denoted by $(\mathcal{G}_t)_{t\geq 0}$. The probability density functions of U_t and S_t are, respectively, denoted by $p_U(t, u) = \frac{\mathrm{d}}{\mathrm{d}\tau}\mathbb{P}(u \leq U_t \leq u+\mathrm{d}u)$ and $g(t, \tau) = \frac{\mathrm{d}}{\mathrm{d}\tau}\mathbb{P}(\tau \leq S_t \leq \tau+\mathrm{d}\tau)$. The survival function of S_t is $\bar{G}(t, s) = \mathbb{P}(S_t > s)$, and by definition, we have

$$\bar{G}(t, s) = \mathbb{P}(U_s < t). \tag{11.15}$$

The next proposition shows that the Laplace transform of the pdf of S_t can be expressed as a function of the Bernstein function $f(\cdot)$.

Proposition 11.2 *The Laplace transform of $g(t, \tau)$ with respect to t is linked to the Laplace exponent of U_t as follows:*

$$\tilde{g}(\omega, \tau) = \int_0^\infty e^{-\omega t} g(t, \tau) \, dt \tag{11.16}$$

$$= \frac{f(\omega)}{\omega} e^{-\tau f(\omega)} \, .$$

Proof From Eq. (10.4) of Chap. 10, the density of S_t is related to the density of U_t through the relation

$$g(t, \tau) = -\frac{\partial}{\partial \tau} \int_0^t p_U(\tau, x) \, dx \, .$$

Integrating by parts allows us to rewrite the Laplace transform of $g(t, \tau)$ as

$$\tilde{g}(\omega, \tau) = -\frac{\partial}{\partial \tau} \int_0^\infty e^{-\omega t} \int_0^t p_U(\tau, x) \, dx \, dt$$

$$= -\frac{\partial}{\partial \tau} \left(\left[-\frac{e^{-\omega t}}{\omega} \int_0^t p_U(\tau, x) \, dx \right]_{t=0}^{t=\infty} + \int_0^\infty \frac{e^{-\omega t}}{\omega} p_U(\tau, t) \, dt \right)$$

$$= -\frac{1}{\omega} \frac{\partial}{\partial \tau} \int_0^\infty e^{-\omega t} p_U(\tau, t) \, dt \, .$$

We recognize in this last equation the derivative with respect to τ of the Laplace transform of U_τ, which is equal to $e^{-\tau f(\omega)}$. □

This relation plays an important role in later developments. The results in Sects. 11.3 and 11.4 are valid for all Lévy subordinators, but we will grant more attention to inverse α-stable and Poisson subordinators because they offer a certain level of analytical tractability.

Inverse α-Stable Subordinators In the α-stable case, the process U_t is a $\frac{1}{\alpha}$-self-similar process and its moment generating function is given by

$$\mathbb{E}_0 \left(e^{-u U_t} \right) = e^{-t \, u^\alpha} \, .$$

Since the Laplace exponent admits the following integral representation,

$$f(\omega) = \omega^\alpha = \frac{\alpha}{\Gamma(1-\alpha)} \int_0^\infty \frac{\left(1 - e^{-\omega x}\right)}{z^{1+\alpha}} \, dz \, ,$$

the Lévy measure is $\bar{\nu}(dz) = \frac{\alpha}{\Gamma(1-\alpha)} \frac{dz}{z^{1+\alpha}}$. As proven in Sect. 10.1 of Chap. 10, the Laplace transform of $g(t, \tau)$ with respect to t is given by

$$\tilde{g}(\omega, \tau) = \omega^{\alpha-1} e^{-\tau \omega^{\alpha}} .$$

We have also seen in the same chapter that the Laplace transform of S_t conditionally to the information available at time zero is given by

$$\mathbb{E}_0 \left[e^{-\omega S_t} \right] = E_\alpha(-\omega t^\alpha) , \qquad (11.17)$$

where $E_\alpha(z) = \sum_{n=0}^{\infty} \frac{z^n}{\Gamma(n\alpha+1)}$ is the Mittag-Leffler function. The moments of S_t are obtained by differentiating and cancelling its Laplace transform:

$$\mathbb{E}_0 \left(S_t^n \right) = \frac{n! t^{n\alpha}}{\Gamma(n\alpha + 1)} .$$

A particular interesting case is when $\alpha = \frac{1}{2}$. Given that $\Gamma(\alpha) = \sqrt{\pi}$, the Lévy measure is equal to

$$\bar{\nu}(dz) = \frac{1}{2\sqrt{\pi}} \frac{dz}{z^{3/2}} .$$

The Laplace transform of the pdf of S_t has a simple expression

$$\tilde{g}(\omega, \tau) = \frac{e^{-\tau \sqrt{\omega}}}{\sqrt{\omega}}$$

that admits an analytical inverse. The probability density function of S_t for $\alpha = \frac{1}{2}$ is given by

$$g(t, \tau) = \frac{e^{-\frac{\tau^2}{4t}}}{\sqrt{\pi t}} \quad t \geq 0 ,$$

which is proportional to the density of an $N(0, \sqrt{2t})$ on the positive real line. If we denote the error function by $\mathrm{erfc}(x) = \frac{2}{\sqrt{\pi}} \int_0^x e^{-t^2} dt$, the Mittag-Leffler exponential for $\alpha = \frac{1}{2}$ is equal to

$$E_{\frac{1}{2}}(z) = \exp(z^2)\mathrm{erfc}(-z) .$$

The first moment of S_t is

$$\mathbb{E}_0 (S_t) = \frac{t^{1/2}}{\Gamma(3/2)} = \frac{t^{1/2}}{\frac{1}{2}\sqrt{\pi}} .$$

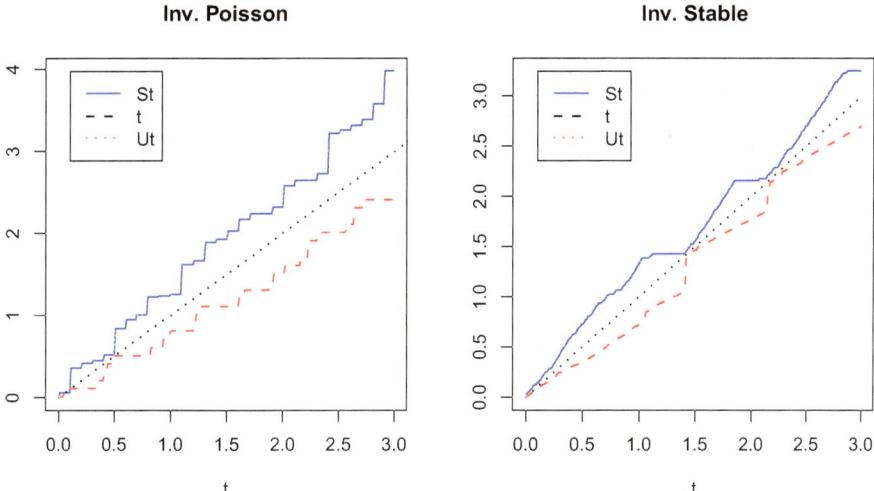

Fig. 11.1 Left plot: sample path of an inverse Poisson subordinator ($\lambda = 10$). Right plot: sample path of an inverse α-stable process ($\alpha = 0.90$)

The right plot of Fig. 11.1 shows a simulated sample path of an α-stable process and of its inverse for $\alpha = 0.90$. We clearly observe periods during which S_t is constant. They are caused by sharp increases of U_t.

Inverse Poisson Subordinators In this case, we invert a subordinator of the form

$$U_t = \eta N_t \quad t, \mu \geq 0 ,$$

where $(N_t)_{t \geq 0}$ is a Poisson process of parameter λ. We choose $\eta = \frac{1}{\lambda}$ in order to ensure that $\mathbb{E}(U_t) = t$. The variance of the subordinator is equal to ηt. The Laplace function of this subordinator is equal to

$$\mathbb{E}\left(e^{-\omega U_t}\right) = \mathbb{E}\left(e^{-\omega \eta N_t}\right)$$
$$= e^{-t f(\omega \eta)} ,$$

where the Laplace exponent is a Bernstein function:

$$f(\omega \eta) = \lambda(1 - e^{-\omega \eta})$$
$$= \int_0^\infty (1 - e^{-\omega z}) \lambda \delta_{\{\eta\}}(z) \mathrm{d}z .$$

We denote by $\delta_{\{\eta\}}(z)$ the Dirac measure at point $z = \eta$. The Lévy measure is therefore $\bar{\nu}(dz) = \lambda \delta_{\{\eta\}}(z)dz$ and the Laplace transform of the pdf of U_t is given by

$$\tilde{g}(\omega, \tau) = \frac{\lambda(1 - e^{-\eta \omega})}{\omega} e^{-\tau(\lambda(1 - e^{-\omega \eta}))}.$$

As N_t is a Poisson process, it takes its values in $\mathbb{N}$ and the survival function of S_t is equal to

$$\mathbb{P}(S_t \geq s) = \mathbb{P}(\eta N_s \leq t)$$
$$= \mathbb{P}(T_{\lfloor \frac{t}{\eta} \rfloor + 1} > s),$$

where $T_{\lfloor \frac{t}{\eta} \rfloor + 1}$ is a gamma (or Erlang) random variable, $\Gamma\left(\lfloor \frac{t}{\eta} \rfloor + 1 ; \lambda\right)$. The pdf of S_t is therefore the following function:

$$g(t, \tau) = \frac{d}{d\tau} \mathbb{P}(\tau \leq S_t \leq \tau + d\tau)$$
$$= \frac{\lambda^{\lfloor \frac{t}{\eta} \rfloor + 1} \tau^{\lfloor \frac{t}{\eta} \rfloor} \exp(-\lambda\tau)}{\lfloor \frac{t}{\eta} \rfloor !} \qquad \tau \geq 0.$$

The Laplace transform is in this case

$$\mathbb{E}_0\left[e^{-\omega S_t}\right] = \int_0^\infty e^{-\omega \tau} g(t, \tau) d\tau \qquad\qquad (11.18)$$
$$= \left(1 + \frac{\omega}{\lambda}\right)^{-\left(\lfloor \frac{t}{\eta} \rfloor + 1\right)}.$$

The variable $T_{\lfloor \frac{t}{\eta} \rfloor + 1}$ is also the sum of $\lfloor \frac{t}{\eta} \rfloor + 1$ exponential random variables, denoted J_k, with parameter λ. The probability density function of J_k is equal to $f_J(x) = \lambda e^{-\lambda x}$. The process S_t^b, defined as the sum of the J_k,

$$S_t^b = \sum_{k=1}^{\lfloor \frac{t}{\eta} \rfloor + 1} J_k, \qquad\qquad (11.19)$$

has the same statistical distribution as S_t for all time $t \geq 0$. Therefore, we can represent S_t by the sum (11.19) and we infer that the cumulative distribution function of $S_{t+\Delta}$, conditionally to the filtration $\mathcal{G}_t$ up to time t, is given by

$$\mathbb{P}\left(S_{t+\Delta} \leq \tau \mid \mathcal{F}_t\right) = \mathbb{P}\left(\sum_{k=1}^{\left\lfloor \frac{t+\Delta}{\eta} \right\rfloor + 1} J_k \leq \tau \mid S_t = \sum_{k=1}^{\left\lfloor \frac{t}{\eta} \right\rfloor + 1} J_k\right)$$

$$= \begin{cases} 0 & S_t > \tau \\ \mathbb{P}\left(T_{\left\lfloor \frac{t+\Delta}{\eta} \right\rfloor - \left\lfloor \frac{t}{\eta} \right\rfloor} \leq \tau - S_t\right) & \left\lfloor \frac{t+\Delta}{\eta} \right\rfloor > \left\lfloor \frac{\Delta}{\eta} \right\rfloor \text{ and } S_t < \tau \\ 1 & \left\lfloor \frac{t+\Delta}{\eta} \right\rfloor = \left\lfloor \frac{t}{\eta} \right\rfloor \text{ and } S_t < \tau. \end{cases}$$

This also means that S_t is a stepwise process defined on a mesh with steps of size η: $S_t = S_{\left\lfloor \frac{t}{\eta} \right\rfloor \eta}$ $\forall t \in \mathbb{R}^+$. The process $(S_t)_{t \geq 0}$ is clearly not Markov since $\mathbb{P}\left(S_{t+\Delta} \leq \tau \mid \mathcal{F}_t\right)$ depends upon the process value at time $\left\lfloor \frac{t}{\eta} \right\rfloor \eta \leq t$. The left plot of Fig. 11.1 shows a simulated sample path of a Poisson process and of its inverse for $\lambda = 10$. S_t is here an increasing stepwise function. Any process time changed by an inverse Poisson subordinator therefore has piecewise constant functions.

11.3   The Dzerbayshan–Caputo Derivatives

In order to introduce motionless phases in the dynamic of a jump-diffusion, we consider a stochastic time scale ruled an inverse Lévy subordinator. We will see in Sect. 11.4 that option prices in this setting are solutions of a PIDE similar to Eq. (11.6) in which the derivative with respect to time is replaced by a convolution-type derivative, called the *Dzerbayshan–Caputo (D–C) derivative*. This section reviews the properties of this convolution-type derivative and its link with Lévy subordinators.

A *Bernstein function* is a function $f : (0, \infty) \to \mathbb{R}$ of class C^∞ such that $f(x) \geq 0$ for all $x > 0$ for which

$$(-1)^k f^{(k)}(x) \leq 0 \qquad \forall x > 0 \, k \in \mathbb{N}.$$

A Bernstein function also admits a similar representation to the Laplace exponent of a Lévy process:

$$f(x) = a + bx + \int_0^\infty \left(1 - e^{-sz}\right) \bar{\nu}(\mathrm{d}z), \tag{11.20}$$

where $a, b \geq 0$. $\bar{v}(\cdot)$ is a positive Lévy measure on $(0, \infty)$. The triplet $(a, b, \bar{v})$ is the *Lévy triplet* of the Bernstein function. We denote by $v(s)$ the tail of the Lévy measure, that is,

$$v(s)\mathrm{d}s = \left(a + \int_s^\infty \bar{v}(\mathrm{d}z) \right) \mathrm{d}s \,.$$

Let us consider $f(\cdot)$, a Bernstein function, and its tail Lévy measure $v(s)$, which is absolutely continuous on $(0, \infty)$. We also need a function $u(t) \in \mathrm{AC}([0, \infty])$, where $\mathrm{AC}([0, \infty])$ is the set of absolutely continuous function on $\mathbb{R}^+$. The generalized D–C derivative according to the Bernstein function $f(\cdot)$ is defined as

$$^f \mathcal{D}u(t) = b \frac{\mathrm{d}}{\mathrm{d}t} u(t) + \int_0^t \frac{\partial}{\partial t} u(t - s) v(s) \mathrm{d}s \quad t \in [0, \infty) \,. \qquad (11.21)$$

This integral is well defined if $|u(t)| \leq M e^{\omega_0 t}$ for some ω_0 and $M > 0$. A direct calculation leads to the following Laplace transform of the D–C derivative:

$$\mathcal{L}\left[^f \mathcal{D}_t u(t) \right](\omega) = f(\omega) \, \tilde{u}(\omega) - \frac{f(\omega)}{\omega} u(0) \quad \mathcal{R}\omega \geq \omega_0 \,, \qquad (11.22)$$

where $\tilde{u}(\omega)$ is the Laplace transform of $u(t)$. We describe the form of this D–C derivative for α-stable and Poisson subordinators.

α-Stable Subordinators Let us consider the α-stable Lévy process, $(U_t)_{t \geq 0}$. The Lévy triplet is in this case $(0, 0, \bar{v})$, where $\bar{v}(\mathrm{d}z) = \frac{\alpha}{\Gamma(1-\alpha)} \frac{\mathrm{d}z}{z^{1+\alpha}}$. The tail of the Lévy measure is given by

$$v(s)\mathrm{d}s = \mathrm{d}s \int_s^\infty \frac{\alpha z^{-\alpha-1}}{\Gamma(1-\alpha)} \mathrm{d}z$$

$$= \frac{s^{-\alpha}}{\Gamma(1-\alpha)} \mathrm{d}s \,.$$

In this case the D–C derivative becomes the Caputo fractional derivative, which we denote by $\mathcal{D}_\alpha u(t)$. Indeed, if we perform the change of variable $s' = t - s$, we infer that

$$\int_0^t \frac{\partial}{\partial t} u(t - s) v(s) \mathrm{d}s = \int_0^t \frac{\partial}{\partial s'} u(s') \, v(t - s') \, \mathrm{d}s' \,,$$

and therefore

$$^f \mathcal{D}u(t) = \mathcal{D}_\alpha u(t) = \frac{1}{\Gamma(1-\alpha)} \int_0^t \frac{u'(s)}{(t-s)^\alpha} \mathrm{d}s \,. \qquad (11.23)$$

Poisson Subordinators If $(U_t)_{t \geq 0}$ is a Poisson subordinator, the Lévy triplet is $(\mu, 0, \bar{\nu})$ with $\bar{\nu}(dz) = \lambda \delta_{\{\eta\}}(z) dz$. The tail of the Lévy measure for $s \geq 0$ is

$$\nu(s) ds = ds \int_s^\infty \lambda \delta_{\{\eta\}}(z) dz$$

$$= \lambda \, \mathbb{I}_{\{s \leq \eta\}} \, ds \,,$$

where $\mathbb{I}_{\{s \leq \eta\}}$ is the indicator variable. The D–C derivative in this case is denoted by $\mathcal{D}_\lambda u(t)$ and given by

$$^f \mathcal{D} u(t) = \mathcal{D}_\lambda u(t) = \lambda \int_0^{\min(t,\eta)} \frac{\partial}{\partial t} u(t - s) \, ds \quad t \in [0, \infty) . \quad (11.24)$$

After a change of variable $s' = t - s$, we infer that

$$\int_0^{\min(t,\eta)} \frac{\partial}{\partial t} u(t - s) \, ds = u(t) - u(t - \min(t, \eta)),$$

and therefore

$$\mathcal{D}_\lambda u(t) = \lambda \left(u(t) - u(t - \min(t, \eta)) \right) \quad t \in [0, \infty) . \quad (11.25)$$

In the next section, we introduce a time-changed version of the financial market presented at the beginning of this chapter.

11.4   Fractional Financial Market

We recall that S_t is an inverse Lévy subordinator defined on $\left(\Omega, (\mathcal{G}_t)_{t \geq 0}, \mathbb{P} \right)$ which is independent from the natural filtration $(\mathcal{F}_t)_{t \geq 0}$ of the asset price $(A_t)_{t \geq 0}$. In this section, we use $(S_t)_{t \geq 0}$ as a stochastic clock and denote by $\mathcal{H}_t$ the augmented filtration $\mathcal{G}_t \vee \mathcal{F}_{S_t}$. This is the smallest filtration at the intersection of $\mathcal{G}_t$ and $\mathcal{F}_{S_t}$. The time-changed risk-free bond has a value at time t equal to

$$B_{S_t} = e^{r S_t} , \quad B_0 = 1. \quad (11.26)$$

Notice that in this framework, the bond return is now stochastic. The time-changed stock price is obtained by replacing the time by S_t in Eq. (11.2):

$$A_{S_t} = A_0 \exp \left(\int_0^{S_t} \mu - \frac{1}{2} \sigma_s^2 ds + \int_0^{S_t} \sigma_s dW_s^P \right) \prod_{k=1}^{N_{S_t}^J} (1 + Y_k) . \quad (11.27)$$

The bond and stock values are also solutions of the time-changed stochastic differential equation:

$$\frac{\mathrm{d}B_{S_t}}{B_{S_t}} = r \, \mathrm{d}S_t \, ,$$

$$\frac{\mathrm{d}A_{S_t}}{A_{S_t}} = \mu \, \mathrm{d}S_t + \sigma_{S_t} \, \mathrm{d}W_{S_t}^P + \mathrm{d}J_{S_t} \, .$$

Figures 11.2 and 11.3 show simulated sample paths of subordinated bond and stock prices when the time change is an inverse α-stable or Poisson subordinator. With an inverse α-stable time change, the sample path of the risky asset alternates between active and motionless phases, whereas for an inverse Poisson time change, the stock path has a piecewise constant sample path. These inverse subordinators describe two different kinds of illiquidity. Inverse Poisson processes are rather adapted to model recurring illiquidity at high or low frequency, depending upon the parameter λ, while the inverse α-stable subordinator replicates temporary illiquidity periods for an asset that is most of the time actively traded.

We have laid down the fractional dynamics under the real measure but need to determine a risk neutral measure in order to evaluate options. For this purpose, let us recall the definition of $\phi\left(\cdot, \kappa\right) = \ln\left(\kappa \frac{f_Y^b(\cdot)}{f_Y(\cdot)}\right)$, where $\kappa \in \mathbb{R}^+$. For an $\mathcal{F}_t$-adapted

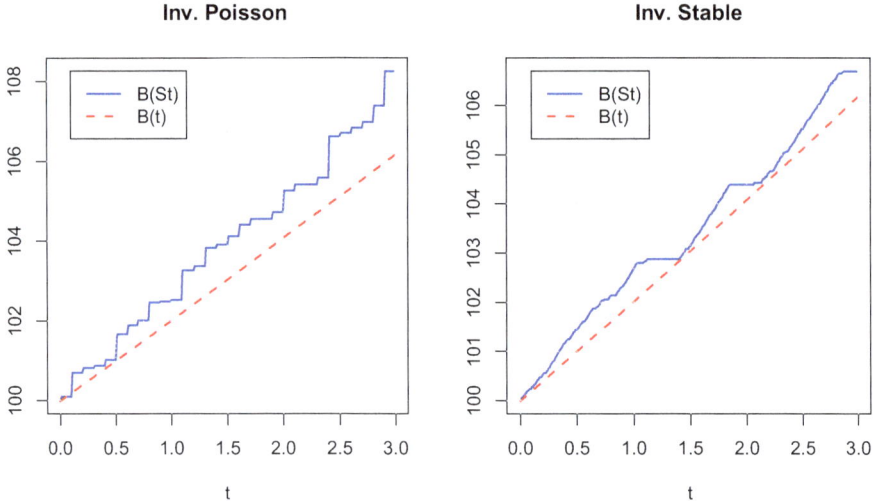

Fig. 11.2 Left plot: sample path of B_{S_t} when S_t is an inverse Poisson subordinator ($\lambda = 10$). Right plot: sample path of B_{S_t} when S_t is an inverse α-stable process ($\alpha = 0.90$). $r = 2\%$

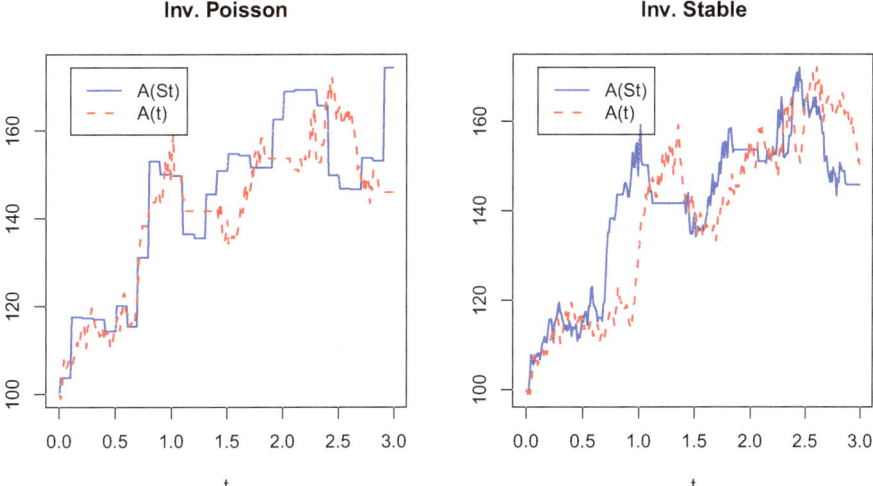

Fig. 11.3 Left plot: sample path of A_{S_t} when S_t is an inverse Poisson subordinator ($\lambda = 10$). Right plot: sample path of A_{S_t} when S_t is an inverse α-stable process ($\alpha = 0.90$). Here, $\mu = 5\%$, $\sigma_s = 15\%$, and no jump

and square integrable process, denoted $(\theta_t)_{t \geq 0}$, we defined the time-changed process $(Z_{S_t})_{t \geq 0}$ as follows:

$$
Z_{S_t} = \exp \left(-\frac{1}{2} \int_0^{S_t} \theta_s^2 ds - \int_0^{S_t} \theta_s dW_s^P + \sum_{k=1}^{N_{S_t}^J} \phi\left(Y_k, \kappa\right) - (\kappa - 1)\lambda_J S_t \right),
$$

$$(11.28)$$

which admits an equivalent infinitesimal representation:

$$
dZ_{S_t} = -Z_{S_t} \theta_{S_t} dW_{S_t}^P - Z_{S_t} (\kappa - 1)\lambda_J dS_t + Z_{S_t} \left(e^{\phi(Y, \kappa)} - 1 \right) dN_{S_t}^J .
$$

We can check that $\mathbb{E}\left(dZ_t | \mathcal{H}_t \vee \mathcal{G}_T\right) = 0$ and therefore $\mathbb{E}\left(Z_T | \mathcal{H}_t \vee \mathcal{G}_T\right) = Z_t + \int_t^T \mathbb{E}\left(dZ_t | \mathcal{H}_t \vee \mathcal{G}_T\right) = Z_t$. Using nested expectations, the process Z_t is then a martingale with $Z_0 = 1$. Since Z_t is $\mathcal{H}_t$-adapted, it is a Radon–Nikodym derivative $Z_{S_t} = \left.\frac{d\mathbb{P}^b}{d\mathbb{P}}\right|_t$, defining a new measure $\mathbb{P}^b$. The next proposition states the dynamics of Brownian motion and jump process under $\mathbb{P}^b$.

Proposition 11.3 *Under the equivalent measure $\mathbb{P}^b$ defined by the Radon–Nikodym derivative (11.28):*

(1) $dW_{S_t} = dW_{S_t}^P + \theta_{S_t} dS_t$ *is a time-changed Brownian motion.*

(2) The process J_{S_t} is a point process with an intensity equal to $\kappa \lambda_J \, dS_t$ and the pdf of jump is $f_Y^b(\cdot)$.

Proof

(1) Let us denote by $(\mathcal{V}_t)_{t \geq 0}$ the subfiltration of $(\mathcal{F}_t)_{t \geq 0}$ carrying exclusively information about the jump process. Using nested expectations, the moment generating function (mgf) of W_t under the equivalent measure is

$$\mathbb{E}^{\mathbb{P}^b}\left(e^{\omega W_{S_t}} \,|\mathcal{H}_0 \right)$$

$$= \mathbb{E}\left(e^{\int_0^{S_t}\left(\omega \theta_s - \frac{1}{2}\theta_s^2 \right)ds - \int_0^{S_t}(\theta_s - \omega)dW_s^P} \, e^{\sum_{k=1}^{N_{S_t}^J}\phi(Y_k,\kappa) - (\kappa-1)\lambda_J S_t} \,|\mathcal{H}_0 \right)$$

$$= \mathbb{E}\left(e^{\sum_{k=1}^{N_{S_t}^J}\phi(Y_k,\kappa) - (\kappa-1)\lambda_J S_t} \mathbb{E}\left(e^{\int_0^{S_t}\left(\omega \theta_s - \frac{1}{2}\theta_s^2 \right)ds - \int_0^{S_t}(\theta_s - \omega)dW_s^P} \,|\mathcal{H}_0 \vee \mathcal{V}_{S_t} \vee \mathcal{G}_t \right) \,|\mathcal{H}_0 \right).$$

Since the Brownian motion is independent from the filtration of the jump process and time change, we have that

$$\mathbb{E}\left(e^{\int_0^{S_t}\left(\omega \theta_s - \frac{1}{2}\theta_s^2 \right)ds - \int_0^{S_t}(\theta_s - \omega)dW_s^P} \,|\mathcal{H}_0 \vee \mathcal{V}_{S_t} \vee \mathcal{G}_t \right)$$

$$= e^{\frac{1}{2}\omega^2 S_t} \mathbb{E}\left(e^{-\frac{1}{2}\int_0^{S_t}(\theta_s - \omega)^2 ds - \int_0^{S_t}(\theta_s - \omega)dW_s^P} \,|\mathcal{H}_0 \vee \mathcal{V}_{S_t} \vee \mathcal{G}_t \right)$$

$$= e^{\frac{1}{2}\omega^2 S_t}.$$

To pass from the second line to the last line, we use the property that the Doléans–Dade exponential of a martingale is a martingale. We recognize the mgf of a Brownian motion, time changed by S_t. Since $e^{\sum_{k=1}^{N_t^J}\phi(Y_k,\kappa) - (\kappa-1)\lambda_J t}$ is a martingale, nesting expectations leads to

$$\mathbb{E}^{\mathbb{P}^b}\left(e^{\omega W_{S_t}} \,|\mathcal{H}_0 \right) = \mathbb{E}\left(e^{\frac{1}{2}\omega^2 S_t} \mathbb{E}\left(e^{\sum_{k=1}^{N_{S_t}^J}\phi(Y_k,\kappa) - (\kappa-1)\lambda_J S_t} \,|\mathcal{H}_0 \vee \mathcal{G}_t \right) \,|\mathcal{H}_0 \right)$$

$$= \mathbb{E}\left(e^{\frac{1}{2}\omega^2 S_t} \,|\mathcal{H}_0 \right).$$

Therefore, we conclude that $W_{S_t} = W_{S_t}^P + \int_0^{S_t}\theta_s ds$ is a time-changed Brownian motion under $\mathbb{P}^b$.

(2) Using nested expectations, the mgf of the time-changed jump process under the measure $\mathbb{P}^b$ may be rewritten as follows:

$$\mathbb{E}^{\mathbb{P}^b}\left(e^{\omega J_{S_t}} \,|\mathcal{H}_0 \right) = \mathbb{E}\left(\mathbb{E}\left(e^{\omega J_{S_t} + \sum_{k=1}^{N_{S_t}^J}\phi(Y_k,\kappa) - (\kappa-1)\lambda_J S_t} \,|\mathcal{H}_0 \vee \mathcal{G}_t \right) \,|\mathcal{H}_0 \right).$$

Given that jump sizes are independent from the number of jumps, we rewrite this last equation as

$$\mathbb{E}\left(e^{\omega J_{S_t}+\sum_{k=1}^{N_{S_t}^J}\phi(Y_k,\kappa)-(\kappa-1)\lambda_J S_t}\,|\mathcal{H}_0\vee\mathcal{G}_t\right)$$

$$=e^{-(\kappa-1)\lambda_J S_t}\mathbb{E}\left(\left(\mathbb{E}\left(e^{\omega Y+\phi(Y,\kappa)}\right)\right)^{N_{S_t}^J}|\mathcal{H}_0\vee\mathcal{G}_t\right).$$

On the one hand, we have that

$$\mathbb{E}\left(e^{\omega Y+\phi(Y,\kappa)}\right)=\int_{\mathbb{R}}\kappa e^{\omega y}f_Y^b(y)\,\mathrm{d}y$$

$$=\kappa\mathbb{E}\left(e^{\omega Y}\right),$$

where Y has the density $f_Y^b(\cdot)$. On the other hand, we know that the moment generating function of a compound Poisson process is equal to

$$\mathbb{E}\left(\exp\left(\omega J_t\right)\right)=\exp\left(\lambda_J t\left(\mathbb{E}\left(e^{\omega Y}\right)-1\right)\right).$$

The mgf of the jump process is therefore equal to

$$\mathbb{E}^{\mathbb{P}^b}\left(e^{\omega J_t}|\mathcal{H}_0\right)=\mathbb{E}\left(e^{-(\kappa-1)\lambda_J S_t}\mathbb{E}\left(\left(\kappa\mathbb{E}\left(e^{\omega Y}\right)\right)^{N_t^J}|\mathcal{H}_0\vee\mathcal{G}_t\right)|\mathcal{H}_0\right)$$

$$=\mathbb{E}\left(e^{-(\kappa-1)\lambda_J S_t}\mathbb{E}\left(e^{N_t^J\,\ln\left(\kappa\mathbb{E}(e^{\omega Y})\right)}|\mathcal{H}_0\vee\mathcal{G}_t\right)|\mathcal{H}_0\right)$$

$$=\mathbb{E}\left(e^{-(\kappa-1)\lambda_J S_t}e^{\lambda_J t\left(\kappa\mathbb{E}(e^{\omega Y})-1\right)}|\mathcal{H}_0\right)$$

$$=\mathbb{E}\left(e^{\kappa\lambda_J S_t\left(\mathbb{E}(e^{\omega Y})-1\right)}|\mathcal{H}_0\right),$$

and we recognize the mgf of a time-changed Poisson process with intensity $\kappa\lambda_J$ and jump density $f_Y^b(\cdot)$.

$\square$

We denote by $\xi_b=\mathbb{E}^{\mathbb{P}^b}(Y)$ the expectation of a jump under the equivalent probability measure. From the previous proposition, we infer that the dynamics of the risky asset under P^b is given by

$$\frac{\mathrm{d}A_{S_t}}{A_{S_t}}=\left(\mu-\theta_{S_t}\sigma_{S_t}\right)\mathrm{d}S_t+\sigma_{S_t}\,\mathrm{d}W_{S_t}+\mathrm{d}J_{S_t}.$$

Using again the independence between the clock and the filtration $(\mathcal{F}_t)_{t\geq0}$, the expectation of the infinitesimal variation of A_t is such that

$$\mathbb{E}^{\mathbb{P}^b}\left(\frac{\mathrm{d}A_{S_t}}{A_{S_t}}|\mathcal{G}_t\right) = \left(\mu + \kappa\,\lambda_J\,\xi_b - \theta_{S_t}\sigma_{S_t}\right)\mathrm{d}S_t\,.$$

the following conditions under which the Radon–Nikodym derivative (11.28) defines an equivalent risk neutral measure.

Corollary 11.4 *The Radon–Nikodym derivative (11.28) defines an equivalent risk neutral measure $\mathbb{Q}$, if ξ_b, κ, and $(\theta_{S_t})_{t\geq0}$ fulfill the following equality:*

$$\theta_{S_t} = \frac{\mu + \kappa\,\lambda_J\,\xi_b - r}{\sigma_{S_t}}\,. \tag{11.29}$$

To lighten the notation, we assume without loss of generality that the frequency and size of jumps are identical under real and risk neutral measures ($\kappa = 1$, $\xi_b = \xi$). The price of a European call option of maturity t, written on the time-changed asset, is the discounted expected payoff under $\mathbb{Q}$:

$$C_S(t, K) = C(S_t, K) \tag{11.30}$$

$$= \mathbb{E}^{\mathbb{Q}}\left(e^{-r\,S_t}\left(A_{S_t} - K\right)_+|\mathcal{F}_0\right),$$

where S_t is a time change. If we remember that the density of S_t is $g(t, \tau)\mathrm{d}\tau = \mathbb{P}(S_t \in [\tau, \tau + \mathrm{d}\tau])$, we can rewrite the call option as an integral:

$$C_S(t, K) = \int_0^\infty C(\tau, K)\,g(t, \tau)\mathrm{d}\tau\,. \tag{11.31}$$

If the non-fractional call and the density of S_t admit closed form expressions, we can evaluate the fractional call price by computing numerically the integral in the above equation. An alternative consists in calculating the price by Monte Carlo simulations. Another solution is provided by the next proposition.

Proposition 11.5 *The call option value in the fractional jump-diffusion setting is the solution of the fractional PIDE equation:*

$$^f\mathcal{D}C_S(t, K) = -(r - \lambda_J\xi)\,K\frac{\partial C_S(t, K)}{\partial K} + \frac{K^2}{2}\mathbb{E}^{\mathbb{Q}}\left(\sigma_{S_t}^2 \mid A_{S_t} = K\right) \tag{11.32}$$

$$\times\frac{\partial^2 C_S(t, K)}{\partial K^2} + \lambda_J\left(\mathbb{E}^{\mathbb{Q}}\left((1 + Y)C_S\left(t, \frac{K}{1 + Y}\right)\right) - (1 + \xi)\,C_S(t, K)\right),$$

with the initial condition $C_S(0, K) = (S_0 + K)_+$. *The fractional put price,* $D_S(t, K) = D(S_t, K)$, *is the solution of the same PIDE but with the initial condition* $D(0, K) = (K - S_0)_+$.

Proof The Laplace transform of $C_S(t, K)$ with respect to time t is equal to

$$\tilde{C}_S(\omega, K) = \int_0^\infty e^{-\omega t} \int_0^\infty C(\tau, K) \, g(t, \tau) \mathrm{d}\tau \, \mathrm{d}t \qquad (11.33)$$

$$= \int_0^\infty C(\tau, K) \, \tilde{g}(\omega, \tau) \, \mathrm{d}\tau ,$$

where $\tilde{g}(\omega, \tau)$ is the Laplace transform, $\int_0^\infty e^{-\omega t} g(t, \tau) \mathrm{d}t$, of the density of S_t. On the other hand, we have that

$$\mathbb{E}\left((1 + Y)\widetilde{C_S \left(t, \frac{K}{1+Y} \right)} \right)$$

$$= \int_0^\infty e^{-\omega t} \int_{-\infty}^{+\infty} (1 + y) C_S \left(t, \frac{K}{1+y} \right) f_Y(y) \mathrm{d}y \, \mathrm{d}t$$

$$= \int_{-\infty}^{+\infty} (1 + y) \int_0^\infty e^{-\omega t} \int_0^\infty C \left(\tau, \frac{K}{1+y} \right) g(t, \tau) \, \mathrm{d}\tau \, \mathrm{d}t \, f_Y(y) \, \mathrm{d}y$$

$$= \mathbb{E}^{\mathbb{Q}} \left((1 + Y)\tilde{C}_S \left(\omega, \frac{K}{1+Y} \right) \right) .$$

Let us momentarily adopt the following notation:

$$h(t, K) := \mathbb{E}^{\mathbb{Q}} \left(\sigma_t^2 \mid A_t = K \right) \frac{\partial^2 C(t, K)}{\partial K^2} ,$$

$$h_S(t, K) := \mathbb{E}^{\mathbb{Q}} \left(\sigma_{S_t}^2 \mid A_{S_t} = K \right) \frac{\partial^2 C(S_t, K)}{\partial K^2} .$$

The Laplace transform of $h(t, K)$ is equal to

$$\tilde{h}_S(\omega, K) = \int_0^\infty e^{-\omega t} \int_0^\infty h(\tau, K) \, g(t, \tau) \mathrm{d}\tau \, \mathrm{d}t$$

$$= \int_0^\infty h(\tau, K) \, \tilde{g}(\omega, \tau) \, \mathrm{d}\tau .$$

We have seen in Sect. 11.2 that the Laplace transform of $g(t, \tau)$ with respect to t is related to the Laplace exponent $f(\cdot)$ of U_t by Eq. (11.16). Combining this expression with Eq. (11.33) gives us

$$\tilde{C}_S(\omega, K) = \frac{f(\omega)}{\omega} \tilde{C}(f(\omega), K),$$

$$\tilde{h}_S(\omega, K) = \frac{f(\omega)}{\omega} h_S(f(\omega), K),$$

and

$$\mathbb{E}^{\mathbb{Q}}\left(\widetilde{(1+Y)C_S}\left(t, \frac{K}{1+Y}\right)\right) = \frac{f(\omega)}{\omega} \mathbb{E}^{\mathbb{Q}}\left((1+Y)\tilde{C}\left(f(\omega), \frac{K}{1+Y}\right)\right).$$

Hence, from the FPE (11.6), we deduce that $\tilde{C}(\omega, K)$ is the solution of the following equation:

$$\omega \tilde{C}(\omega, K) - C(0, K) = -(r - \lambda_J \xi) K \frac{\partial}{\partial K} \tilde{C}(\omega, K) + \frac{K^2}{2} \tilde{h}(\omega, K)$$

$$+ \lambda_J \left(\mathbb{E}^{\mathbb{Q}}\left((1+Y)\tilde{C}\left(\omega, \frac{K}{1+Y}\right)\right) - (1+\xi)\, \tilde{C}(\omega, K) \right).$$

As $\tilde{C}_S(\omega, K) = \frac{f(\omega)}{\omega} \tilde{C}(f(\omega), K)$, replacing ω by $f(\omega)$ leads to

$$f(\omega)\tilde{C}(f(\omega), K) - C(0, K) =$$

$$- (r - \lambda_J \xi) K \frac{\partial}{\partial K} \tilde{C}(f(\omega), K) + \frac{K^2}{2} \tilde{h}(f(\omega), K)$$

$$+ \lambda_J \left(\mathbb{E}^{\mathbb{Q}}\left((1+Y)\tilde{C}\left(\omega, \frac{K}{1+Y}\right)\right) - (1+\xi)\, \tilde{C}(f(\omega), K) \right).$$

Multiplying this last equation by $\frac{f(\omega)}{\omega}$ and since $C_S(0, K) = C(0, K)$, we infer that

$$f(\omega)\tilde{C}_S(\omega, K) - \frac{f(\omega)}{\omega} C_S(0, K) =$$

$$- (r - \lambda_J \xi) K \frac{\partial}{\partial K} \tilde{C}_S(\omega, K) + \frac{K^2}{2} \tilde{h}_S(\omega, K)$$

$$+ \lambda_J \left(\mathbb{E}^{\mathbb{Q}}\left((1+Y)\tilde{C}\left(\omega, \frac{K}{1+Y}\right)\right) - (1+\xi)\, \tilde{C}_S(\omega, K) \right),$$

which is the Laplace transform of Eq. (11.32).   □

Notice that if $\sigma_t = \sigma(t, A_t)$ is a function of time and of the asset value, the conditional expectation in Eq. (11.32) becomes

$$
\mathbb{E}^{\mathbb{Q}}\left(\sigma_{S_t}^2 \mid A_{S_t} = K\right) = \mathbb{E}^{\mathbb{Q}}\left(\mathbb{E}^{\mathbb{Q}}\left(\sigma_{S_t}^2 \mid A_{S_t} = K, \mathcal{G}_t\right)\right) \tag{11.34}
$$

$$
= \mathbb{E}^{\mathbb{Q}}\left(\sigma^2(S_t, K)\right) .
$$

In the Black and Scholes (B&S) framework, the Brownian volatility is constant $\sigma^2(S_t, K) = \bar{\sigma}^2$ and there are no jumps. In this case, the fraction PIDE (11.32) can be rewritten as

$$
{}^f\mathcal{D}C_S(t, k) = -r\frac{\partial}{\partial k}C_S(t, k) + \frac{\bar{\sigma}^2}{2}\frac{\partial^2 C_S(t, k)}{\partial k^2}, \tag{11.35}
$$

where $k = \ln(K)$. In the case of an inverse α-stable process, ${}^f\mathcal{D}C_S(t, k) = \mathcal{D}_\alpha C_S(t, k)$ and the fractional B&S call price is the solution of

$$
\frac{1}{\Gamma(1-\alpha)}\int_0^t \frac{\frac{\partial}{\partial s}C_S(s, K)}{(t-s)^\alpha}ds = -r\frac{\partial}{\partial k}C_S(t, k) + \frac{\bar{\sigma}^2}{2}\frac{\partial^2 C_S(t, k)}{\partial k^2}. \tag{11.36}
$$

If the stochastic clock is an inverse Poisson subordinator, ${}^f\mathcal{D}C_S(t, k) = \mathcal{D}_\lambda C_S(t, k)$, the fractional B&S call price can be computed by iterating the following recursion:

$$
C_S(t, k) = C_S(\max(0; t - \eta), k) - \frac{r}{\lambda}\frac{\partial}{\partial k}C_S(t, k) + \frac{\bar{\sigma}^2}{2\lambda}\frac{\partial^2 C_S(t, k)}{\partial k^2} \quad t \geq 0.
$$

In the constant elasticity volatility (CEV) model, the expectation (11.34) is constant and equal to

$$
\mathbb{E}^{\mathbb{Q}}\left(\sigma(S_t, K)^2\right) = \sigma_0^2 K^{2\gamma-2}.
$$

If the Brownian volatility depends upon time as in Eq. (11.13), the conditional expectation of the variance in Eq. (11.32) is a function of the Laplace transform of the time change

$$
\mathbb{E}^{\mathbb{Q}}\left(\sigma(S_t, K)^2\right) = \beta_0 + \beta_1 \mathbb{E}^{\mathbb{Q}}\left(\left(\frac{K}{F_{S_t}}\right)^\gamma\right) + \beta_2 \mathbb{E}^{\mathbb{Q}}\left(\left(\frac{K}{F_{S_t}}\right)^{2\gamma}\right) \tag{11.37}
$$

$$
= \beta_0 + \beta_1\left(\frac{K}{A_0}\right)^\gamma \mathbb{E}^{\mathbb{Q}}\left(e^{-\gamma r S_t}\right) + \beta_2\left(\frac{K}{A_0}\right)^{2\gamma} \mathbb{E}^{\mathbb{Q}}\left(e^{-2\gamma r S_t}\right).
$$

If the clock is an inverse α-stable process, this local volatility is a continuous function of time:

$$\mathbb{E}^{\mathbb{Q}}\left(\sigma(S_t, K)^2\right) = \beta_0 + \beta_1 \left(\frac{K}{A_0}\right)^{\gamma} E_\alpha(-\gamma\,r\,t^\alpha) + \beta_2 \left(\frac{K}{A_0}\right)^{2\gamma} E_\alpha(-2\,\gamma\,r\,t^\alpha),$$

where $E_\alpha(\cdot)$ is the Mittag-Leffler function. For an inverse Poisson subordinator, the local volatility (11.13) is a stepwise function of time:

$$\mathbb{E}^{\mathbb{Q}}\left(\sigma(S_t, K)^2\right) = \beta_0 + \beta_1 \left(\frac{K}{A_0}\right)^{\gamma} \left(1 + \frac{\gamma\,r}{\lambda}\right)^{-\left(\left\lfloor \frac{t}{\eta} \right\rfloor + 1\right)}$$

$$+ \beta_2 \left(\frac{K}{A_0}\right)^{2\gamma} \left(1 + \frac{2\,\gamma\,r}{\lambda}\right)^{-\left(\left\lfloor \frac{t}{\eta} \right\rfloor + 1\right)}.$$

The next section proposes a numerical method for solving Eq. (11.32).

11.5   Numerical Framework

Andersen and Brotherton-Ratcliffe [3] have demonstrated the reliability of a finite difference approach for solving the Dupire equation in a Brownian setting. We extend their framework to the fractional jump-diffusion and opt for an implicit method. We need to specify the distribution of jumps and the form of the local volatility. We make the common assumption that jumps are exclusively negative and defined on $[-1, 0]$. Furthermore, we consider a continuous pdf for Y. We also assume that the volatility is a function of time and asset value: $\sigma_t = \sigma(t, A_t)$. Let us consider a domain $[0, t_{\max}] \times [0, K_{\max}]$ on which we wish to estimate fractional call prices. We choose two steps of discretization, denoted Δ_t and Δ_K, in order to define pairs (t_k, K_j) where

$$t_k = k\,\Delta_t \quad , \quad K_j = K_0 + j\,\Delta_K$$

for $k = 0, \dots, n_t$ and $j = 0, \dots, n_K$. The numbers n_t and n_K are integers equal to $n_t = \left\lfloor \frac{t_{\max}}{\Delta_t} \right\rfloor$ and $n_K = \left\lfloor \frac{K_{\max} - K_0}{\Delta_K} \right\rfloor$. To lighten the notation, we denote by $C_S(k, j)$ the approached value of $C_S(t_k, K_j)$. Under the assumption that $K_0 \ll A_0$, we have the following boundary conditions:

$$C_S(0, j) = \left(A_0 - K_j\right)_+ ,$$

$$C_S(k, 0) = \mathbb{E}^{\mathbb{Q}}\left(e^{-r S_{t_k}} A_{S_{t_k}}\right) = A_0.$$

We write $\sigma^2(k, j) = \mathbb{E}^{\mathbb{Q}} \left(\sigma^2_{S_{t_k}} \mid A_{S_{t_k}} = K_j \right)$, while the $(n_t + 1) \times (n_K + 1)$ matrix of variances is $\Sigma = (\sigma^2(k, j))_{k \in \{0,\ldots,n_t\}, j \in \{0,\ldots,n_K\}}$.

The first-order derivative of $C_S(t, K)$ with respect to K in the right-hand side of Eq. (11.32) is approached by a central finite difference approximation:

$$\frac{\partial C_S(k, j)}{\partial K} \approx \frac{C_S(k, j + 1) - C_S(k, j - 1)}{2\Delta_K} \quad \text{for } 0 < j \leq n_K. \tag{11.38}$$

On the boundary, we set

$$\frac{\partial C_S(k, 0)}{\partial K} = \frac{C_S(k, 1) - C_S(k, 0)}{\Delta_K} \quad \text{and}$$

$$\frac{\partial C_S(k, n_K + 1)}{\partial K} = \frac{C_S(k, n_K + 1) - C_S(k, n_K)}{\Delta_K}.$$

The second-order derivative is approached in the same way:

$$\frac{\partial^2 C_S(k, j)}{\partial K^2} \approx \frac{C_S(k, j + 1) - 2C_S(k, j) + C_S(k, j - 1)}{\Delta_K^2} \quad \text{for } 0 < j \leq n_K.$$

On lower and upper boundaries, we set $\frac{\partial^2 C_S(k,0)}{\partial K^2} = \frac{\partial^2 C_S(k,1)}{\partial K^2}$ and $\frac{\partial^2 C_S(k,n_K+1)}{\partial K^2} = \frac{\partial^2 C_S(k,n_K)}{\partial K^2}$, respectively. In order to rewrite these partial derivatives in matrix form, we introduce $(n_K + 1) \times (n_K + 1)$ matrices R_1 and R_2 defined by

$$R_1 = \frac{1}{\Delta_K} \begin{pmatrix} -1 & 1 & 0 & \ldots & 0 \\ -\frac{1}{2} & 0 & \frac{1}{2} & \ldots & 0 \\ 0 & -\frac{1}{2} & 0 & \frac{1}{2} & \vdots \\ \vdots & \ddots & \ddots & \ddots & \vdots \\ 0 & \ldots & 0 & -1 & 1 \end{pmatrix} \quad R_2 = \frac{1}{\Delta_K^2} \begin{pmatrix} 1 & -2 & 1 & 0 & \ldots & 0 \\ 1 & -2 & 1 & 0 & \ldots & 0 \\ 0 & 1 & -2 & 1 & 0 & \vdots \\ \vdots & \ddots & \ddots & \ddots & \ddots & 0 \\ 0 & \ldots & 0 & 1 & -2 & 1 \\ 0 & \ldots & 0 & 1 & -2 & 1 \end{pmatrix}.$$

The vector partial derivatives with respect to the strike at time t_k are then equal to following matrix products:

$$\frac{\partial C_S(k, \cdot)}{\partial K} \approx \bar{K} R_1 C_S(k, \cdot)^\top, \quad \frac{\partial^2 C_S(k, \cdot)}{\partial K^2} \approx R_2 C_S(k, \cdot)^\top, \tag{11.39}$$

where $C_S(k, \cdot)$ is the kth line of the $(n_t + 1) \times (n_K + 1)$ matrix of call prices, denoted by C_S. The next step consists in discretizing the continuous pdf of jumps. We denote by $y_m^{(j)} < 0$ the size of the jump for transiting from $K_m = \dfrac{K_j}{1+y_m^{(j)}} > K_j$ to K_j. By definition, these $y_m^{(j)}$ are equal to

$$y_m^{(j)} = \frac{K_j}{K_m} - 1 = -\frac{(m-j)\,\Delta_k}{K_0 + m\,\Delta_k} \qquad j = 0, ..., n_K\,, \ m = j, ..., n_k\,.$$

Notice that $y_m^{(j)}$ are ordered as follows: $y_{n_K}^{(j)} < ... < y_{m+1}^{(j)} < y_m^{(j)} < ... < y_j^{(j)} = 0$. The associated discrete probabilities of observing such jumps are

$$p_m^{(j)} = \mathbb{P}\left(Y \in \left[\, y_m^{(j)} - \frac{y_m^{(j)} - y_{m+1}^{(j)}}{2}\,;\, y_m^{(j)} + \frac{y_{m-1}^{(j)} - y_m^{(j)}}{2}\right)\right) \qquad j < m < n_K\,,$$

$$p_{n_K}^{(j)} = \mathbb{P}\left(Y \in \left(-\infty\,;\, y_{n_K}^{(j)} + \frac{y_{n_K-1}^{(j)} - y_{n_K}^{(j)}}{2}\right)\right),$$

$$p_j^{(j)} = \mathbb{P}\left(Y \in \left[\frac{y_{j+1}^{(j)}}{2}\,;\, 0\right]\right).$$

For a given $j \in \{0, ..., n_K\}$, we approximate the expectation related to the jump part in Eq. (11.32) by the following sum:

$$\mathbb{E}^{\mathbb{Q}}\left((1+Y)C_S\left(t_k, \frac{K_j}{1+Y}\right)\right) \approx \sum_{m=j}^{n_K} p_m^{(j)}\,(1+y_m^{(j)})\,C_S\,(k, m)\,. \qquad (11.40)$$

To rewrite this last expectation in matrix form, we denote by Y the $(n_K+1) \times (n_K+1)$ matrix of $y_m^{(j)}$:

$$Y = \begin{pmatrix} 0 & y_1^{(0)} & \dots & \dots & y_{n_K}^{(0)} \\ 0 & 0 & y_2^{(1)} & \dots & y_{n_K}^{(1)} \\ \vdots & & \ddots & \ddots & \vdots \\ \vdots & & & 0 & y_{n_K}^{(n_K-1)} \\ 0 & \dots & \dots & \dots & 0 \end{pmatrix}$$

and by T the $(n_K + 1) \times (n_K + 1)$ matrix of probabilities $T_{j,m} = p_m^{(j)}$ for $j \in \{0, ..., n_t\}$ and $m \in \{0, ..., n_K\}$:

$$
T = \begin{pmatrix}
p_0^{(0)} & p_1^{(0)} & \cdots & & \cdots & p_{n_K}^{(0)} \\
0 & p_1^{(1)} & p_2^{(1)} & \ddots & & p_{n_K}^{(1)} \\
\vdots & \ddots & \ddots & \ddots & & \vdots \\
\vdots & \ddots & 0 & p_{n_K-1}^{(n_K-1)} & p_{n_K}^{(n_K-1)} \\
0 & \cdots & \cdots & 0 & p_{n_K}^{(n_K)} = 1
\end{pmatrix}.
$$

The elementwise product (also called the Hadamard product) of the matrix T and $(1 + Y)$ is denoted by $T \bullet (1 + Y)$. Using this notation, we infer that

$$
\left[\mathbb{E}^{\mathbb{Q}} \left((1 + Y) C_S \left(t_k, \frac{K_j}{1 + Y} \right) \right) \right]_{k \in \{0, ..., n_t\}, j \in \{0, ..., n_K\}}
$$
$$
= T \bullet (1 + Y) \; C_S(k, \cdot)^{\top}. \tag{11.41}
$$

The Dzerbayshan–Caputo (D–C) derivative depends upon the chosen Bernstein function $f(\cdot)$. Therefore there is no general procedure for approaching it numerically. In the next two subsections, we focus on D–C derivatives obtained with inverse α-stable and Poisson subordinators.

Inverse α-Stable Subordinators   In this case, the D–C derivative becomes the Caputo derivative and is numerically approached by the finite difference sum:

$$
\mathcal{D}_\alpha C_S(k, j) \approx \frac{(\Delta_t)^{-\alpha}}{\Gamma(1 - \alpha)} \sum_{m=0}^{k-1} (k - m)^{-\alpha} \left(C_S(m + 1, j) - C_S(m, j) \right)
$$
$$
= \frac{(\Delta_t)^{-\alpha}}{\Gamma(1 - \alpha)} \left(C_S(k, j) - C_S(k - 1, j) \right) \tag{11.42}
$$
$$
+ \frac{(\Delta_t)^{-\alpha}}{\Gamma(1 - \alpha)} \sum_{m=0}^{k-2} (k - m)^{-\alpha} \left(C_S(m + 1, j) - C_S(m, j) \right).
$$

In order to rewrite this derivative in matrix form, we define a matrix $D(k)$ of dimension $(k - 1) \times k$ as follows:

$$
D(k) = \frac{(\Delta_t)^{-\alpha}}{\Gamma(1 - \alpha)} \begin{pmatrix}
0 & \cdots & \cdots & 0 & -(2)^{-\alpha} & (2)^{-\alpha} \\
\vdots & & \ddots & 0 & -(3)^{-\alpha} & (3)^{-\alpha} & 0 \\
\vdots & & \ddots & \ddots & \ddots & & \vdots \\
-(k)^{-\alpha} & (k)^{-\alpha} & \cdots & \cdots & & \cdots & 0
\end{pmatrix}. \tag{11.43}
$$

Therefore, the Caputo derivative at time t_k admits the following representation:

$$\mathcal{D}_\alpha C_S(k, \cdot)^\top =$$

$$\frac{(\Delta_t)^{-\alpha}}{\Gamma(1-\alpha)} \left(C_S(k, \cdot) - C_S(k-1, \cdot)\right)^\top + \left(D(k)\, C_S(0:k-1, \cdot)\right)^\top \mathbf{1}_{k-1},$$

where $\mathbf{1}_{k-1}$ is a $(k-1)$-vector of ones. We denote by $\bar{K} = \mathrm{diag}\left(K_j\right)_{j\in\{0,1,\ldots,n_K\}}$ the diagonal matrix of strikes and by $\bar{\Sigma}(k) = \mathrm{diag}\left(\Sigma(k, \cdot)\right)$ the diagonal matrix of $\sigma(k, \cdot)^2$. If we insert expressions (11.43) and (11.39) in the fractional Dupire equation (11.32), we obtain its finite difference approximation:

$$\frac{(\Delta_t)^{-\alpha}}{\Gamma(1-\alpha)} \left(C_S(k, \cdot) - C_S(k-1, \cdot)\right)^\top + \left(D(k)\, C_S(0:k-1, \cdot)\right)^\top \mathbf{1}_{k-1}$$

$$= \left(-(r-\lambda_J\xi)\,\bar{K}\,R_1 + \frac{\bar{K}^2}{2}\,\bar{\Sigma}(k)\,R_2\right) C_S(k, \cdot)^\top$$

$$+ \lambda_J \left(T \bullet (1+Y) - (1+\xi)\,I_{n_K+1}\right) C_S(k, \cdot)^\top,$$

where I_{n_K+1} is the identity matrix. Finally, fractional call prices are computed iteratively from t_0 to t_{n_K} with the following recursion:

$$C_S(k, \cdot)^\top = \left[\frac{(\Delta_t)^{-\alpha}}{\Gamma(1-\alpha)}\,I_{n_K+1} + (r-\lambda_J\xi)\,\bar{K}\,R_1 - \frac{\bar{K}^2}{2}\,\bar{\Sigma}(k)\,R_2 \right. \qquad (11.44)$$

$$\left. - \lambda_J \left(T \bullet (1+Y) - (1+\xi)\,I_{n_K+1}\right)\right]^{-1} \times$$

$$\left[\frac{(\Delta_t)^{-\alpha}}{\Gamma(1-\alpha)} C_S(k-1, \cdot)^\top - \left(D(k)\, C_S(0:k-1, \cdot)\right)^\top \mathbf{1}_{k-1}\right].$$

Inverse Poisson Subordinators If the stochastic clock is an inverse Poisson subordinator and if η is a multiple of Δ_t (η/Δ_t in $\mathbb{N}$), then the D–C derivative is equal to

$$\mathcal{D}_\lambda C_S(k, \cdot) = \begin{cases} \lambda\left(C_S(k, \cdot) - C_S(0, \cdot)\right) & k\,\Delta_t < \eta \\ \lambda\left(C_S(k, \cdot) - C_S\left(k - \frac{\eta}{\Delta_t}, \cdot\right)\right) & k\,\Delta_t \geq \eta. \end{cases}$$

If η is not a multiple of Δ_t, $C_S\left(k - \frac{\eta}{\Delta_t}, \cdot\right)$ is computed by linear interpolation of nearest call prices. The finite difference version of the fractional Dupire equation (11.32) is in this case

$$
\lambda \left(C_S(k, \cdot) - C_S\left(k - \frac{\eta}{\Delta_t}, \cdot\right) \right)^{\top}
$$

$$
= \left(-(r - \lambda_J \xi) \, \bar{K} \, R_1 + \frac{\bar{K}^2}{2} \, \bar{\Sigma}(k) \, R_2 \right) C_S(k, \cdot)^{\top}
$$

$$
+ \lambda_J \left(T \bullet (1 + Y) - (1 + \xi) \, I_{n_K + 1} \right) C_S(k, \cdot)^{\top} .
$$

Fractional call prices are computed iteratively from t_0 to t_{n_K} with the following recursion for $k \, \Delta_t < \eta$:

$$
C_S(k, \cdot)^{\top} = \left[\lambda I_{n_K + 1} + (r - \lambda_J \xi) \, \bar{K} \, R_1 - \frac{\bar{K}^2}{2} \, \bar{\Sigma}(k) \, R_2 \right. \tag{11.45}
$$

$$
\left. - \lambda_J \left(T \bullet (1 + Y) - (1 + \xi) \, I_{n_K + 1} \right) \right]^{-1} \times \lambda \, C_S(0, \cdot) ,
$$

and for $k \, \Delta_t \geq \eta$,

$$
C_S(k, \cdot)^{\top} = \left[\lambda I_{n_K + 1} + (r - \lambda_J \xi) \, \bar{K} \, R_1 - \frac{\bar{K}^2}{2} \, \bar{\Sigma}(k) \, R_2 - \right. \tag{11.46}
$$

$$
\left. \lambda_J \left(T \bullet (1 + Y) - (1 + \xi) \, I_{n_K + 1} \right) \right]^{-1} \times \lambda \, C_S\left(k - \frac{\eta}{\Delta_t}, \cdot \right) .
$$

We conclude this section by presenting the recursion to estimate call prices in the non-fractional case. The approached solution of the Dupire Equation (11.6) may be obtained by the following recursion:

$$
C_S(k, \cdot)^{\top} = \left[I_{n_K + 1} + (r - \lambda_J \xi) \, \Delta_t \bar{K} \, R_1 - \Delta_t \frac{\bar{K}^2}{2} \, \bar{\Sigma}(k) \, R_2 \right.
$$

$$
\left. - \lambda_J \Delta_t \left(T \bullet (1 + Y) - (1 + \xi) \, I_{n_K + 1} \right) \right]^{-1} \times C_S(k - 1, \cdot)^{\top} ,
$$

for $k = 0, ..., n_t$ and $j = 0, ..., n_K$. In the next section, we test these numerical approximations and compare option prices in each of these cases.

11.6 Numerical Illustration

We need to specify the statistical distribution of jumps. The size of a jump cannot exceed the current stock price; otherwise, the price would become negative. Therefore, we assume that $Y = e^Z - 1$, where Z is a negative exponential random variable of parameter $\rho > 0$, under the risk neutral measure. Clearly, $Y \in [0, 1]$. The pdf of Z is $f_Z(z) = \rho\, e^{\rho z} I_{\{z \leq 0\}}$. A direct calculation yields that $P(Y \leq y) = (1 + y)^\rho I_{\{y \in [0,1]\}}$ and its expectation is $\xi = \mathbb{E}(Y) = -\frac{1}{1+\rho}$. As the purpose of this section is to emphasize the impact of time changing the dynamics of asset prices, we consider a constant volatility $\sigma_t = \bar{\sigma}$, but the numerical schemes proposed in Sect. 11.5 are also applicable with local volatility functions (11.12) and (11.13).

Figure 11.4 presents 1 year call prices obtained with inverse α-stable and Poisson subordinators for a range of strikes from $K = 50$ to 150. Market parameters are $r = 1\%$, $\lambda_J = 20$, $\bar{\sigma} = 25\%$, $\rho = 49$, and $A_0 = 100$. The numbers of steps of

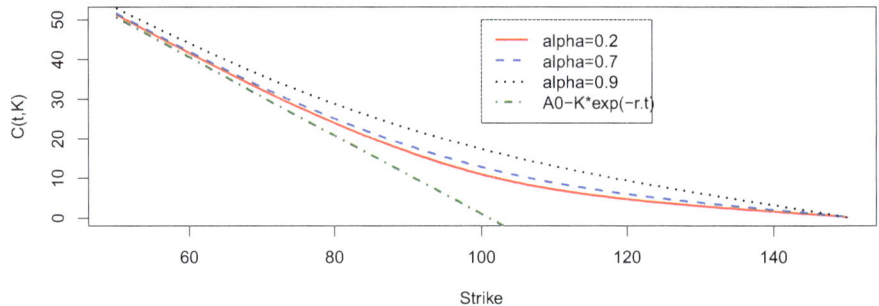

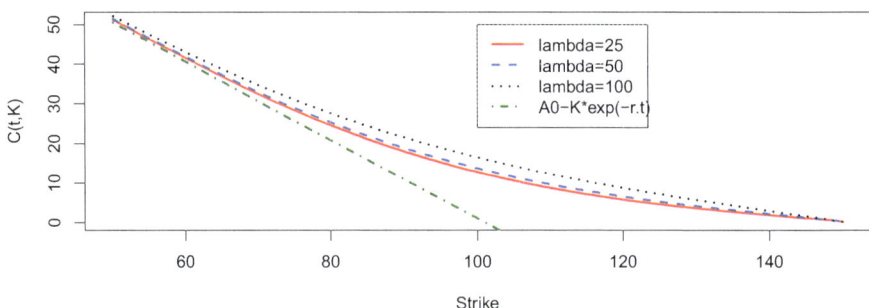

Fig. 11.4 Prices of 1 year call option for varying strikes. Upper plot: inverse α-stable subordinator. Lower plot: inverse Poisson subordinator

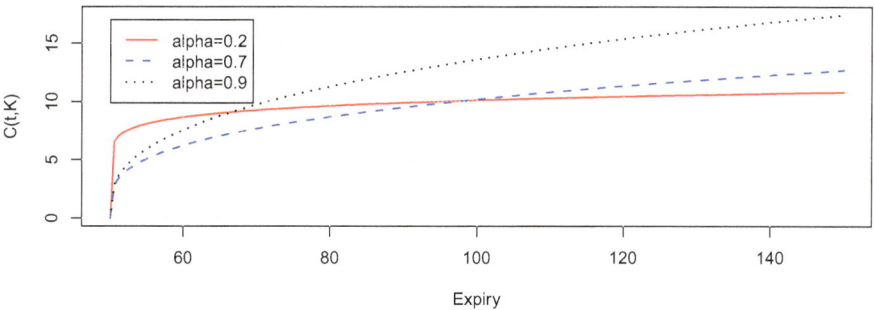

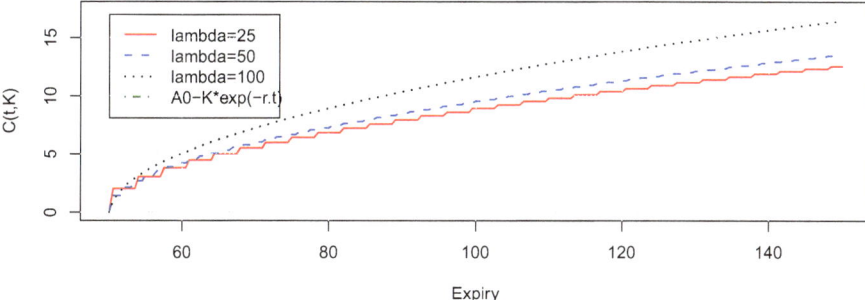

Fig. 11.5 Prices of "at the money" (ATM, $K = S_0$) call options for varying maturities. Upper plot: inverse α-stable subordinator. Lower plot: inverse Poisson subordinator

discretization are $n_t = 200$ and $n_K = 200$. The upper plot reveals that prices rise when α increases. The lower plot emphasizes that prices increase with λ.

Figure 11.5 shows ATM call prices ($K = S_0$) for varying maturities. The upper plot reveals that increasing α reduces the concavity of the curve of prices with respect to time. The lower plot emphasizes firstly that call prices form a stepwise increasing function of expiry times, for the inverse Poisson subordinator. The length of steps is inversely proportional to λ. Furthermore, increasing λ globally raises option prices.

To conclude this section, Fig. 11.6 compares the implied volatility surfaces (obtained by inverting the Black & Scholes formula) in fractional and non-fractional cases. Here, we consider an inverse α-stable subordinator with $\alpha = 0.8$. These plots reveal that time changing the jump-diffusion leads to higher implicit volatilities than those obtained with a pure jump-diffusion. We also observe that implied volatilities are rather flat for ATM options, whereas they decrease with the expiry in the fractional model.

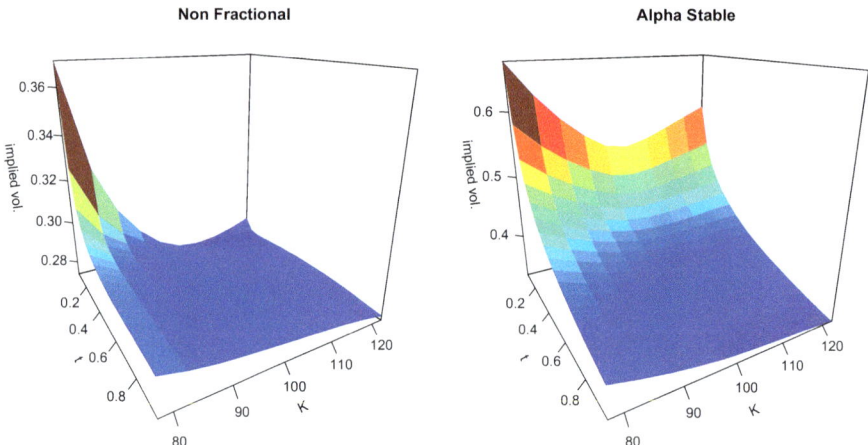

Fig. 11.6 Left plot: implied volatility surface for a jump-diffusion. Right plot: implied volatility surface for the jump-diffusion subordinated by the inverse α-stable subordinator

11.7   Further Reading

Dupire's approach has enjoyed a certain popularity among practitioners, at least partly because of its simplicity. It has subsequently been extended by many authors, notably to Markov process with jumps as in Andersen and Andreasen [2] and Friz et al. [10]. Carr et al. [6] and Cont and Voltchkova [7] develop a Dupire equation for options when the underlying asset is ruled by a general Lévy process. Bentata and Cont [4] propose forward equations for option prices in a semi-martingale model.

We refer the reader to Schilling et al. [12] for a detailed exposition of Bernstein functions and their properties. Toaldo [14] considers subordinators with Laplace exponent that are Bernstein function. He shows that the governing equations of their inverse processes are presented by means of convolution-type integro-differential operators similar to the Dzerbayshan–Caputo derivative. In the same spirit, Kochubei [11] develops a kind of fractional calculus with operators in the time variable, of the form $\left(D_{(k)}f\right)(t) = \frac{d}{dt}\int_0^t k(t-\tau)f(\tau)d\tau - k(t)f(0)$, where $k(\cdot)$ is a non-negative locally integrable function.

References

1. Acay, B., Bas, E., Abdeljawad, T.: Fractional economic models based on market equilibrium in the frame of different type kernels. Chaos Solitons Fractals **130**, 109438 (2020)
2. Andersen, L., Andreasen, J.: Jump diffusion models: volatility smile fitting and numerical methods for pricing. Rev. Deriv. Res. **4**, 231–262 (2000)
3. Andersen, L., Brotherton-Ratcliffe, R.: The equity option volatility smile: a finite difference approach. J. Comput. Finance **1**(2), 5–38 (1998)

4. Bentata, A., Cont, R.: Forward equations for option prices in semimartingale models. Preprint (2010). Available at https://arxiv.org/abs/1001.1380
5. Bergomi, L.: Stochastic Volatility Modeling. CRC Financial Mathematics Series. Chapman & Hall, London (2016)
6. Carr, P., Geman, H., Madan, D.P., Yor, M.: From local volatility to local Lévy models. Quant. Finance **4**, 581–588 (2004)
7. Cont, R., Voltchkova, E.: Integro-differential equations for option prices in exponential Lévy models. Finance Stoch. **9**, 299–325 (2005)
8. Cox, J.C.: Notes on option pricing I: constant elasticity of variance diffusions, reprinted in J. Portf. Manag. **23**, 15–17 (1996)
9. Dupire, B.: Pricing with a smile. Risk **7**, 18–20 (1994)
10. Friz, P.K., Gerhold, S., Yor, M.: How to make Dupire's local volatility work with jumps. Quant. Finance **14**(8), 1327–1331 (2014)
11. Kochubei, A.N.: General fractional calculus, evolution equations, and renewal processes. Integr. Equ. Oper. Theory **71**, 583–600 (2011)
12. Schilling, R.L., Song, R., Vondracek, Z.: Bernstein Functions. Theory and Applications. Walter de Gruyter, Berlin (2010)
13. Shreve, S.E.: Stochastic Calculus for Finance II: Continuous-Time Models. Springer Finance. Springer, Berlin (2004)
14. Toaldo, B.: Convolution-type derivatives, hitting-times of subordinators and time-changed C_0-semigroups. Potential Analy. **42**, 115–140 (2015)